Solar Power Generation

Solar Power Generation

Technology, New Concepts & Policy

P. Jayarama Reddy

CRC Press
Taylor & Francis Group
Boca Raton   London   New York   Leiden

CRC Press is an imprint of the
Taylor & Francis Group, an **informa** business

A BALKEMA BOOK

Front cover illustration: Courtesy of Dyesol Ltd.: A dye-sensitized solar panel.
Back cover illustration: Courtesy of Professor Paul Yu, source Novotny et al 2008,
Copyright © 2008 American Chemical Society: SEM image of n-type InP nanowire
growth on ITO taken at a 45° tilt.

CRC Press/Balkema is an imprint of the Taylor & Francis Group,
an informa business

© 2012 Taylor & Francis Group, London, UK

Typeset by MPS Limited, Chennai, India
Printed and Bound by CPI Group (UK) Ltd, Croydon, CR0 4YY

Library of Congress Cataloging-in-Publication Data

Reddy, P. Jayarama.
  Solar power generation : technology, new concepts & policy / P. Jayarama Reddy.
    p. cm.
  Includes bibliographical references and index.
  ISBN 978-0-415-62110-6 (hardback)
    1. Photovoltaic power generation.   I. Title.
  TK1087.R424 2012
  621.31'244–dc23

                                                                    2012001741

Published by: CRC Press/Balkema
                P.O. Box 447, 2300 AK Leiden, The Netherlands
                e-mail: Pub.NL@taylorandfrancis.com
                www.crcpress.com – www.taylorandfrancis.com

ISBN: 978-0-415-62110-6 (Hbk)
ISBN: 978-0-203-10941-0 (eBook)

Dedicated to SAI

Table of Contents

Foreword

The amount of solar energy reaching the Earth is more than 10,000 times the current energy consumption by man. This resource is larger than all the alternative power sources currently available. With the rapidly growing population of mankind, the anticipated depletion of fossil fuels, the dangers associated with pollution such as acid rain and global warming and the hazards associated with using nuclear power generation, this resource can no longer be ignored.

The heating effect of solar energy can of course be used to directly heat water for domestic applications and passively heat buildings. Indirect differential heating of the Earth produces the wind which in turn can produce wave energy. In this book the direct conversion of solar energy into electricity using "photovoltaic (PV) solar cells" is discussed. PV, the direct conversion of the sunlight into electricity is the most efficient route for power generation simply because the number of steps needed for conversion is a minimum. Also electricity is the most useful form of energy available.

PV generation has many other advantages over other forms of power generation. It is freely and conveniently available everywhere needing no mains supply, it is silent in operation and can be visually unobtrusive, it can be planned and installed in a matter of months rather than taking the ten or more years needed to build a conventional power plant, the technology is non-polluting (any toxic materials used in manufacture can be controlled using existing industrial methods) and it is modular, such that the generation capacity can be expanded easily and any breakages easily replaced.

India is fortuitous in being blessed by an abundance of solar energy. It is also fortunate in having some of the best solar energy scientists in the world. India has an opportunity to use this vast potential to lead the world in supplying its population with clean, non-polluting, affordable and renewable energy.

Professor Jayarama Reddy is one of the best known, accomplished and inspiring scientists and teachers of PV in India. This exciting book includes the wisdom and knowledge of a master scientist and teacher who has devoted his life to developing PV for use not just in India but throughout the world. Both young scientists, new to the field, and experienced scientists will find its insights useful and inspiring as we move forward to provide the world with the energy it needs.

The topics in this book include discussions on crystalline, multicrystalline, microcrystalline and amorphous silcon solar cells, polycrystalline copper gallium indium diselenide and cadmium telluride thin film devices, organic and dye-sensitized solar cells, high efficiency III-V multijunction solar cells and new concept design solar cells including quantum well, quantum dot and other nanostructured devices.

The book is completed with a chapter on the policies and incentives used/needed to develop photovoltaics in India and other parts of the world including Japan, the European Union, the USA, Taiwan and Canada.

R. W. Miles
BSc ARCS MSc DIC PhD MInstP CPhys
Northumbria Photovoltaics Applications Centre,
Northumbria University,
Newcastle upon Tyne, UK.

Acknowledgements

No book, generally, is the work of its author(s) alone. Scores of scientific discoverers contribute through their tireless efforts over time to the growth of knowledge in an area of study. The stupendous contributions of such research groups in the Solar Photovoltaics over several decades and the encouragement of the subject experts, my former students, friends and family have facilitated me to complete this book. I am indebted to all of them.

I wish to express my personal thanks to: Professor Robert Miles, Northumbria Photovoltaics Application centre, Northumbria University, UK for his kind support from the start and for writing the foreword; Professors Charles Lieber (Harvard), Harry Atwater (CalTech), Peidong Yang (UCBerkeley), Martin Green and Gavin Conibeer (UNSW), Marc Brongersma (Stanford), Di Gao (Pittsburgh), Paul Yu (UCSanDiego), René Janssen (Eindhoven Univ), Makoto Konagai (Tokyo), Robert Miles (Northumbria), V.D. Rumyantsev (Russian Ioffe Inst), Ryne Raffaelle (Rochester Inst Tech), Ayodhya Tiwari (EMP, Switzerland), Andreas Bett (Fraunhofer ISE), Rommel Noufi and Martha Symko-Davies (NREL), Daisuke Kanama (Hokkaido Univ), Kyotaro Nakamura and Tatsuo Saga (Sharp), Luc Feitknecht (U Neuchatel), L. Tsakalakos (GE Global Res), Janne Halme (Aalto Univ), Robert McConnell (Amonix), Sam Carter (Solar Systems), Ezri Tarazi and Roy Segev (Zenith Solar), several scientific societies and Hindawi Publishing Corporation for permitting to reproduce figures and use material from their publications; Dr. Janne Halme, University of Aalto, Finland for preparing a figure especially for the book; Prof. K.T. Ramakrishna Reddy, my former student and active researcher, for his personal and professional support throughout the preparation of the book, Prof. C. Suresh Reddy and his students for preparing figures and other odd works, My family members Sreeni and Geetanjali for drawing figures, preparing the manuscript in the desired format, and computer-related and diverse works, Jagan and Vijayalakshmi for collecting and organizing references, and Hitha for her suggestions in the design of the cover page, in addition to their involvement, understanding and ceaseless support; My wonderful granddaughters Divi, Diya, Tanvi and Hitha for providing cheer with their endless enthusiasm; and Dr. Germaine Seijger, Senior Editor, CRC Press for persuading me to take up this work and her regular back-up, and the Editorial staff, particularly Alistair Bright and José van der Veer, for their wonderful support.

P. Jayarama Reddy

Chapter 1

Introduction

*'Solar radiation (energy) received within one hour on the earth is sufficient
to meet the total energy demand of the world for more than one year!'*

Energy production is among the top problems that humanity will face over the next century because of growing energy demand, particularly in developing countries. A major
part of electric power is currently produced by burning fossil fuels – coal, oil or gas. The
increasing energy demand currently as well as in future, however, will possibly not be
met by fossil fuels for two reasons: (a) their limited reserves and (b) their significant contribution to greenhouse gases which have adverse effects on the climate (Solomon *et al.*
2007; Pearce 2008). Consequently, many organizations from around the world have
been searching for alternatives to fossil fuels that are low-cost, sustainable and clean.

In recent years, usage of nuclear energy, which has the potential to satisfy a great
part of overall electrical energy demand, is being discussed in the context of growing
global warming because its contribution to the greenhouse effect is minimal compared
to other sources of energy. The amount of nuclear fuel, however, is also limited and not
renewable. The recent nuclear disaster involving the Japanese reactors at Fukushima
due to a massive earthquake caused an incomparable tragedy and has triggered global
discussion and questions the future of nuclear power. Therefore nuclear energy, in spite
of its huge potential, is surrounded by some controversy though no scientific evidence
forbids their installation and use.

Increasing demand for electrical energy as well as environmental concerns related
to fossil fuel usage are the driving forces behind the development of new energy sources,
which are renewable and ecologically safe. The energy sources, which include energy
from wind, water and biomass, geothermal and solar energy are renewable and envi-
ronmentally friendly. Among these clean energy sources, solar energy is one of the most
promising and fastest growing renewable energy sources worldwide. The predicted
potential and role of each source in the coming decades is presented in Figure 1.1
(Feltrin & Freundlich 2008). This forecast is based on the assumption that the amount
of CO_2 will remain below 450 ppm/year. According to the figure, the renewable sources
will contribute significantly to overall electrical energy production; and in the long-
term, solar energy has the greatest potential among the renewable sources, and is
predicted to play a major role after 2050.

Converting solar energy into electricity or hydrogen fuel using PV cells is one of
the most attractive solutions to modern energy issues because solar energy is produced
with almost zero carbon-emission (Jayarama Reddy 2011).

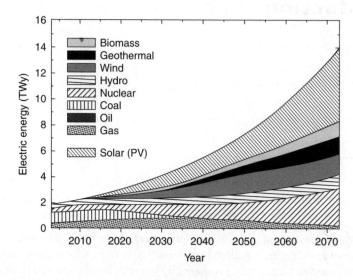

Figure 1.1   Predicted production of electrical energy in the next 60 years (under assumption that the emission of the CO_2 will be held under 450 ppmv)
(*Source*: Feltrin & Freundlich 2008, Copyright © 2008 Elsevier, reproduced with permission)

Solar energy comes from an abundant source, and is available in direct form as solar radiation and in indirect form as wind, biomass, etc. The sun deposits 120,000 TW of radiation on the surface of earth. So, there is clearly enough solar power available if an efficient means of harvesting solar energy is developed. The total energy needed by the humans in 2020 is projected to be 20 TW. The current biggest drawback of wider usage of photovoltaic (PV) is the higher price of the electricity produced – about ~4 to ~8€/Wp (Sinke *et al.* 2007) – which is not competitive with the commercial grid-electricity. However, it is expected to fall progressively and reach around 2–4€/Wp by 2015 (Sinke *et al.* 2007), and the competitive price for the whole of Europe and 'grid parity' will be attained in 2020. The current high price of electricity from solar cells is, however, compensated by the positive effects achieved by this ecologically friendly source as well as the necessity to replace fossil fuel. Although reliable PV systems are commercially available and widely deployed, further breakthrough development of solar PV technology is crucial for it to become a major source of electricity.

Current status of PV technology: Different semiconductor materials with suitable optoelectronic properties have been proposed for photovoltaic applications. Among them, silicon has been the most widely accepted and utilized in the production of PV modules. The average annual growth rate in PV cell production over the last decade has been more than 40% (Dorn 2007). In 2008, the world annual production of photovoltaic (PV) cells reached more than 7.9 GWp [Photon Intl. 2009]. In 2009, the cell manufacturers produced a record 10.7 GWp of PV cells globally – an impressive increase from the year before. While growth in 2009 slowed from the remarkable 89-percent expansion in 2008, it continued the rapid rise of an industry that first reached 1,000 megawatts of production in 2004. By the end of 2009, nearly 23,000 megawatts of PV had been installed worldwide.

If one considers the installation of power-related photovoltaic plants globally, around 80% of all large photovoltaic plants are installed in Europe (700 MWp). The share of the USA accounts for about 16% (142 MWp) and that of Asia for 4% (34 MWp). At present Germany hosts nearly 50% of the world's installed photovoltaic power, but its market share was decreasing slowly very recently (Lenardic and Hug 2008). Primary PV world markets are still Germany with about 45% of the installed power, followed by Spain (28%) and the USA with a 16% market share. The rest of the world has small markets.

Spain proved to be the most dynamic PV market with an impressive growth, albeit it one that slightly declined in 2008. The average installed capacity of a single large commercial power plant has increased from 400 kWp in 1997 to 1.64 MWp in 2007. The average capacity of sole commercial PV plants accounts for 1.14 MWp. Germany remains the market leader but Spain and the USA are catching up (Lenardic and Hug 2008).

Solar PV is considered the world's fastest-growing power technology, and is used to generate electricity in more than 100 countries. Despite this tremendous growth, however, solar power still accounts for a share of less than 0.1% of the global energy generation because of its high cost of production (Photon Intl. 2009; BP's Report 2010). Nevertheless, the strong growth in PV cell production is expected to continue for many years.

The basic advantage of silicon is that it is abundant in nature and can be processed at relatively low costs, resulting in a low energy payback time, i.e., the time needed for a photovoltaic system to produce electrical energy until it pays back the energy costs associated with its production. The raw silicon, after purification, is presented in various crystallization forms: mono-crystal, multi-crystal, micro-crystal and amorphous. These phases of silicon are the most frequently used in the production of photovoltaic cells. Almost 90% of the share of module production belongs to the mainstream wafer-based photovoltaic technologies based on mono-crystal (33%) and multi-crystal (53%) silicon. The prevalence of wafer-based silicon technology has its origin in the development of silicon wafers as well as a compatibility of solar cell technological processes with the microelectronics industry. However, the major drawback of wafer-based silicon technology is the necessity to use silicon wafers 200–300 μm thick, which represents a huge amount of silicon for large scale production of solar modules. Due to the high cost of silicon wafers, its cost reduction potential seems to be limited.

In 2006, for the first time, more than half of the silicon production went into PV instead of computer chips. The production of silicon wafers has not been growing at the same rate as the PV industry, which results in shortage of silicon wafers as well as increase of their price. Today's silicon wafer price is about 60% of the overall price of a wafer-based silicon solar cell (Rogol 2008). The silicon wafer shortage was one of the reasons to research new technologies that do not need huge amounts of expensive silicon. The two commercial technologies, namely, (i) thin-film Si solar cells, i.e., solar cells with hetero junction a-Si: H/c-S (hetero junction with intrinsic thin layer, HIT) and (ii) solar cells based on polycrystalline thin films are the result of these research efforts. Due to greatly reduced semiconductor material consumption and the ability to fabricate the solar cells on inexpensive large area flexible substrates and to monolithically series-connect the fabricated solar cells, thin-film PV technology has the potential of achieving module costs well below €1/Wp.

Technology Overview and Forecasts: The solar cell technology based on crystalline silicon wafers (first generation Si technology) is the dominant technology for terrestrial applications today. Single- and multi-crystalline wafers used in commercial production allow power conversion efficiencies up to 25%, although the manufacturing techniques at present limit them to about 15 to 20%. Recent progress in mono- and multi-Si modules and concepts are presented by leading manufacturers at Intersolar Europe (2011). For example, *SunPower Corp.* presented two models, 327 W and 333 W of rated peak power, the first panels worldwide with conversion efficiency rates above the 20% level. The record efficiency is attributed to a low temperature coefficient, the use of anti-reflective glass and exceptional low-light performance. *LG* has come up with a new module of 260 W fabricated with 60 mono-Si cells with boosted efficiency compared to their earlier cells. *Canadian Solar* has presented new cells with 19.5% efficiency, manufactured using the metal-wrap-through (MRT) process. Their modules with outputs between 245 and 265 W will be on the market soon. Plans are to increase the production capacity to 2.3 GW/year based on their new 600 MW cell production facility in China. The Chinese manufacturer *Yingli* has planned to increase production capacity to 1.7 GW from the current 1 GW by the year's end. *Sanyo* has presented new generation HIT modules: 250 W output with an efficiency of 18%, and 240 W output with an efficiency of 19%. While the major Chinese manufacturers are offering the standard crystalline Si modules at 1.0€/W, the established manufacturers in the West are offering them at 1.20 to 1.30€/W (approximately 1.75 to 1.90$/W).

The second generation (2nd G) of photovoltaic materials is based on the thin-film deposition of semiconductors such as amorphous silicon (a-Si), cadmium telluride (CdTe), copper indium gallium diselenide/disulphide (CIGS/CIS). The efficiencies of thin-film solar cells tend to be lower compared to Si-wafer solar cells, around 6% to 10%; but manufacturing costs are also lower, so that a price in terms of $/watt of electrical output can be reduced. Besides, there is an advantage of less material usage, and facility to deposit films on panels of light or flexible materials. The impressive recent developments in thin-film module production, output power and efficiency, and costs are demonstrated at Intersolar Europe (2011). The established thin-film module manufacturers, Sharp, Kaneka, and Schott Solar have displayed new, full-surface coated modules with double glazing. Schott has offered a 10-year product warranty and a 30-year performance warranty. Global Solar Energy Inc. has presented flexible, lightweight CIGS modules suitable for roofs with limited load-bearing capacities produced from their two plants at Arizona (40 MW) and Berlin (35 MW). Xunlight Corporation has started releasing their flexible and light-weight tandems consisting of a-Si layer and two layers of a-SiGe alloy from early 2011. Several manufacturers of a-Si modules such as Masdar PV GmbH, Sun Well Solar, and Sunner Solar have planned to introduce micromorphous Si tandem modules towards the end of 2011. The CIGS module producers, Soltecture, Avanics GmbH, the Q-cells subsidiary Solibro GmbH have planned to substantially increase the capacity to bring down costs. The Japanese pioneer Solar Frontier K.K. is producing $1.25 \times 1.0 \, m^2$-size modules with efficiencies up to 12.2%; their small $30 \times 30 \, cm$ modules have already achieved 17.2% efficiency. Another CIGS manufacturer, PVNext Corp., with 30 MW production capacity, has produced modules with 10% efficiency and is trying to push up the efficiency to 12.5%. The CdTe thin-film leader First Solar, with a capacity of 1500 MW, has achieved production costs of 0.75$/W at the end of 2010. With the turnkey 120-MW line supplied by the equipment manufacturer Oerlikon Solar AG, micro-morphous modules are produced in China for 0.42€/W, and the Company believes that with an efficiency of 12%, the costs can be reduced to 0.35€/W (Sun & Wind Energy, July/2011, p. 117).

Efficiency and cost potential of silicon and thin-film modules given by Solarpeq (2009) are shown in the Table (Source: EU Platform, research by Solarpeq, 2009).

	mono-Si	multi-Si	CdTe	CIS	a-Si	a-Si/μ-Si
η achieved by industry (%)	19.6	18.5	11.1	12.0	7.0	9.0
η achievable (%)	>20	20	18	18	10	15
Manufacturing cost/Watt (€)	2	1.5–2	0.67	2	1	1
Expected costs as from 2020 (€)	<0.5	<0.5	<0.3	<0.3	<0.3	<0.3

One may find evidently small variations in the current data as well as projections related to production and installation volumes and costs given by the different institutions. But the conclusion is that more research and development effort to find novel approaches and innovations in production processes, and supportive policies by the governments can bring down the costs of the established PV technologies in future.

The concept of third generation (3rd G) of photovoltaic cells is developed by Martin Green, University of New South Wales (UNSW) (Figure 1.2) with a twofold object of achieving high conversion efficiencies and reduction in costs of PV technology (Green 2004).

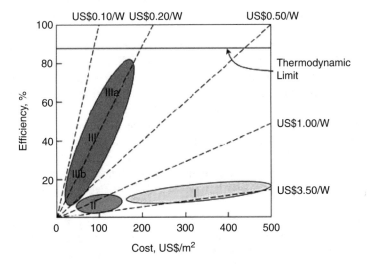

Figure 1.2 Overview of the different PV technologies – Efficiency and Cost projections: first generation (I) are thick film Si devices and second generation (II) are inorganic thin-film technologies; third-generation technologies (III) are described in the text
(*Source:* Green 2004, reproduced with permission of the author)

The research efforts are already showing the potential: a remarkable increase in efficiency that maintains the cost advantage of second-generation materials. The approaches include development of dye-sensitized nano crystalline solar cells (Graetzel solar cells), organic polymer-based photovoltaics, multi-junction solar cells based on III-V semiconductors, solar cells using Quantum Dots and Nanowires, and multi-band, Hot Carrier and Plasmonic solar cells. This third generation currently comprises two categories. The first (IIIa) consists of novel approaches mentioned above that strive to achieve very high efficiencies. All these concepts have theoretical maximum efficiencies well above the 31% limit for single-junction devices. Hence, these high efficiency cells can afford higher costs and still show a favorable $/Wp balance. In the second type of third-generation devices (IIIb), a low $/Wp balance would be achieved via moderate efficiencies (15–20%), but at very low cost. This requires inexpensive semiconductor materials, packaging solutions, and production processes (low-temperature atmospheric routes), as well as high fabrication through-put, low or no investment into the production facility and a production-on-demand scenario. Organic PVs (OPVs) are promising low-cost PV technologies currently available. In the different types of OPVs, at least one key function for PV energy conversion is handled by the organic semiconductor or conductor. The most prominent and mature technology is dye-sensitized nanostructured oxide solar cells (DSCs). DSCs use an organic dye to absorb light and undergo a rapid electron transfer to a nanostructured oxide such as anatase TiO_2. The mesoscopic structure of the TiO_2 allows processing of rather thick, nanoporous films. At an active-layer thickness of several micro-meters, the entire light is absorbed, and these devices reach external quantum efficiencies of over 80%. The hole transport is achieved by a redox couple, such as iodide/tri iodide (I^-/I_3^-). There is a lot of interest in replacing the liquid electrolyte with a solid-state hole transporter; however, current progress is limited by the transport properties of the solid-state system. Solid-state bulk-hetero junction devices (BHJ) based on conjugate polymers and molecules is other very successful recent technology among OPVs.

Large scale production or utilization is possible in any country if the industry and the customer are provided with appropriate policy and incentive programmes. This has been emphasized in discussions at many international forums in the recent years. Hence, the policies and incentives offered in countries such as Japan, China, member states of the European Union (especially Germany, Spain, France and Italy), the United States, India, Taiwan, and Canada where PV activity has been significant have helped the large scale growth and utilization of solar electricity in those countries. Since deploying renewable energies, particularly solar energy, has been recognized as the most affordable path to combat global warming/climate change, the environment/energy policies of several countries have provided incentives to develop solar energy technologies.

The ultimate goal of any R&D effort today is to bring down the cost of stable solar electric power generation to the level of the utility grid in the near future.

This book is designed to draw attention to (i) the matured PV technologies currently utilised in industry, (ii) R&D activity for the development of low cost processes, and to improve the designs and overall performance of PV cells, (iii) development status of PV technologies based on 'new' concepts (third-generation technologies) which have high potential for mass production and (iv) Policy status in selected countries for promoting the manufacture and use of Solar power. The book thus comprises:

- The mono- and multi-crystalline Si wafer-based and amorphous & micro-crystalline Si-based solar cell technologies: Various processes and equipment currently used to produce standard mono- and poly-Si PV cells are discussed with respect to the corresponding material technologies, such as silicon ingot and wafer production. An overview of more advanced solar cells with a higher efficiency potential produced by a few companies, and a few of the latest concepts/technologies that could lead to efficiencies of greater than 25% are outlined. The development of amorphous silicon thin-film solar cells and modules, and micro-crystalline silicon solar cells and the suggested technological aspects for improving their performance are covered (chapter 2)
- Polycrystalline CdTe- and CIGS-based thin-film PV technologies, and novel multi-junction concepts based on them (chapter 3)
- Third generation cell technologies: polymer solar cells (DSCs and Organic cells), high efficiency solar devices [III-V Multi-junction solar cells, and Concentrator Photovoltaics (CPV)] (chapters 4 & 5)
- New solar cell concepts: Progress in solar cells based on quantum dots and nanowires; concepts still under study (hot carrier solar cells, plasmonic photovoltaics, nanostructure material-based thin-film solar cells); and CSG technology (chapter 6)
- Policies and incentives for large scale production/utilization of solar electricity in different countries (Japan, China, European Union countries, United States, India, Taiwan etc) (chapter 7).

The research in the areas covered in chapters 4, 5 and 6 is so intense that several publications appear regularly. Due to space constraints, all could not be covered, and the reader is advised to refer to them for more details.

REFERENCES

BP (2010): 'Statistical Preview of World Energy Full Report, 2010'

Bernreuter, J. (2011): Taking the bull by the horns, *Sun & Wind Energy*, 7/2011, p. 117

Buddenseik, V., Iken, J., & Uphoff, V. (2011): Reliable forecast required urgently, *Sun & Wind Energy*, 7/2011, p. 110

Dorn, J. (2007): Earth Policy Institute, *Solar Cell Production Jumps 50% in 2007*, [Online] Available from: http://www.earth-policy.org/Indicators/Solar/2007.htm (accessed 2nd May 2011)

EU PV Platform (2009): Research by Solarpeq. [Online] available from www.solarserver.com/solarmagazin/solar-report_0809_e.html

Feltrin, A., & Freundlich, A. (2008): Material considerations for terawatt level deployment of photovoltaics. *Renewable Energy*, 33, 180–185

Green, M.A. (2004): Third-generation Photovoltaics: Theoretical and Experimental Progress, *Proc. 19th EU PV Solar Energy Conf*, 7–11 June, 2004, Paris, France, pp. 3–8

Hoffert, M.I., Caldeira, K., Jain, A.K., Haites, E.F., Harvey, L.D.D., Potter, S.D., Schlesinger, S.M., Schneider, S.H., Watts, R.G., Wigley, T.M.L., & Wuebbles, D.J. (1998):

Energy implications of future stabilization of atmospheric CO2 content, *Nature*, **395**, 881–884

Jayarama Reddy, P. (2010): *Science and Technology of Photovoltaics*, 2nd Edn, CRC Press, Leiden, The Netherlands, ISBN 13: 978-0-415-57363-4

Jayarama Reddy, P. (2011): *Pollution and Global Warming*, BS Publications, Hyderabad, India, ISBN: 978-93-81075-09-8

Lenardic, D. & Hug, R. (2008). Large PV Power Plants – Average growth by almost 100% since 2005. [Online] Available from Solarserver.com

Miroslav, Mikolasek (2009): Current status and progress in New generation Silicon based solar cells, *Posterus*, 6th July 2009. [Online] Available from http://www.posterus.sk/?p=1247 (Accessed on 2nd May 2011)

Pearce, J. (2008): Industrial symbiosis of very large-scale photovoltaic manufacturing, *Renewable Energy*, **33**, 1101–1108

Photon International (2009), p. 176, March 2009

Rogol, M. (2008): Refining Benchmarks and Forecasts – Rogol's Monthly Market Commentary, *Photon International, 2008*

Roney, J.M. (2010): Solar cell production climbs to another record in 2009, *Earth Policy Institute*, Washington, D.C.

Solomon, S. *et al.* (2007): Climate change 2007: The Physical Science basis – The Intergovernmental Panel on Climate Change, Cambridge University Press, ISBN 978-052188009-1

Sinke, W. (2007): A Strategic Research Agenda for Photovoltaic Solar Energy Technology. *The European Photovoltaic Technology Platform*, Luxembourg: Office for Official Publications of the European Communities, ISBN 978-92-79-05523-2.

Chapter 2

Silicon solar cells

2.1 INTRODUCTION

Crystalline silicon photovoltaic (PV) cells have very wide and largest usage among all types of solar cells the market offers. The highest energy conversion efficiency reported so far for research crystalline silicon PV cells is 25%. Standard industrial cells, however, remain limited to 15–18% with the exception of certain high-efficiency cells capable of efficiencies greater than 20%. High-efficiency research PV cells have advantages in performance but are often unsuitable for low-cost production due to their complex structures and the extended manufacturing processes. Even so, both high conversion efficiency and low processing cost can concurrently be achieved through the development of superior manufacturing technologies and equipment.

Silicon is one of the most abundant and safe resources on Earth, representing 26% of crustal material. The environment friendly silicon solar cell has a record of over 60 years of development and longest period of production. World annual solar cell production of 100 GWp is expected to be achieved by around 2020, and the silicon PV cell is the most viable candidate to meet this demand.

The crystalline silicon PV cell is essentially a semiconductor diode. In the early years of solar cell production, many technologies for crystalline silicon cells were proposed on the basis of silicon-based devices. The technologies and equipment developed for already existing silicon-based semiconductor devices, such as large-scale integrated circuits and several other kinds of silicon semiconductor applications, and those developed for PV solar cells supported growth in both fields. For example, process technologies such as photolithography helped to increase energy conversion efficiency in solar cells, and mass-production technologies such as wire-saw slicing of silicon ingots developed for the PV industry were also readily applicable to other silicon-based semiconductor devices. However, the value of a PV cell per unit area is much lower than that for other silicon-based semiconductor devices.

Production technologies such as silver-paste screen printing and firing for contact formation are therefore needed to lower the cost and increase the volume of production for crystalline silicon solar cells. To achieve parity with existing mains grid electricity prices, achieving lower material and process/production costs are as important as higher solar cell efficiencies. The realization of high-efficiency solar cells with low process cost is currently the most important technical issue before solar cell manufacturers.

2.2   FEATURES OF STANDARD C-SILICON CELLS

The structure of typical commercial crystalline-silicon PV cells is shown in Figure 2.1. Standard cells are produced using one of two different boron-doped p-type silicon substrates, either mono crystalline or polycrystalline. The cells of mono-Si type are typically 125 mm (5 inches), and that of poly-Si, 156 mm (6 inches) square. Mono-crystalline solar cells are produced from pseudo-square silicon wafer substrates cut from column ingots grown by the Czochralski (CZ) process. In contrast, poly-crystalline cells are made from square silicon substrates cut from polycrystalline ingots grown in quartz crucibles. Details on the CZ process and poly-Si ingots growth are given in several articles, for example, author's earlier book (Jayarama Reddy 2010).

The dominant technology for p-type Si solar cells involves screen-printing based metallization (Figure 2.1). The front surface of the cell is covered with micrometer-sized pyramid structures (textured surface) to reduce reflection loss of incident light. An anti-reflection coating (ARC) of silicon nitride (SiN_x) or titanium oxide (TiO_x) is overlayed on the textured silicon surface to further reduce reflection loss. Crystalline silicon solar cells have highly phosphorous-doped n^+ (electron-producing) regions on the front surface of boron-doped p-type (electron-accepting) substrates to form p–n junctions. Back-surface p^+ field (BSF) regions are formed, usually by firing screen-printed aluminum paste in a belt furnace, on the back surface of the silicon substrate which serves as a back surface passivation layer and suppress recombination of minority carriers (photo-generated electrons). However, the Al-BSF may not be appropriate for thinner substrates because its formation introduces high stress, which can bend thin wafers. In addition, the Al-BSF provides only low-to-moderate quality surface passivation and exhibits low internal surface reflection, which is not sufficient for excellent electrical and optical confinement, respectively (Meemongkolkiat 2008).

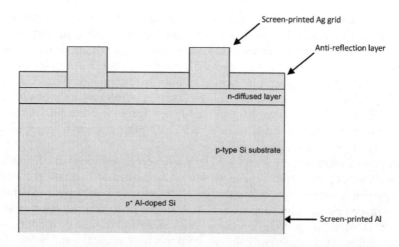

Figure 2.1   Schematic of a commercial mono-crystalline silicon solar cell

The carriers (electrons) generated in the silicon bulk and diffusion layers are collected by the front and back silver contacts (electrodes). The front contact consists of gridlines connected by a busbar to form a comb-shaped structure. The back contact is usually a series of silver stripes connected to the front bus bar of the adjacent cell via soldered copper inter-connects. The contacts are usually formed by firing of screen-printed silver paste at the same time as firing for formation of the BSF regions. The front contact is similarly formed using screen-printed silver paste applied on top of the ARC layer. The contact between the front electrode and the n^+ region of the silicon substrate is achieved by firing, facilitating silver to penetrate through the ARC layer.

The screen printing equipment for the formation of the front and rear contacts of solar cells is commercially available. This equipment is robust, simple, and can easily be automated. The screen printed front contact has to have (i) low contact resistance, (ii) no junction shunting, (iii) low specific resistance, (iv) high aspect ratio, (v) good adhesion to Si, (vi) firing through SiN, and (vii) good solderability for series interconnection with tabbing ribbons within the module.

The screen-printed front silver contact prepared by firing to penetrate the ARC is one of the most important techniques for large-volume fabrication of modern standard crystalline silicon cells. Other techniques, such as using boron-doped BSF and nickel–copper plating contacts, are used by a few cell manufacturers. The efficiencies of typical commercial crystalline silicon solar cells with standard cell structures are in the range of 16–18% for mono-crystalline substrates, and 15–17% for (poly) multi-crystalline substrates.

The substrate thickness used in most standard crystalline cells is 160–240 μm. The solar cells are assembled into modules by soldering and laminating to a front glass panel using ethylene vinyl acetate as an encapsulant. The energy conversion efficiency of modules of standard solar cells is roughly 2% less than the individual cell efficiency, falling in the range, 12–15%.

The value chain for crystalline silicon solar cells and modules is longer than that for thin-film solar cells as we see later. Three kinds of industries are associated with crystalline silicon solar cell and module production: (1) metallurgical and chemical plants for producing raw silicon material, (2) mono-crystalline and polycrystalline ingot preparation, and wafer slicing by multi-wire saw, and (3) solar cell and module production.

The fabrication cost for solar cell modules includes the cost of the silicon substrate (50%), cell processing (20%) and module preparation (30%). The cost share is therefore strongly affected by the market price for poly-silicon feedstock. Hence, reducing the cost of the silicon substrate remains one of the most important issues in the PV industry. The energy conversion efficiency of solar cells is another important issue because the cell efficiency influences the entire value-chain cost of the PV system, from material production to system installation. The solar cell efficiency is limited by the three main loss mechanisms: (1) photon losses due to surface reflection, silicon bulk transmission and back contact absorption; (2) minority carrier (electrons in the p-region and holes in the n-region) loss due to recombination in the silicon bulk and at the surface; and (3) heating joule loss due to series resistance in the gridlines and busbars, at the interface between the contact and silicon, and in the silicon bulk and diffusion region. In the design of solar cells and processes, these losses are minimized without lowering the productivity of the solar cells.

The physics of crystalline Si solar cell is dealt in many publications (for example, Green 1982, 1995; Partain, 1995; Goetzberger *et al.* 1998). The current-voltage characteristics and optical parameters are measured for each cell to determine (i) the optical quality, (ii) its current at the maximum power point and to sort the cells into current classes to minimize mismatch losses in the module consisting of series-connected solar cells (Neuhaus *et al.* 2005), (iii) its reverse breakthrough characteristics to avoid hot spot heating within the module (Neuhaus *et al.* 2006), and (iv) the solar cell parameters (short-circuit current, I_{sc}, open-circuit voltage, V_{oc}, current at the maximum power point, I_{mp}, voltage at the maximum power point, V_{mp}, maximum power, P_{max}, fill factor, FF, and energy conversion efficiency, η) as a final process control.

In research and development, short-circuit current density (J_{sc}) is also used. An air mass 1.5 (AM1.5) spectrum condition ($1000\,W/m^2$) is the standard test condition for terrestrial solar cells. The AM1.5 condition is defined as 1.5 times the spectral absorbance of Earth's atmosphere; in contrast, the spectral absorbance for space is zero (air mass zero, AM0). The solar energy under the AM1.5 condition is used as the input energy for calculation of solar cell efficiency. The solar cell fill factor and efficiency are calculated using the following equations.

$$FF = P_{max}/I_{sc} \times V_{oc} \quad \text{where } P_{max} = I_{mp} \times V_{mp} \tag{2.1}$$

$$\text{Efficiency} = P_{max}/1000\ (W/m^2) \times \text{Cell area } (m^2) \tag{2.2}$$

2.3 PROGRESS IN CELL EFFICIENCY

The first crystalline silicon solar cells were fabricated by Bell laboratory in 1953 achieving 4.5% efficiency followed in 1954 with 6% efficiency devices (Pearson 1985; Chapin *et al.* 1954). In the next ten years since the first demonstration, the efficiency of crystalline silicon cells improved to around 15%, and was sufficiently efficient to be used as electrical power sources for spacecraft, special terrestrial applications such as lighthouses, and consumer products such as electronic calculators. The improvements in research-cell efficiencies achieved for various kinds of solar cells over the past 30 years are shown in Figure 2.2 (NREL graph, Ullal 2009). The basic cell structure used in current industrial crystalline solar cells, which includes features such as a lightly doped n^+ layer (0.2–$0.3\,\mu m$) for better blue-wavelength response, a BSF formed by a p/p^+ low/high junction on the rear side of the cell, a random pyramid-structured light-trapping surface, and an ARC optimized with respect to the refractive index of the glue used to adhere it, were developed for space and terrestrial use in the 1970s. The efficiency of mono-crystalline cells for space use is in the range of 14–16% under '1 sun' AM0 test conditions, equivalent to 15–17% at AM1.5. These standard structures for crystalline silicon cells are still used in the production of standard industrial crystalline cells, which offer efficiencies in the range, 14–17%.

The key technologies needed to realize efficiencies of higher than 20% were developed in the 1980s and '90s. Most of the features presented in Table 2.1 are utilized in the production of latest high-efficiency crystalline silicon cells (Saga 2010).

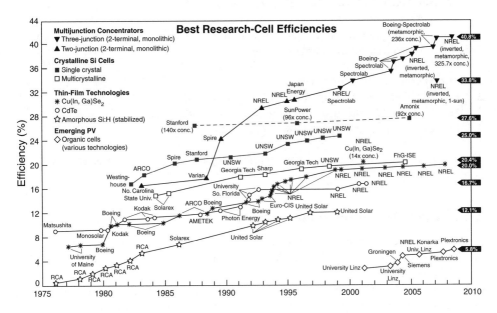

Figure 2.2 Best Research – Cell efficiencies (Personal communication from Ullal, NREL, 2009)

Table 2.1 Key technologies for high-efficiency c-Si solar cells

(a) Minimising Photon loss:
Front textured surface of random pyramid or inverted pyramid structures to reduce surface reflection loss, Simple- or double-layer ARC to reduce surface reflection loss, Back contact cell structure to reduce front contact shadow loss, Flat back surface by chemical etching of Si to improve back reflectivity and reduce photon absorption, Back surface reflector consisted of a dielectric layer and high reflectivity thin metal layer to reduce photon absorption.

(b) Minimising carrier loss:
Passivation of front electrode (partly in contact with highly doped Si layer) to reduce carrier combination under front electrode, Shallow doped p-n junction with front surface dielectric passivation layer to reduce carrier recombination in the n^+-doped region and at the surface; hetero junction with thin amorphous layers on a c-Si base in a hetero junction cell; front surface field and surface passivation for back contact in a back-junction cell, Locally p^+-doped back surface field and point contact structure to reduce carrier recombination in highly doped p^+ back region, Back surface passivation by a dielectric layer or hetero junction structure to reduce back surface recombination.

(c) Minimising electrical loss:
Fine grid line front contact to reduce series resistance of n^+-doped region, Selective emitter (deep and highly doped emitter under the contact) to reduce contact resistance of front contact with Si surface, n-type or p-type Si substrates with minority carrier diffusion lengths longer than the base thickness.

2.4 MONO-CRYSTALLINE Si SOLAR CELLS

Typical high-efficiency mono crystalline silicon PV cell designs that have been applied to industrial production are explained here:

1 Buried contact (BC) Solar cell
2 Passivated emitter rear localized (PERL) Solar cell
3 Hetero junction with intrinsic thin layer (HIT) Solar cell
4 Interdigitated Back contact (IBC) Solar cell.

These PV cells feature many of the technologies that provide high efficiency in this type of PV cell.

2.4.1 Buried contact solar cells

Based on 'passivated emitter solar cells' the concept of more industry-oriented technology of buried contact solar cell was developed and patented by Green and co-workers from UNSW (Green 1985; Wenham & Green 1985). The emphasis during the development was on evolving simple and low cost processes and methods suitable for large area cells and for mass production. This research group recognised that the screen printed contacts decrease the device efficiency. Factors such as high shading losses, the high resistivity of the screen printed silver grids (compared to pure silver), a high contact resistance between the grid and silicon and poor aspect ratio also affect device efficiency. This has led to using photolithography/laser scribing for the development of cell contacts. In the buried contact silicon solar cell, the metal is buried in a laser-formed groove inside. This is an important high efficiency feature of the buried contact solar cell which allows for a large metal height-to-width aspect ratio. A large metal contact aspect ratio in turn allows a large volume of metal to be used in the contact finger, without having a wide strip of metal on the top surface. Therefore, a high metal aspect ratio allows a large number of closely spaced metal fingers, while still retaining a high transparency. For example, on a large area device, a screen printed solar cell may have shading losses as high as 10 to 15%, while in a buried contact structure, the shading losses are only 2 to 3%. These lower shading losses allow low reflection and therefore higher short-circuit currents. The original design for this solar cell has the buried metal contacts on the top surface as shown in Figure 2.3a although more recently the interdigitated rear contact design applied to n-type wafers has become mainly popular.

Figure 2.3b shows the partially plated laser groove. In this BC-design the contact metals, nickel, copper and then silver, are deposited using electroless methods. The improved design permits shallower phosphorus doping at the surface of the device without degrading the V_{oc}, improving the short wavelength response of the cell. The higher doping concentration of this n^+ region also reduces the contact resistance to the grid contact.

The contact resistance of a buried contact solar cell is lower than that in screen printed solar cells due to the formation of a nickel silicide at the semiconductor-metal interface and the large metal-silicon contact area. These reduced resistive losses basically allow large area solar cells with high fill factors. Efficiencies of 20% have

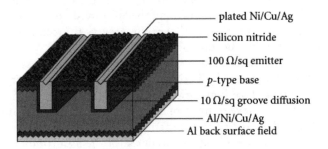

Figure 2.3a Schematic representation of a Buried contact (BC) solar cell
(*Source*: Neuhaus & Munzer 2007, Hindawi Publishing Corporation; Copyright © 2007 D.-H. Neuhaus and A. Munzer)

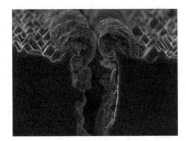

Figure 2.3b Cross section of a partially plated laser groove (Courtesy: PVEducation.org)

been demonstrated on small area devices; however, 21% efficiency is assumed to be achievable on large area CZ-silicon ultimately (UNSW PVC Annual Report 2008).

In industrial production, BP Solar cells have reached average efficiencies of 17% and the best cells have demonstrated 18.3% efficiency (Mason *et al.* 2004). Mason and Jordon (1991) and Hartley *et al.* (2002) have described the steps for the manufacture of BP Solar's buried contact solar cells. The latest developments in the fabrication of industrial solar cell are incorporating buried contact grid on the front side and laser fired contacts on the rear side of the cells (Mason *et al.* 2006; Schultz *et al.* 2006).

2.4.2   Passivated emitter, rear locally diffused (PERL) cell

The next advancement was PERL cell where local diffusion has been used in the area of rear point contact (Zhao *et al.* 1995, 1999). This is a research cell developed at UNSW (Figure 2.4a). The back structure of the PERL cell includes local boron BSF around the local back contacts, while thermally grown SiO_2 passivates the other area of the back surface. The incorporation of a local BSF minimizes the recombination at the back contact and allows an ohmic contact on high-resistivity substrates. The fabrication of this local boron BSF cell, however, requires lengthy procedure with several masking and high-temperature process steps, including (a) growth of SiO_2 that serves as a boron diffusion mask, (b) photo-lithography process to define BSF windows, (c) boron-diffusion using a BBr$_3$ source, (d) removal of the SiO$_2$ mask, (e) re-growth of SiO$_2$ for surface passivation, and (f) a photo-lithography process to open the SiO$_2$ layer for a contact formation within the boron BSF region.

The key to success of such a complex structure is excellent process control to diffuse boron without degrading the bulk lifetime. Despite its high complexity, this structure demonstrates the potential for high efficiency of Si solar cells. Further, it provides the basis and strategy for the development of high-efficiency cells with simpler structures and processes. The bulk minority carrier lifetime in PERL cells is longer than 1 ms. This cell approaches the limit of existing technologies for the absorption of solar photons and the collection of carriers generated in the cell emitter and base.

A PERL cell record efficiency of 24.7% was reported almost ten years ago (Zhao et al. 1999), and the record of 25.0% reported by research group at UNSW was obtained after re-measurement of the same cell using newer solar reference spectrum. The jump in efficiency resulted from new knowledge about the composition of sunlight that led to the adoption of a new reference solar spectrum by the International Electrotechnical Commission in April 2008. As shown in Figure 2.4b, the new reference spectrum has more energy at both blue and red ends of the spectrum, with relatively less energy at intermediate green wavelengths. The exceptionally strong response of the PERL cell at both blue and red wavelengths increases its performance margin over less highly performing devices. Referenced to the new spectrum, the independently confirmed parameters for this 'milestone' cell are $V_{oc} = 706$ mV, $J_{sc} = 42.7$ mA/cm^2, FF $= 0.828$ and $\eta = 25.0\%$ for a 4 cm^2 laboratory cell (Green et al. 2009).

The PERL cell has remained the most efficient type of mono-silicon PV cell for the past several years (Zhao et al. 1999), and has been the most popular laboratory structure of all the high-efficiency crystalline silicon PV cells. However, *the full PERL design is not easy to apply to low-cost industrial production* because of the necessity for multiple photolithography steps, similar to complex designs of semiconductor devices. Expensive silicon PV cells for space applications have a similar structure to the PERL cell (Washio et al. 1993).

Boron BSF solar cells: It is seen that boron BSF has a better surface passivation; and also avoids wafer bowing, especially when the wafers become thinner. Since the application of boron involves more technological steps, Siemens Solar (Munzer et al. 1995), using boron coating on the rear side and subsequent boron drive-in at elevated temperatures, has developed one-sided fabrication process. A low surface reflection was obtained by the combination of a surface texture with an extra SiN AR coating. The emitter formation and metallization on the front and rear are done by conventional methods. These steps have resulted in relative efficiency increase of over 10%.

Siemens Solar and later, Shell Solar and Solar World (the succeeding companies) have utilized this process for mass production of mono-Si modules of 60 MW/yr. Schematic of the BSF solar cell developed by Solar World is shown in Figure 2.4c. A selective emitter structure was developed to improve efficiency and an average cell efficiency of 18.4% was obtained on more than 500 cells (Neuhaus & Munzer 2007), and the best solar cell efficiency recorded was 18.8% (Munzer et al. 2006). These improved cells exhibit a high red response due to boron BSF and almost constant blue response due to shallow emitter. The next development is PERF (Passivated emitter

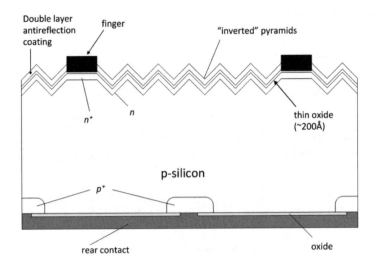

Figure 2.4a Schematic representation of PERL Cell developed by UNSW

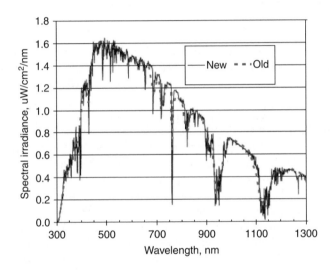

Figure 2.4b Comparison of the spectral content of the new and old solar reference spectrum
(IEC60904-3 Ed.2 and Ed.1, respectively)
(*Source*: ARC – UNSW PV Annual Report 2008, reproduced with the permission of
Prof. Martin Green)

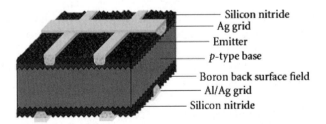

Figure 2.4c Schematic drawing of a boron BSF solar cell of Solar World
(*Source*: Neuhaus & Munzer 2007, Hindawi Publishing Corp.; Copyright © 2007
D.-H. Neuhaus and A. Munzer)

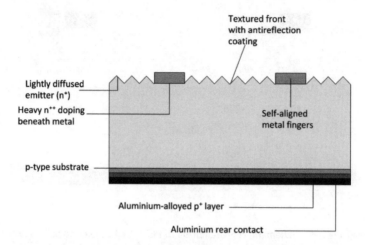

Textured front
with antireflection
coating

Lightly diffused
emitter (n⁺)

Heavy n⁺⁺ doping
beneath metal

Self-aligned
metal fingers

p-type substrate

Aluminium-alloyed p⁺ layer

Aluminium rear contact

Figure 2.5 Schematic of simplified PLUTO cell developed by SunTech Power

rear floating junction) solar cell which offers the best long term potential for high performance. This structure is used to produce mono-silicon cells with highest V_{oc} to date with $720\,mV$ under standard test conditions (Wenham *et al.* 1994) and efficiencies $>23\%$.

The PLUTO cell (Figure 2.5) developed for industrial production by SunTech Power was also based on the design (PESC) developed at the UNSW in 1985 (Green *et al.* 1985). It has a *simpler* passivated emitter solar cell (PESC) structure and provides efficiency of up to 19.2% in a 4 cm-square cell (Shi & Wenham 2009).

The PESC features front passivation, a selective emitter, and a plated front contact with fine gridlines. The production sequence for the simplified Pluto cells is: Texture and clean, Inline diffusion, Wet edge isolation, Anti-reflection & surface passivation coating, Patterning of anti-reflection coating, Selective emitter formation, Screen print Al and formation of BSF, and Self aligned metallization. The simplified version of Pluto technology can be easily retrofitted onto existing screen-printing production lines using essentially the same equipment, wafers and materials, and can achieve a performance advantage of around 10–15% (Shi *et al.* 2009).

2.4.3   HIT solar cell

This cell was developed by SANYO (Taguchi *et al.* 1990) combining a-silicon and mono-silicon. It is a unique hetero junction structure consisting of very thin, amorphous p- and n-doped layers and intrinsic amorphous layers on the front and rear surfaces of a CZ n-type mono-silicon substrate. The current trend in wafer-based technology is to decrease wafer thickness; for example, reducing wafer thickness by $70\,\mu m$ brings 10–15% savings of silicon material resulting in decrease of overall price. However, when the wafer thickness is decreased below $150\,\mu m$, the high temperature processes used during fabrication may bend the wafer. This aspect is taken care of in the fabrication of HIT cell.

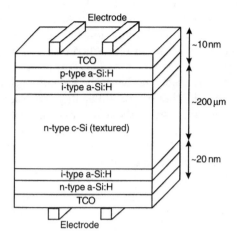

Figure 2.6 SANYO HIT solar cell structure
(*Source*: Jayarama Reddy 2010)

In HIT solar cell, a single thin crystalline silicon wafer ($\sim$120 μm) is sandwiched between two ultra-thin intrinsic silicon layers and n-type and p-type doped amorphous silicon layers, deposited at temperature below 300°C (Figure 2.6).

The low temperature deposition enables use of thin wafers. The p/n junction of the p-type substrate HIT solar cell is formed by the p-type amorphous silicon (a-Si:H) layer and the n-type crystalline silicon substrate. The impurity-free intrinsic amorphous silicon thin layer is inserted between the highly doped a-Si:H layer and the substrate in order to reduce recombination of charge carriers at the interface. The intrinsic amorphous layer has ability to decrease the defect density at the interface because this layer has lower defect density by around three orders of magnitude (Jensen *et al.* 2002).

Fabrication process of HIT cell is simple with relatively low manufacturing cost. HIT solar cells demonstrate numerous advantages (Cervantes, 2008): good surface passivation, simple and low temperature process enabling the use of thin wafers ($<$100 μm), and high conversion efficiency. Since the base material of the structure is crystalline silicon, the typical degradation (Staebler-Wronski effect) observed in amorphous silicon solar cells does not take place in HIT solar cells. In addition, HIT solar cells exhibit better temperature characteristics than conventional c-Si ones – a low temperature coefficient of about 0.30% K^{-1} at P_{max} compared to about 0.45% K^{-1} for standard industrial crystalline silicon PV cells – which means more power generated in outdoor conditions for the same nominal conversion efficiency. This hetero junction structure improves V_{oc} considerably by the effects of the large energy band gap of the front amorphous silicon layer and the excellent quality of the interface between the amorphous layer and the crystalline substrate. This cell has a transparent conductive oxide (TCO) ARC, which reduces the sheet resistivity of the front amorphous layers. The J_{sc} is distinctly lower compared to other high-efficiency PV cells. This appears to be due to reduced photocurrent collection by the front a-silicon layers and the bulk silicon by the effects of the lesser transparency of the TCO layer compared to other ARCs, and/or the lower internal quantum efficiency of the amorphous layers.

The main dissimilarity between HIT cells and wafer-based Si solar cells is the presence of hetero junction in case of HIT cells. The presence of different conduction and valence band offsets at a-Si/c-Si hetero junction brings about the inherent superior performance of the HIT cell with the n-type silicon wafer (Stangl *et al.* 2001). The simulation studies carried out by Maydell *et al.* (2006) indicate that the quality of the a-Si:H/c-Si interface, which is associated with the recombination, has serious impact on the performance of the cell. Obtaining good quality intrinsic a-Si:H layer as well as the interface of the a-Si:H/c-Si is one of the main challenges in order to achieve good efficiency.

This Sanyo product fabricated with state-of-the-art manufacturing techniques provides industry-leading performance and value. Sanyo has reported world's highest conversion efficiency level of 22.3% for HIT solar cells in 2007, mainly due to the excellent a-Si:H/c-Si hetero-interface properties obtained by optimized surface cleaning process. The other cell parameters achieved are, open circuit voltage, 725 mV, short circuit current, 39.09 mA/cm^2 and FF, 79.1%.

Sanyo (Tsunomura *et al.* 2008) projects further increase in the performance of HIT cells by improving the a-Si:H/c-Si hetero junction and the grid electrodes and by reducing the absorption in the a-Si:H and TCO layers. The improving of the hetero junction will be attainable by optimization of cleaning technology of c-Si surface before a-Si:H deposition and by lower plasma and/or thermal damage to the c-Si surface during a-Si:H, TCO and conductive electrode fabrication. The best output parameters for the Sanyo HIT cell developed for industrial use are recently reported: $V_{oc} = 729$ mV, $J_{sc} = 39.5$ mA/cm^2, FF $= 0.800$ and $\eta = 23.0\%$ for a large 100.4 cm^2 cell (Taguchi *et al.* 2009). In an effort to decrease the price of fabricated modules, experiments with wafers upto 70 μm thick are carried out. A conversion efficiency of 22.8% has been achieved with a 98 μm thick Cz-Si (100.3 cm^2) cell with $P_{max} = 2.290$ W (Taguchi *et al.* 2009). At Intersolar Europe 2011, Sanyo presented two module versions of new generation HIT modules: one has a 250 W output and an efficiency of 18%; the second has a 240 W output with an efficiency of 19% (Sun & Wind Energy 7/2011).

2.4.4   Interdigitated back contact (IBC) solar cells

Back contact solar cells achieve potentially higher efficiency by putting all or part of the front contact grids on the rear of the device (Figure 2.7). This configuration reduces shading on the front of the cell, thereby resulting in higher efficiency. These designs are especially useful in high current cells such as concentrator PVs. There are three distinct configurations of back contact solar cells and IBC is one. Emitter wrap through (EWT) and Metalization wrap through (MWT) are the other two configurations of back contact solar cells discussed under polycrystalline Si solar cells.

Originally called the front surface field (FSF) cell, the interdigitated back contact (IBC) cell was first studied in the late 1970s (Lammert & Schwartz 1977; Roos *et al.* 1978; Cheng & Leung 1980). The point contact (PC) cell developed by Stanford University in the 1980s gave efficiencies of more than 20% from the beginning (Sinton *et al.* 1985, 1986). Also called BJ-BC solar cells, these back junction solar cells have a collecting junction on the rear of the solar cell whereas the front surface is passivated. The minority carriers that generate at the front surface have to diffuse all the way to back surface. As a result, these cells require a high ratio of bulk diffusion length to cell thickness.

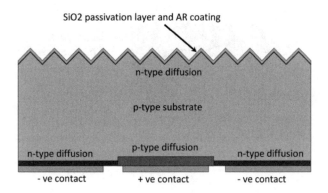

SiO2 passivation layer and AR coating

n-type diffusion

p-type substrate

n-type diffusion

p-type diffusion

n-type diffusion

- ve contact   + ve contact   - ve contact

Figure 2.7 Schematic representation of a Back contact cell with contact grids on the rear side

The IBC cell has front and rear surface passivation layers, a random-pyramid light-trapping surface, FSF, interdigitated n- and p-doped regions on the back surface, n and p contact gridlines on n- and p-doped regions, a single-layer ARC and CZ n-type mono-crystalline silicon substrate. Because the minority carriers must diffuse through the entire wafer thickness to reach the junction at the rear, the IBC cell design requires extraordinary high lifetime silicon material. SunPower Corp. which has commercialized this cell uses Si wafers with lifetimes greater than 1 milli-second and a thickness of 200 μm (Mulligan *et al.* 2004; Rogol & Conkling 2007).

Of all the crystalline Silicon PV modules in the market, those based on BC-BJ cells provide the possibility of module efficiencies >20%. Schematic drawing of an IBC cell manufactured by SunPower is shown in Figure 2.8a.

These cells were first fabricated for unmanned aircraft and solar race cars by Sun-Power in the 1990s, and were extended to large-scale production for PV generation systems in the 2000s. To reduce the fabrication cost, several low cost processes are introduced by SunPower (Verlinden *et al.* 1994; Mulligen *et al.* 2004; Cudzinovic & McIntosh 2006). The best conversion efficiency reported so far for a large-area industrial cell is 23.4% (Swanson 2008).

Several laboratories and manufactures have investigated for improving the design and processing of BC-BJ cells (Kluska *et al.* 2008; Granek *et al.* 2008; Munzer *et al.* 2008). BC-BJ cells have several advantages compared to the conventional front-contact cell structure: no gridline (sub-electrode) or busbar (main electrode) shading, a front surface with good passivation properties due to the absence of front electrodes, freedom in the design of back contacts (electrodes), and improved appearance with no front electrodes. They also provide advantages in module assembly, allowing the simultaneous interconnection of all cells on a flexible printed circuit. The low series resistance of interconnection formed by this type of technology results in a high FF of 0.800 compared with around 0.75 for standard silicon PV cell modules (Bultman *et al.* 2003). Figure 2.8b shows the schematic representation of the module structure that Nakamura *et al.* newly investigated. The IBC cell is directly mounted on the PWB or FPC and the cell electrodes are connected to the wiring lines directly. This method is called surface-mount technology (SMT). The SMT concept single-cell module made

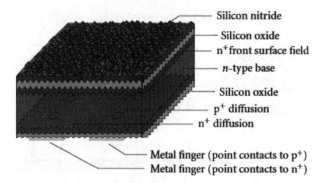

Silicon nitride
Silicon oxide
n^+ front surface field
n-type base
Silicon oxide
p^+ diffusion
n^+ diffusion
Metal finger (point contacts to p^+)
Metal finger (point contacts to n^+)

Figure 2.8a   Schematic representation of IBC solar cell
(*Source*: Neuhaus & Munzer 2007, Hindawi Publishing Corp.; Copyright © 2007 D.-H. Neuhaus & A. Munzer)

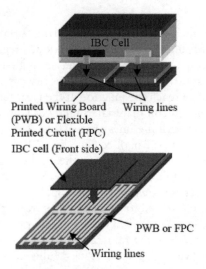

IBC Cell
Printed Wiring Board   Wiring lines
(PWB) or Flexible
Printed Circuit (FPC)
IBC cell (Front side)
PWB or FPC
Wiring lines

Figure 2.8b   The Surface mount technology concept of IBC solar cell
(Courtesy: Nakamura *et al.* 2008, reproduced with the permission of the author)

of 126 mm quasi-square IBC cell shows high FF nearly equal to the FF of the cell. The module parameters are $J_{sc} = 38.6$ mA/cm^2, $V_{oc} = 0.643$ V, FF $= 0.80$, and $\eta = 19.8\%$. Thus it appears that SMT module structure is effective to reduce the series resistance caused by the interconnection (Nakamura *et al.* 2008). Further, the studies on thickness dependence on efficiency have shown that the thickness can be reduced to 120 μm.

2.4.5   Industrial mono-Si solar cells

p-type mono crystalline Si substrates sliced from boron-doped CZ ingots are used for standard industrial PV cells for many years. In the early years of terrestrial PV cell production, small 2–5-inch-diameter CZ ingots were used. Hence, cost reduction for mono crystalline cells with small size and high cost was a difficult issue. With great deal of research and development over the past 20 years, CZ wafers with side lengths of 125 and 156 mm, sliced from 6- and 8-inch-diameter ingots, respectively, are now available for mono-silicon PV cell fabrication. The fabrication of mono crystalline cells and modules using wafers of the same size as those used for polycrystalline cell production has improved the competitiveness of mono crystalline cells against their polycrystalline counterparts in terms of manufacturing cost per output watt. Mono crystalline cells represented 38% of all solar cells manufactured in 2008 (Photon Intl. 2009).

There are large differences between the efficiencies of the best research crystalline silicon PV cells and the corresponding industrial cells. The efficiencies of standard industrial mono-Si PV cells remain in the range of 16–18%, considerably lower than the 25% efficiency levels of the best research cells. Due to economics of production, industrial cells are restricted to simple cells that are suitable for high-speed, automated production using low-cost materials. Simple design features, such as front surface texturing and BSF similar to those developed for terrestrial crystalline-silicon PV cells in the early 1980s are still adopted in most current industrial crystalline cells. To improve cell efficiencies, many cell manufactures are systematically attempting to introduce high-efficiency features, such as finer gridlines, selective emitters or more shallowly doped n^+ regions, into existing production processes. The IBC (BC-BJ) cells and HIT cells have exceptionally high efficiencies for industrial mono crystalline Si PV cells, but have complex cell structures that require a much longer production process and more specialized equipment compared with the other industrial cells. As a result, it is difficult for these advanced cell types and modules to compete commercially in terms of production cost per output watt. The challenge lies in balancing efficiency enhancement and cost reduction for solar cells and modules using existing manufacturing technologies. Innovative and simple manufacturing processes and equipment for the fabrication of high-efficiency crystalline silicon solar cells are therefore needed in order to realize significant cost reductions.

Another drawback of the mono-Si cell technologies is that mono-crystalline cells based on p-type CZ silicon substrates are susceptible to light-induced degradation (LID) caused by the recombination of reactive boron–oxygen complexes (B s–O $2i$). Many studies have been undertaken in an attempt to eliminate LID effects in mono-silicon PV cells, and permanent deactivation of the complex at high temperature ($>170°C$) has been reported (Lim $et\ al.$ 2008). Boron-doped magnetic-field CZ wafers and gallium-doped CZ wafers also show promise for eliminating LID effects in mono-Si solar cells. CZ-silicon cells based on phosphorous-doped n-type CZ wafers are also free of LID effects. The high-efficiency PV cells of Sun Power and Sanyo are fabricated using n-type CZ-silicon wafers. Recently, the Fraunhofer ISE (2009) has developed prototype n-type silicon solar cells with conversion efficiency exceeding a record 23.4%. A $2 \times 2\ cm^2$ n-type cell was demonstrated achieving 23.4% conversion efficiency; and a bigger cell ($12.5 \times 12.5\ cm^2$), using much simpler process including a screen-printing application of the aluminum alloy emitter, showed 18.2% efficiency. The research groups are working on the process technology to push the commercially-viable silicon solar cells to exceed 20% efficiency.

It is also difficult to expect further price decrease of the mono-crystalline silicon solar cells. Firstly, the power generation conversion efficiency of the material is approaching the theoretical value. Therefore, in recent years, the progress in the conversion efficiency of the solar cells available in the market has not been very significant. Secondly, the price of the materials has soared. The extent of price decreases for solar cells has been declining compared with the price decreases of its auxiliary equipments.

2.5   MULTI (POLY)CRYSTALLINE SILICON SOLAR CELLS

The study of multi-crystalline silicon ingots and wafers has started with an object of using them for solar cell fabrication since the production process of mono-silicon crystals is laborious and the costs are very high. The science and production of poly-Silicon have been studied since the mid-1970s (Lindmeyer 1976; Fischer & Pschunder 1976). The literature on the preparation of poly-Si ingots and wafers is vastly available. Tremendous progress is achieved in getting the required equipment, and modern poly-crystalline furnaces are designed for maximum productivity to cast ingots of around 450 kg.

Polycrystalline cells are currently the most widely produced cells, making up about 48% of world solar cell production in 2008 (Photon Intl. 2009). Standard poly-crystalline industrial cells offer efficiencies of 15–17%, roughly 1% lower than for mono-crystalline cells fabricated on the same production lines. The efficiencies of poly-crystalline cell modules, however, are almost the same as those for mono-crystalline cells (14%) due to the higher packing factor of the square polycrystalline cells.

Mono-crystalline cells are fabricated from pseudo-square CZ wafers and have relatively poor packing factors. The efficiencies of both mono-crystalline and poly-crystalline PV cells are likely to become better in the future through the introduction of high-efficiency structures. But, the variation in efficiency between mono- and multi-crystalline cells may become wider due to the difference in crystal quality (*i.e.* minority carrier lifetimes) resulting from high-efficiency structures. The best of the current research polycrystalline silicon cells, a PERL cell developed by Fraunhofer ISE (Schultz *et al.* 2004), provides an energy conversion efficiency of 20.3%. This PERL cell has a laser-fired contact back structure that gives a V_{oc} of as high as 664 mV. The efficiency of this polycrystalline cell, however, remains about 5% lower than that for the best research mono-crystalline PERL cells, mainly attributable to the quality difference between mono- and polycrystalline substrates. Polycrystalline substrates are subject to high electrical loss due to higher recombination rates of minority carriers, both at active grain boundaries and within crystalline grains due to high dislocation and impurity densities in comparison with FZ or CZ mono-crystalline substrates.

Extensive research activity over years has led to progress in the efficiency of poly-Si solar cells, and the recent high-efficiency solar cells have the features listed in Table 2.2 (Saga 2010). The honeycomb-structured polycrystalline solar cells demonstrated recently by Mitsubishi Electric exhibit efficiencies of over 19.3% and come in large size, 15 cm × 15 cm (Hamamoto *et al.* 2009). These polycrystalline cells with screen-printed and fired silver-paste electrodes have a distinct front honeycomb-textured surface to reduce light reflection, resulting in a high J_{sc}, 37.5 mA/cm^2. This cell also has the PERL structure of front surface passivation, rear surface passivation with local BSF, and a selective emitter for improved cell efficiency.

Table 2.2 Key Technologies for high efficiency poly-Si solar cells

(a) Minimising Photon loss:
Front textured surface by acid etching (including honeycomb texturing) or reactive-ion etching; Chemical polishing (etching) of back surface; Back surface reflector consisting of a dielectric layer and highly reflective thin metal layer.

(b) Minimising carrier loss:
Shallowly doped n^+ regions with SiN_x surface passivation layer; p-type Si substrates with minority carrier diffusion lengths longer than base thickness; Local BSF and point contacts, such as the laser fired contacts, Back surface passivation with SiO_2 and/or SiN_x layers.

(c) Minimising electrical loss:
Fine grid lines for front electrode by advanced screen printing techniques such as stencil printing; Selective emitter (deep, highly doped emitter beneath front electrode); Deposition of seed layers, then copper plating for metallization instead of screen printed silver paste.

These features are common to those of the recently produced mono-crystalline solar cells.

Several approaches to reduce cell costs including usage of thinner silicon wafers are investigated. High-efficiency (18.1%) polycrystalline silicon cells using 100 μm-thick wafers were fabricated by Sharp in 2009 (Ooka *et al.* 2009). High-efficiency polycrystalline cells with SiNx passivation layer and thin aluminum reflector on the back silicon surface display less performance degradation with decreasing substrate thickness for substrates of 100–180 μm thick. Cells with rear passivation and local BSF on 100 μm-thick substrates provide the additional advantage of less cell bowing compared with the standard aluminum alloyed BSF cells on substrates of the same thickness.

2.5.1 EWT and MWT polycrystalline Si cells

New types of back-contact polycrystalline cells, such as Emitter wrap through (EWT) cells and Metallization wrap through (MWT) cells (Figure 2.9), have been developed by ECN, Kyocera and Advent Solar (Mewe *et al.* 2009; Gee *et al.* 2008; Inoue *et al.* 2008). These back-contact cells are suitable for use with lower quality crystalline Si material including polycrystalline Si having relatively short minority carrier lifetimes (related to cell thickness). These cells have a fraction of the collector emitter on the rear side and an additional second carrier collecting junction at the front side leading to a higher current collection. In EWT solar cell, all metal contacts are moved on to the rear side and use a front side emitter for additional current collection (Hall & Soltys 1980). These cells have laser-drilled through-holes which have heavy phosphorous diffusion that can wrap through front n-electrodes and/or n-doped regions to the back surfaces. The EWT cells have a larger number of close-spaced through-holes, which direct photogenerated electrons to the back surface solely through n-doped emitters. EWT solar cells are designed similar to IBC on the cell front side and holes for connecting the front to the rear side emitter. Haverkamp *et al.* at University of Konstanz have discussed these designs, identified the challenges that arise for industrial production and suggested

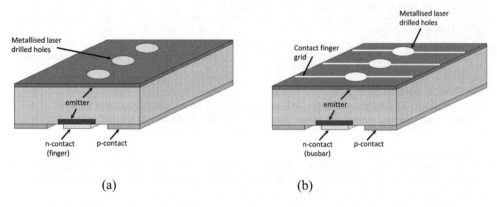

Figure 2.9 Schematic diagrams of (a) EWT and (b) MWT back-contact solar cell structures studied at University of Konstanz (Redrawn from Haverkamp *et al.*)

new technologies that could be implemented. An excellent review on EWT and MWT back contact solar cells was published by van Kerschaver and Beaucarne (2005).

Best cell efficiencies of 21.4% have been achieved on a small area, 4 cm², and FZ silicon (Glunz *et al.* 2001) using sophisticated fabrication processes. Applying industrial processes, 16.1% efficiency has been reported on 100 cm² size solar cells on CZ silicon (Kress *et al.* 2000). Hacke *et al.* of Advent Solar (2005) has discussed fabrication steps and reported production of EWT solar cells with efficiencies of 15.2% and $V_{oc} = 600$ mV, $J_{sc} = 35.4$ mA/cm², and FF = 71.3% with screen-printed Ag metallization. By improving the series resistance, the same group could achieve 15.6% efficiency (Hacke *et al.* 2006). The EWT cells produce higher photocurrents by eliminating the both busbar (main electrode) and gridline (sub-electrode) shading on the front surface. A high J_{sc} of 37.5 mA/cm² and efficiency of 17.1% were reported recently for EWT cells by Q-Cells. A new efficiency record of 19.5% (confirmed by the Fraunhofer ISE) for multi crystalline solar cells of six-inch size, 180 μm thick, was achieved by Q-Cells (Sun & Wind Energy 2011). During the process, the wafer was mirrored and passivated using functional nanolayers on the back. The back of the cell consists of dielectric layers in combination with local contacts which improves the electrical and optical characteristics.

The MWT cells require only a relatively small number of through-holes to direct photo-generated electrons to the back surface through the metal electrodes and n-doped emitters, and produce higher collection photocurrents due to absence of a bus bar (main electrode) on the front surface as in conventional cells.

A high J_{sc} of 37.3 mA/cm² and a cell efficiency of 18.3% were reported for a recent MWT cell by Kyocera (Inoue *et al.* 2008). The ECN's metal-wrap-through technology has a number of advantages over standard H-pattern cells: current gain due to reduced cell front metallization coverage, higher fill factor for larger cells due to the unit cell design, higher packing density in the module, less resistance losses in the module and less cell breakage during module manufacture as the cell is fully backcontacted. The substrate foil concept was used to reach a high packing density of cells in the module

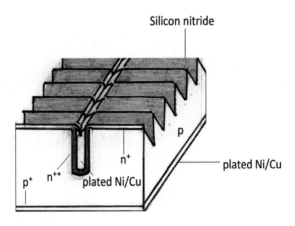

Silicon nitride

p

plated Ni/Cu

n⁺

p⁺   n⁺⁺   plated Ni/Cu

Figure 2.10 Schematic diagram of a V-textured Back Contact solar cell with selective emitter (Redrawn from Jooss *et al.* 2002)

and to minimize resistance losses. Four modules were manufactured with the advanced MWT cells in the ECN's module pilot line, all with 100% mechanical yield. The best module, made with 160 µm thin cells, showed a module efficiency of 16.4% on an aperture area of 8904 cm², as independently measured and certified by JRC-ESTI. This module efficiency for MWT cell modules by ECN is the highest reported to date (Mewe *et al.* 2009).

2.5.2 Polycrystalline buried contact solar cells

Buried contact solar cells using CZ-Si are being produced by BP Solar with efficiencies, 16–17%. Jooss *et al.* (2002) have investigated the development of poly-Si cells incorporating the advantages of buried-contact concept to reduce the cost of production. The functional process includes mechanical V-texturing for the reduction of reflection losses as well as bulk passivation by a remote hydrogen plasma source. The front surface is coated by SiN_x as single layer ARC, the contact grooves are perpendicular to the V-texture and metallisation is performed by electroless plating of Ni/Cu. The defect density in multi-crystalline Si being high, bulk passivation has been done to improve the quality of the Si material. The schematic structure is shown in Figure 2.10. Record efficiencies of 17.5% with $V_{oc} = 628$ mV, $J_{sc} = 36.3$ mA/cm² and FF = 76.8% (independently confirmed at FhG-ISE) have been obtained on Polix mc-Si on a cell area of 144 cm². The high J_{sc} is a result of low shadowing and reflection losses, high bulk diffusion lengths and a selective emitter structure. Hydrogenation was investigated for Baysix multi-Si and led to an increase in V_{oc} of 5–11 mV and in J_{sc} of 0.3–0.6 mA/cm², which were caused by an increase in the effective diffusion length of 40–50 µm. It has been shown, that hydrogenation from a PECVD SiN_x layer applied to screen printed solar cells could be more effective than remote plasma-hydrogenation in buried contact solar cells.

The target for industrial poly-Si PV cells is to realize average cell efficiencies of 17% in large-scale production (Mewe *et al.* 2009). Many approaches have been investigated

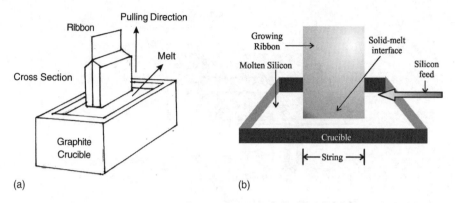

Figure 2.11   (a) Principle of EFG Process. (b) String-Ribbon process from Evergreen: The molten silicon is drawn between two parallel wires (strings) and solidifies into a continuously growing
ribbon
(*Source*: Jayarama Reddy 2010)

to improve the quality of polycrystalline substrates to match that of the more expensive
CZ mono-crystalline wafers. The dendritic casting method is one such approach that
allows the grain orientation and size to be controlled, resulting in high-quality dendritic
crystals with parallel twinning. Solar cells based on dendritic polycrystalline wafers
show efficiencies of as high as 17%, comparable to the efficiencies provided by CZ
mono-crystalline cells using the same cell fabrication process (Nakajima *et al.* 2008).

2.5.3   EFG and SRG techniques

Among several alternative growth methods proposed over the past four decades for the
production of polycrystalline silicon substrates directly from molten silicon, include
edge-defined film-fed growth (EFG), string ribbon growth (SRG), and ribbon growth
on substrate (RGS) (Ravi 1977; Sachs 1984; Lange 1990) shown in Figures 2.11(a) &
(b). These methods potentially make it possible to reduce the amount of silicon used
in PV cell fabrication. The details are discussed by the author elsewhere (Jayarama
Reddy 2010).

The EFG and SRG methods are used on industrial production scales by SCHOTT
Solar and Evergreen Solar, respectively (Seren *et al.* 2006).

These technologies have the advantages of low silicon consumption per Wp and
high cell efficiencies in comparison with the RGS method. In the improved EFG, silicon
is shaped into a hollow octagonal tube during the growth itself using a special die, with
flat polycrystalline silicon sheets forming the eight vertical sides. Each octagonal tube
is 5.3 meters long; each side is about 10 cm wide; and sheet thickness is ~300 μm.

Wafers of 10 cm × 10 cm are cut from the sheets with Nd:YAG laser. These wafers
have enabled to fabricate large area cells with efficiencies above 14% (Sarasin basic
report 2001). These methods afford inexpensive and environment-friendly wafer
production process.

Smart Solar Fab of Schott Solar produces octagonal tubes with drawing height
of 7 meters, edge length of 125 mm, with a wall thickness of 300 μm (Rosenblum

et al. 2002). Recently manufactured cells based on direct-grown substrates have almost the same efficiencies as those of standard cast-silicon polycrystalline cells. However, the smaller EFG- and SRG-based cells, which are roughly half the size of standard industrial cells, incur higher cell and module processing costs. A crystallization on dipped substrate method, which can be used to produce standard-sized wafers (156 mm × 156 mm) directly from molten silicon in a crucible, was recently proposed by Sharp (Takakura *et al.* 2008).

2.5.4 PLUTO multi-crystalline Si modules

Mass production of the new high efficiency solar cells by PLUTO technology with selective emitter has recently commenced by SunTech Power. The PLUTO solar cell technology was developed specifically for low quality and low cost multi-crystalline silicon wafers *for consistency with the industry trend* towards this type of wafer. It retains the majority of the fabrication processes in common with conventional screen-printing technology, making it relatively simple to retrofit onto existing screen-printed production lines. The Pluto technology was mainly developed to overcome the fundamental performance limitations of standard screen-printing approaches through achieving narrow metal lines (typically 30 microns wide making them only about 25% of the width of standard screen-printed lines), low metal/silicon interface area, selective emitter and well passivated surfaces. Production on 30MW production lines for the two technologies (Pluto as well as screen-printed) simultaneously shows that the Pluto technology has a 12–13% performance advantage over screen-printing technology while the production costs for the two technologies are the same per unit area. Efficiencies in the range of 17–17.5% in large scale production have been routinely achieved during the first two months of operation for standard commercial multi-crystalline silicon material, with 17.2% efficiency independently confirmed by the Fraunhoffer Institute. The PLUTO modules are believed to be the most efficient multi-crystalline silicon PV modules world-wide (Shi *et al.* 2009).

2.6 MATERIALS AND PROCESSING

The raw, high-purity poly silicon material used for the fabrication of crystalline silicon solar cells is generally prepared by the Siemens method. The market price for raw silicon is affected by the demand–supply balance for solar cell and semiconductor fabrication, and can fluctuate markedly. In 2006–2008, for example, the cost of raw silicon as a proportion of total solar cell module cost jumped from 20–30% to more than 50% due to a market shortage of silicon. Reducing the cost of silicon in a cell by reducing the substrate thickness is therefore an important aspect of achieving overall cost reductions for solar cell modules.

Wire-saw wafer slicing is one of the key production technologies for industrial crystalline silicon PV cells; and improvements in wafer slicing technology have brought down raw wafer thickness from 370 μm in 1997 to 180 μm by early 2006 for industrial polycrystalline-silicon cells fabricated by Sharp (Saga 2010). To introduce wafers thinner than 150 μm, sophisticated manufacturing processes suitable for ultrathin wafers will be needed.

The front emitter layer of crystalline Silicon PV cells is formed by phosphorus diffusion techniques in a quartz tube or belt furnace. Solid P_2O_5 or liquid $POCl_3$ is used as the phosphorus diffusion source. Phosphorous diffusion methods reduce the concentration of impurities in Si wafer by gettering resulting in improved minority carrier lifetime. This procedure has been found to be effective, provided diffusion is conducted under phosphorus supersaturation conditions (doping level above the solid solubility in silicon) (McHugo et al. 1996; Wenham & Green 1996; Lossen et al. 2005). It was also shown that a double-sided diffusion gives better efficiencies than a single-sided diffusion (Schneider et al. 2005).

The BSF layers in industrial cells are formed by alloying of screen-printed aluminum paste in a belt furnace. This process provides high productivity and relatively low process cost for BSF formation. Al-paste alloying has the additional advantage of inducing wafer gettering effects in both poly- and mono-silicon PV cells similar to the phosphorus diffusion technologies (Cheek et al. 1984; Lolgen et al. 1993). Metal impurities, such as iron or copper, can be eliminated from bulk silicon by aluminum gettering effects, which can improve the minority carrier diffusion length.

The screen printing and firing of silver paste to make contact with the bulk silicon surface by penetrating the ARC is a well-established, simple and fast process for forming front and rear electrodes. Being low-cost additionally, it is the most widely used method for forming electrodes in industrial crystalline silicon PV cells. The front gridlines are designed so as to optimize the trade-off between shadow loss and series resistance. As an alternative to the screen-printed silver paste approach, plated electrodes of layered nickel, copper and silver have been developed at UNSW for use in buried-contact cells (Wenham et al. 1988a, 1988b). Crystalline silicon PV cells with plated electrodes have excellent electrical characteristics due to their low series resistance and fine gridlines, which result in a much smaller shadow area. However, plated electrodes, which are formed by a wet process, are not yet widely used as the screen-printed silver paste electrodes. The silver used as the electrode material in crystalline silicon cells might become a critical material resource in future when the production reaches the predicted large volumes. Copper and aluminum have therefore been considered as substitutes for silver in silicon PV cells.

Silicon nitride, SiN_x:H with upto 40 at% of hydrogen is deposited onto the front side as an anti-reflection coating using PECVD. This layer not only minimizes optical losses but also serves as a good surface passivation to reduce recombination losses of the emitter (Aberle & Hezel 1997; Aberle 2000; Cuevas et al. 2003). Further, the hydrogen released from SiN:H layer during post-deposition anneal reduces bulk recombination in multi-crystalline silicon (Cuevas et al. 2003; Nagel et al. 1997).

2.6.1 Al_2O_3 deposition for surface passivation

In order to achieve simultaneously the twin object of using thinner silicon wafers and achieving higher efficiencies in the industrial Si solar cell production, the control of surface recombination losses become very important. In recent years, aluminium oxide (Al_2O_3) films deposited by various techniques such as atomic layer deposition (ALD), plasma-enhanced chemical vapour deposition (PECVD) and reactive sputtering have proven capable of providing an effective surface passivation on low-resistivity p-type and n-type silicon wafers as well as on boron- and aluminium-doped p^+-emitters

(Agostonelli *et al.* 2006; Hoex *et al.* 2008; Schmidt *et al.* 2008, 2009; Benick *et al.* 2008; Cesar *et al.* 2010; Miyajima *et al.* 2008; Saint-cast *et al.* 2009; Li & Cuevas 2009).

Schmidt *et al.* (2010) have systematically studied the passivation quality of Al_2O_3 films deposited by (a) Plasma-assisted and thermal atomic ALD which are unsuitable for industrial solar cell production due the low rates (<2 nm/min), and (b) high-rate spatial ALD, PECVD, and reactive sputtering which are shown to have a vast potential for industrial cell production.

Using spatial ALD and PECVD on 1 cm p-type Si, surface recombination velocities (SRVs) below 10 cm/s are obtained whereas Sputtered Al_2O_3 layers have provided rather high SRV of 35–70 cm/s. Despite their lower passivation quality, the sputtered Al_2O_3 layers are found to be still suitable for the fabrication of 20.1% efficient PERC (passivated emitter and rear cells) solar cells.

Schmidt *et al.* (2010) have implemented Al_2O_3 rear passivation layers deposited by plasma ALD, thermal ALD and sputtering into PERC-type solar cell. They achieved independently-confirmed efficiencies of 21.4% for cells with plasma ALD, and 20.7% for cells with thermal ALD-Al_2O_3.

2.6.2   Inkjet technology for cell fabrication

Inkjet technology has been extensively investigated in the recent years at UNSW (ARC PV Report 2008, 2009) for a range of solar cell fabrication processes that include texturing, grooving, patterning of dielectric layers for metal contacting, localized diffusions, etc. Two approaches are being studied: Resist based method and Direct etching through aerosol jetting. The techniques developed to carry out these processes are distinctively different to those used before.

In the resist based approach, a non-corrosive plasticizer is inkjet printed onto a low cost resist layer, altering the chemical properties of the resist layer in these localized regions. These restricted regions then become permeable to etchants such as hydrofluoric acid (HF). This facilitates the patterning or etching of the primary dielectric or semiconductor material to facilitate a range of semiconductor processes. The change in resist permeability, interestingly, is a reversible process making it feasible to return the resist to its original state after carrying out processes on the underlying material. The main objective at UNSW was to develop inkjet printing techniques for patterning low cost resist layers as a simple, much *cheaper* alternative to expensive photolithographic based processing so that high efficiency PERL solar cell technology could be commercialised at low cost and extend the usage to terrestrial applications. These new approaches appear capable of achieving similar device performance levels; but the greatest challenge is to achieve the dimensions of features in the resist patterning comparable with photolithography. So far, test devices have achieved feature dimensions such as holes of 30–40 μm dia. which need to be reduced to about 10 microns dia. to fully match the performance levels achieved by photolithographic based processing. New and innovative inkjet printing techniques have been recently developed for further reducing the resist patterning dimensions.

The dielectric patterning facility via inkjet printing was also experimented for the formation of textured surfaces and light trapping scheme. Figure 2.12(a) shows the

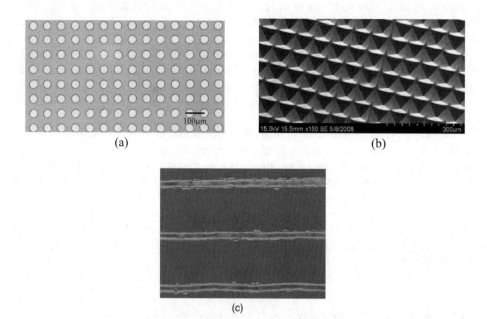

(a)   (b)

(c)

Figure 2.12 (a) Optical Microscope photograph of ink jet patterned holes in a SiO_2 layer; (b) SEM
of the inverted pyramids formed in the silicon surface following KOH etching of (a); and
(c) Optical image of grooves, 200 nm deep and $\sim$10 μm wide at Si/SiO$_2$ interface
(*Source:* ARC – UNSW PV Annual report 2009, reproduced with the permission of
Prof. Martin Green)

inkjet patterned holes in a SiO_2 layer and (b) SEM of the inverted pyramids formed in
the Si surface with KOH etching of (a).

The other method uses Optomec's M3D aerosol jetting device to deposit a solution
containing fluoride ions, based on an etching pattern, onto an acidic water-soluble
polymer layer formed over the dielectric layer (SiO_2 or SiN_x). This solution reacts
with the polymer layer, at the deposited areas, to form an etchant that etches the SiO_2
and SiN_x under the polymer layer to form a pattern of openings in the dielectric layer.
After the formation of the pattern, the polymer and the etch residue are removed by
rinsing in water. The method uses inexpensive chemicals and involves fewer steps than
photolithography and is safer than existing immersion etching techniques. The aerosol
etching method has been used to etch groove structures in SiO_2 grown thermally on
polished silicon wafers. By varying the aerosol and sheath gas flow rates the geometry
and depth of etching can be varied.

Figure 2.12(c) is an optical image of grooves etched in a 260 nm thick thermal
oxide grown on a polished Si wafer by deposition of 25 layers of aerosolised 10%
(w/v) NH_4F solution onto a polyacrylic acid polymer layer of thickness $\sim$2.2 μm. The
aerosol and sheath gas flow rates were 18 and 55 cm^3/min, respectively. The process
velocity of the stage was 10 mm/s and the platen was heated to 45°C. Etched grooves
which are 200 nm deep and $\sim$10 μm wide at the Si/SiO$_2$ interface are formed. The
current etched feature sizes are sufficiently small for use in many current front-contact

and rear-contact silicon solar cells. However, much more reduction in size is expected to apply to high efficiency solar cell technologies.

2.7   FUTURE FOR CRYSTALLINE SILICON SOLAR CELLS

Industrial solar cells module must reach a price level of \$1/Wp with a total system price level of \$2/Wp to reach grid parity. To become competitive with thermal (coal) or nuclear power generation, the Cells need to be mass produced at a total system cost of less than \$1/Wp. Achieving even a module price of \$1/Wp will require modules to be produced at a cost of less than 0.7\$/Wp. Although such low costs remain very challenging for crystalline silicon solar modules, cost reductions to such a level are considered to be possible based on the technologies presented above. The cost reduction must be achieved while public incentives for PV systems remain in effect.

The annual production volume for all kinds of solar cells is expected to exceed 100 GWp/year by around 2020. Crystalline silicon cell modules have a long history of proven field operation and offer high efficiencies while presenting fewer resource issues than many competing technologies. As such, crystalline silicon PV cells are expected to be strongly represented in the future solar cell market.

To reach these future price levels, new technologies listed in Table 2.3 (Saga 2010) need to be developed and used for the fabrication of c-Si solar cells and modules. New technologies to break through the efficiency barrier of 25% for c-silicon PV cells are being studied by many researchers and institutes around the world, but any practical improvements in cell efficiency are yet to be realised. The peak theoretical efficiency in a crystalline silicon solar cell based on a single homo junction and a bulk silicon energy band gap of 1.1 eV is 30% under 1 sun $AM_{1.5}$ illumination. To break through this ideal efficiency limit based on existing Schockley and Queisser solar cell theory, novel technologies based on quantum dot (QD) and quantum well structures have been proposed and studied by many researchers.

Multi-junction designs have been attempted in many forms for improving solar cell efficiency beyond that of a single-junction cell. For example, a triple-junction solar cell with a silicon bottom cell is expected to give efficiencies of more than 40%. Researchers at the UNSW have also proposed a silicon-based tandem junction solar cell incorporating silicon QD technology (Green et al. 2008). An effective bandgap of up to 1.7 eV has been demonstrated for 2 nm-diameter silicon QDs embedded in SiO_2 (Park et al. 2008).

Photon management, such as up- and down-conversion and plasmonic effects, are other potential approaches that could add extra efficiency based on existing high-efficiency silicon cells (Schweizer et al. 2008; Mataki et al. 2009; Schaadt et al. 2005). These technologies aim at shifting the photon energy of sunlight to match the sensitivity of the solar cell by adding special optical features (e.g. a fluorescent coating layer including rare-earth elements for up-and down-conversion) to the front and/or rear surface of the cells without modifying the structure of the solar cell itself. These aspects are further discussed in chapter 6.

Developments in light trapping architectures applied to thin-film (5 micron) silicon solar cells are recently investigated by Mutitu et al. (2009). These structures have the potential of reducing costs as well as finding applications in emerging areas such

Table 2.3 New Technologies for high-efficiency c-Si solar cells and modules

Novel technologies to break through efficiency barrier of 25%;
Wafer slicing technologies and equipment for ultrathin (50 μm) wafers;
Direct slicing technologies and equipment for ultrathin wafers without kerf loss;
Production technologies and equipment for solar cells and modules based on ultrathin wafers;
High-quality polycrystalline ingot technologies providing performance comparable to mono-Si cells;
Low-cost contact forming technologies and materials to replace screen-printed and fired silver paste;
Low concentration (around 10×) and high efficiency module technologies to minimize total
  PV system cost.

as BIPV. The new designs implement photonic bandgap engineering and diffractive optical concepts to increase the optical path length of light within the active solar cell region. To this end, a diffraction grating integrated one dimensional photonic crystal dielectric stack is incorporated to the back surface of silicon solar cell design structures. A metallic layer is added to the base of the stack to increase the reflective capacity of the integrated structures. The combination of a dielectric and metallic structure serves to mitigate material losses that arise due to the metal and hence increase the overall reflective properties of the integrated back structure. The increased reflectivity combined with diffractive optical engineering enhances the light trapping ability of the new design structures. Electromagnetic modeling and optimization tools such as the scattering matrix method and the particle swarm optimization algorithm are employed in the analysis and realization of the design structures. Simulation results show short circuit current characteristic improvements of over 10% in the new structures when compared to corresponding structures without the extra metallic layers.

The high-efficiency technologies, however, generally incur higher production costs compared to standard silicon cells. Cell and module manufacturing technologies that satisfy both high efficiency and low cost will be essential for industrial production in the near future. The impacts of novel technologies such as QDs and photon management will be interesting to watch as research and development on crystalline silicon solar cells continues. These aspects are further discussed in chapter 6.

2.8   AMORPHOUS SILICON SOLAR CELLS

Amorphous silicon technology is the oldest and best established thin-film silicon technology. Thin film solar cells based on hydrogenated amorphous silicon (a-Si:H) technology can potentially lead to low cost through lower material costs than conventional modules, and also do not suffer from some critical drawbacks of other thin-film technologies, such as limited supply of basic materials or toxicity of the components.

Hydrogenated amorphous silicon (a-Si:H) differs from crystalline silicon by the lack of long-range lattice order and the high content of bonded hydrogen (typically around 10% in device-quality a-Si:H).

2.8.1   Overview of technology development

The first experimental a-Si:H solar cell was made at RCA Laboratory in 1970s (Carlson & Wronski 1976). This single junction *p*-i-n a-Si:H solar cell deposited on a glass substrate coated with TCO and aluminium back contact exhibited 2.4% conversion efficiency. To increase the output voltage of a-Si:H solar cells, stacked (also called multi-junction) solar cell structure was introduced (Hamakawa *et al.* 1979). A key step to industrial production was the development of a monolithically integrated type of a-Si:H solar cell (Kuwano *et al.* 1980). Using the monolithic series integration of a-Si:H solar sub-cells, a desired output voltage from a single substrate can be easily achieved. In 1980 the integrated type a-Si:H solar cells were commercialised by Sanyo and Fuji Electric for small applications, such as in calculators and watches, so on.

In 1980s, hectic research activity in the field of a-Si:H solar cell was devoted to developing and optimising a-Si:H based alloys. A p-type hydrogenated a-silicon carbide (a-SiC:H) was implemented as a low-absorbing layer, usually denoted as a window layer (Tawada *et al.* 1981); and a hydrogenated a-Silicon Germanium (a-SiGe:H) alloy became an attractive low band gap material for stacked solar cells (Nakamura *et al.* 1980). Surface-textured substrates for optical absorption enhancement were introduced (Deckman *et al.* 1983). The laboratory cells reached an initial efficiency of 11 to 12%. Commercial single junction a-Si:H modules with efficiencies up to 5% were available by the end of 1980s, with annual production capacity of a-Si:H cells and modules reaching about 15 MWp.

In 1990s, research and manufacturing effort was aimed at achieving 10% stabilized module efficiency and a high throughput process. Several companies optimised and implemented an a-SiGe:H alloy in tandem [BP Solar (Arya *et al.* 2002; Sanyo (Okamoto *et al.* 2001; Fuji Electric (Ichikawa *et al.* 2001)] and triple-junction [United Solar (Guha *et al.* 2000a)] solar cell structures. The annual total production capacity for TF Si single- and multi-junction modules reached around 30 MWp at the end of 20th century.

Hydrogenated microcrystalline silicon (μc-Si) emerged in this period as a new candidate for the low band gap material in multi junction a-Si:H based solar cells. Low temperature PECVD was used to deposit these films. The University of Neuchâtel introduced a micromorph tandem solar cell in 1994, which comprised an a-Si:H top cell and a μc-Si:H bottom cell (Meier *et al.* 1994).

A promising potential of the micromorph cell concept was soon demonstrated by fabricating micromorph tandem and triple solar cells with stabilized efficiencies in the range of 11 to 12% (Meier *et al.* 1998; Yamamoto *et al.* 2000) and Kaneka Corporation soon started the development of micromorph module production (Yamamoto *et al.* 2000). The introduction and implementation of μc-Si:H in TF Si solar cells focused the attention on higher deposition rates. Several new deposition methods (Schropp & Zeman 1998) have been investigated for fabrication of solar cells at high deposition rates (0.5 to 1.0 nm/s), such as very high frequency (VHF) and microwave PECVD, hot wire CVD, and expanding thermal plasma CVD. Of these, VHF and hot wire CVD techniques have potential for future high throughput production for PV. The a-Si and μc-Silicon based typical cell structures are now briefly discussed.

2.8.2 a-Si:H thin film solar cells

The research and development of a-Si solar cells broadly involve three areas: (1) technology development for improving conversion efficiency, (2) development of low-cost manufacturing process, and (3) technology development for evaluating the

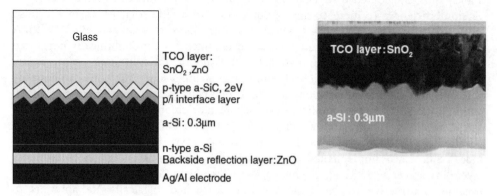

Figure 2.13 Schematic structure and cross-sectional TEM image of single junction a-Si solar cell
(*Source*: Konagai 2011, Copyright © 2011 The Japan Society of Applied Physics, reproduced
with permission)

performance of solar cells. Area (1) can be further divided into development of technology for manufacturing high-quality a-Si films, for suppressing light induced degradation, and the device technology for increasing conversion efficiency. The area (2) can be classified into the development of high rate deposition technology and technology for manufacturing large-area a-Si solar cell modules (Konagai 2011). Many areas of research are, thus, coordinated, and the results integrated to achieve reliable high efficiency and low cost a-silicon based solar modules of long durability.

Generally, a-Si films are fabricated by RF plasma CVD at 13.56 MHz. However, VHF plasma CVD at about 60 MHz is considered effective for high rate deposition and has already reached the stage of practical application. Several novel deposition methods are also investigated, e.g., hot filament assisted CVD (hot wire-CVD) technique was used at Utrecht University, and the expanding thermal plasma CVD (ETP-CVD) technique was developed at Eindhoven University of Technology. To increase the quality of films, it is important to control the proportion of SiH_3 and atomic hydrogen (H) on the growing surface.

There are two types of a-Si solar cell structures: the single junction structure and the multi junction (double- and triple-junction) structures. The single junction a-Si:H solar cell structure consists of three layers, a p-type a-SiC:H layer, an intrinsic a-Si:H layer, and an n-type a-Si:H layer, forming p-i-n junction (Figure 2.13). The doped layers are usually very thin, a p-type a-SiC:H layer is ~10 nm thick and an n-type a-Si:H is ~20 nm thick. The active layer in the a-Si:H solar cell is the intrinsic layer, typically 300 to 500 nm thick, which is sandwiched between the doped layers so that an internal electric field is present across the intrinsic layer. The electron-hole pairs that are generated in the intrinsic a-Si:H layer immediately experience the internal electric field that separates electrons and holes from each other. The separated carriers drift under the influence of the internal electric field towards the doped layers (electrons towards the n-type layer and holes towards the p-type layer) and are collected by the electrodes. Since the dominant transport mechanism of the photo-generated carriers is drift in the internal electric field, an a-Si:H solar cell is called a *drift device*.

The thickness of the a-Si:H solar cell is just equal to the thickness of the depletion region in the crystalline silicon solar cell, i.e. about 0.5 μm, and this small thickness is sufficient for absorption of the visible part of solar spectrum. The small thickness implies a large reduction in material as well as in energy consumption during production compared to crystalline silicon solar cells. Furthermore, when deposited on a light substrate such as a flexible foil the weight of a-Si:H solar modules is strongly decreased becoming appropriate for many applications, specifically in space and building-integrated applications.

The absence of long-range order in a-Si:H causes a relaxation in the transition rules; hence, it behaves as a direct gap semiconductor with a band gap around 1.75 eV. This feature of a-Si:H also enables to absorb larger part of visible solar spectrum with much less material. Active (intrinsic) layer thicknesses below 0.5 μm are common in a-Si:H solar cell fabrication, whereas nearly 400 times thick layers (~200 μm) are required for crystalline silicon PV cells.

Before comparing the performance of crystalline silicon and amorphous silicon cells, it is necessary to understand the main differences in the band gap of these materials. While in case of crystalline silicon, almost no density of states is present in the band states, a significant density of states can be observed in the bandgap of a-Si:H. The presence of defects in a-Si:H is attributed two mechanisms: (i) the lack of order in the material causes a broadening of the valence and conduction bands leading to band tails, where localized energy states exist, and (ii) unpassivated dangling bonds create defect states near the centre of the band gap.

Amorphous silicon PV technology has several advantages: (a) low deposition temperatures (typically 200–300°C), which permit the use of low-cost substrates; (b) easy integration of the modules into facades, roofs, and other structures due to use of flexible substrates; (c) relatively less energy and less material consumption in the fabrication of modules; (d) abundance of all raw materials involved; and (e) potential for environment-friendly large-scale manufacturing operation.

But, the main disadvantage noticed in early stage is degradation of its performance under illumination, called Staebler-Wronski effect (Staebler & Wronski 1977). The SW effect has been found to be due to the creation of new defects (dangling bonds) that act as additional recombination centers. Subsequent annealing at 100–250°C could restore the original efficiency. Due to this effect, the electivity of the solar cell usually decreases by 15–35%. The presence of the light-induced degradation can be overcome by combining thin individual cells into a tandem and triple junction cells. Thinner p-i-n cells suffer less from collection problems, even if the defect density is increased by the SW effect.

2.8.3 Microcrystalline (μc)-Silicon thin film solar cells

μc-Si is a material containing an amorphous-Si phase and a polycrystalline Si phase. The band gap of μc-Si is around 1.2 eV (equivalent to that of mono-Si), and its absorption coefficient is higher than that of mono-Si due to the presence of amorphous component whose fraction decides the magnitude of the bandgap. At IMT Neuchatel (Feitknecht *et al.* 2003), μc-Si:H solar cell in the n-i-p configuration was fabricated using VHF PECVD deposition process. The fabrication of substrate n-i-p solar cells is substantially different from that of superstrate p-i-n solar cells, because of the change in the deposition order of the layers and its technological consequences (like initial n- or p-type nucleation layer and optically transparent and doped window-layers). The interface of the back TCO to the n-type Si film (back-interface) is very vital for the deposition of n-i-p-type cells because it can influence the structure and performance of the entire cell. Most importantly, nucleation of silicon grains and electronic and optical properties are key issues which depend on the nature of the back TCO material (Fujiwara et al. 2003; Bailat et al. 2001). The study of the interface of the back TCO/n-layer of the n-i-p solar cell is also not simple. Since most of the light is absorbed within the first 2 µm of the absorber, only the longer wavelengths of the solar spectrum reach the back interface under white-light illumination. The external quantum efficiency (EQE) measurement performed from both sides of the cell (double-sided illumination), only provides better characterization of the cell. The schematic of the cell structure with a surface area of 0.3 cm^2 and thickness, 2.5 µm is shown in Figure 2.14a. The cell exhibited an AM 1.5 conversion efficiency exceeding 9% with $V_{oc} = 520$ mV, FF $= 73\%$, and $J_{sc} = 24.2$ mA/cm^2.

Figure 2.14b shows the cell characteristics and EQE. This remarkable performance was possible due to (i) fine-tuning of the silane concentration in hydrogen feedstock gas used for deposition of the intrinsic absorber layer, (ii) incorporation of anoptimized back reflecting substrate into the cell after investigating four different back-contacts; and (iii) the ideal combination of each of these key components. Compared to earlier results with n-i-p-type µc-Si:H solar cells, a substantial increase in V_{oc} was obtained, while maintaining reasonable J_{sc} values. Earlier investigations on the role of i-layer material had revealed a trade-off between cells with high J_{sc} but low V_{oc}, or cells of low J_{sc} and high V_{oc}.

The present study shows the successful combination of a cell with an acceptable V_{oc} and good J_{sc} generation in the wavelength region above 700 nm. This is mainly due to suitable light-diffusing back-reflectors whose optical and electrical performance is excellent (Feitknecht et al. 2001). For industrially relevant solar cell design, the introduction of highly scattering back-reflectors is necessary. The only drawback of this cell is the low deposition rate of 2.6 Å/sec mainly due to rather low VHF-plasma frequency of 70 MHz. Nonetheless, this is a hopeful result of a n-i-p solar cell approaching 10% conversion efficiency.

High-performance µc- and a-silicon solar cells are vital for a successful combination to form the 'micromorph' tandem cell (Feitknecht et al. 2001; Meier et al. 2002).

Ever since Meier and colleagues reported high conversion efficiency for an a-Si/µc-Si tandem solar cell (Fischer et al. 1996), the focus of the research on bottom cell materials used in tandem solar cells has changed from a-SiGe to µc-Si (for example, Yamamoto et al. 1998). Along with µc-Si layer in solar cells, a light absorbing layer with a thickness of 1–3 µm is required to allow sufficient amount of photons to be absorbed because its absorption ability is low in the visible part of the solar spectrum. This value is approximately 10-fold greater than the layer thickness in the regular a-Si solar cells; thus, there is need for the development of high rate deposition technology (deposition rate of 1–5 nm/s) to increase the throughput of the fabrication process.

In addition, to enable even thin absorbing layers to absorb maximum photons, light confinement technologies using textured TCO layers, and a photonic structure to allow the reflection of light with a specific wavelength are considered key development

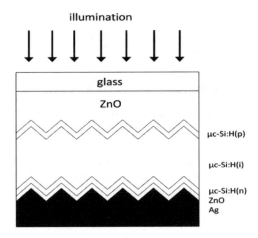

Figure 2.14a Schematic view of the layer structure of μc-Si:H thin film solar module developed by IMT, University of Neuchatel (Redrawn from Feitknecht et al. 2003)

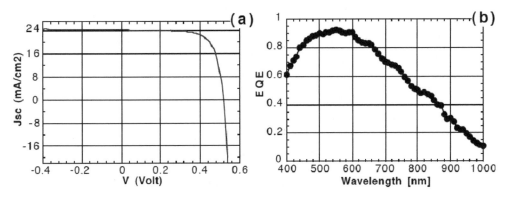

Figure 2.14b The best μc-Si:H solar cell in the n-i-p configuration fabricated at IMT, Neuchatel (Source: Feitknecht et al. 2003, Copyright © 2003 Materials Research Society. Reprinted with the permission of Cambridge University Press)

areas. During the μc-Si film formation by plasma CVD, the film structure is determined by the ratio of SiH_3 molecules impinging on the growing surface to atomic hydrogen. As the fraction of atomic H increases, crystallization starts occurring, and finally in the μc-Si film, each microcrystal is surrounded by a-Si tissue. The relative fraction of the crystalline and amorphous phases can be obtained by Raman scattering studies. A μc-Si film with a crystallization rate of ~50–60% is used as i-layer in μc-Si solar cells.

A p-i-n-type structure similar to that of a-Si single-junction solar cell is used because the electric field in i-layer assists to accumulate photo-generated carriers as in the case of a-Si solar cells. To rapidly prepare i-layer of required thickness (1–3 μm), VHF plasma CVD at ~60 MHz is normally used. The high-pressure depletion method (Konagai 2011) is effective for fabricating high-quality μc-Si films using VHF plasma CVD and, in particular, a phase modulation method is being developed to create uniform plasma over a large area (Goya *et al.* 2003). From the perspective of high rate deposition, the work of Matsui's group who achieved conversion efficiency of 8.2% with deposition rate of 2.1 nm/s has received attention (Matsui *et al.* 2003).

2.8.4 Tandem and multi-junction a-Si:H cells

The multi-junction solar cell has thickness similar to a conventional single junction solar cell, but each component (sub) cell is thinner and therefore less sensitive to light-induced defects. In a multi-junction cell structure, each component cell can be adjusted to a specific range of the solar spectrum, thereby better utilizing the solar spectrum and increasing the solar cell conversion efficiency. Intrinsic a-Si:H has an optical band gap of $\sim$1.70 eV and absorbs only photons with energies above 1.70 eV efficiently. Part of the photons with energies lower than 1.70 eV can be absorbed by a-SiGe:H layer that has an optical band gap of 1.45 eV. By varying the content of germanium in a-SiGe:H, its optical band gap can be lowered to achieve tuning of the response to the solar spectrum. This type of realization of materials with different optical band gaps for the active layers in multi-junction solar cells is called *multi-band gap* approach. Most of current high efficient solar cells are based on the multi junction and multi-band gap approach. For successful operation of the multi junction solar cell, the current generated at the maximum power point has to be equal in each component cell, and a *tunnel-recombination junction* (TRJ) between the component cells has to feature low electrical and optical losses. Tandem solar cells typically have a-Si/a-SiGe structure. a-Si/a-SiGe/a-SiGe triple-junction tandem solar cells are also being developed to further increase the conversion efficiency.

An a-Si:H/a-SiGe:H tandem cell fabricated at Delft University of Technology is presented in Figure 2.15. The a-Si:H component cell absorbs photons with energies above 1.70 eV, and photons with lower energies, which pass through the a-Si:H top cell, are absorbed in the bottom cell, a-SiGe:H. The observed parameters of the Cell are $V_{oc} = 1.49$ V, $J_{sc} = 8.74$ mA/cm^2, FF = 0.67, and efficiency, 8.7%. The stabilized conversion efficiencies of the above a-Si solar cells are 9 to 10% for small-area single-junction cell.

The highest stabilised efficiency for a-Si:H based solar cells to date has been demonstrated by USSC with a triple junction structure. This solar cell structure is shown in Figure 2.16. The complexity of the solar cell structure is clear from the graded layers and the utilization of different materials for the active layers. The top cell uses a-Si alloy with a band gap of $\sim$1.8 eV for the i-layer and captures blue photons. a-SiGe alloy with about 10–15% Ge, is the i-layer for the middle cell with an optical gap of $\sim$1.6 eV and is well suited for absorbing green photons. The bottom cell uses an i-layer of a-SiGe alloy with about 40–50% Ge and captures the red and infrared photons corresponding to optical gap of $\sim$1.4 eV. Light that is not absorbed gets reflected from the back reflector (silver/zinc oxide) which is textured to enable light trapping. The cells are interconnected by the heavily doped layers that form tunnel junctions between adjacent cells (Guha 2005).

USSC has reported an initial efficiency of 14.6% and the stabilized efficiency of 13.0% for a small area cell (0.25 cm^2). Later, an initial efficiency of 15.2% was reported signifying the viability of further improvement in stabilised efficiency.

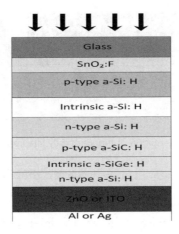

Figure 2.15 Schematic structure of an a-Si:H/a-SiGe:H tandem cell fabricated by Delft University

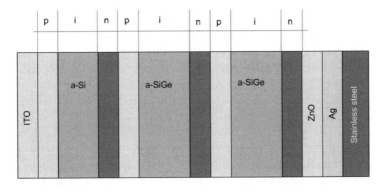

Figure 2.16 Schematic of Triple junction cell demonstrated by USSC

Comparing the external parameters of the single junction cell ($J_{sc} = 15.7 \, mA/cm^2$, $V_{oc} = 0.74 \, V$, FF $= 0.64$, $\eta = 6.3\%$) with tandem cell ($J_{sc} = 8.74 \, mA/cm^2$, $V_{oc} = 1.49$, FF $= 0.67$, $\eta = 8.7\%$), it is noticed that the J_{sc} is lower while the Voc is higher in case of a tandem cell. This observation is even more pronounced for triple-junction cells. For practical applications, the external parameters of multi-junction solar cells are advantageous in comparison to single junction cells, because the lower current means less loss in the electrodes, especially in the TCO, and the higher output voltage allows more flexible design of a module with required voltage.

2.8.5   Fabrication of a-Si thin film modules

The development of the monolithically integrated type a-Si:H solar cell by Kuwano *et al.* (1980) was the beginning for the industrial production of a-Si:H solar cells/modules. In this process, several subcells deposited on one substrate are connected in series as shown in Figure 2.17.

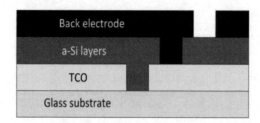

Figure 2.17 Representation of Monolithically integrated a-Si PV modules

The integration of the subcells can be obtained by using conventional masking techniques or by including two or more laser scribing steps between depositions of the respective layers. The laser cutting selectively removes narrow lines (width 50 to 150 μm) of material to limit the area losses. The laser scribing is also a speed limiting step in the production line. Pulsed Nd:YAG lasers are generally used for scribing the films. Encapsulation is done to protect the product against the atmospheric impacts to extend its lifetime. As a-Si:H solar modules are becoming an attractive component for building applications, the framing is done very carefully.

Using the monolithic integration of a-Si:H solar cells, a desired output voltage from a single substrate can be easily achieved. The process steps for the integration of the subcells can be simply implemented in the whole fabrication process of a module. This gives a-Si:H technology a great advantage over mono-silicon module technology, where the subcells are mechanically connected with each other.

The first commercial a-Si:H solar cells, typically a few square centimeters in size, were introduced to the market in 1980. Since then the manufacturing technology of a-Si:H solar modules has advanced rapidly in terms of module size and efficiency. Several mass-production lines are in operation such as Kaneka in Japan, BP Solarex and USSC in USA that manufacture large area a-Si:H modules. Several companies work on the basis of a pilot line such as Phototronics in Europe, Fuji Electric and Sanyo in Japan.

Presently, glass-plates or stainless steel sheets are used as the substrate materials. The choice of the substrate material influences the production technology, though the principal steps of the production process remain the same for both. In case of glass substrates the p-i-n deposition sequence is used, in case of stainless steel the n-i-p sequence process is followed. The PECVD reactor is vital to the production units because its design determines throughput, flexibility to cell design variations, quality and uniformity of the layers and defect density levels.

There are several configurations of the PECVD reactors that are used for production: single chamber system, multi-chamber system (in-line configuration and cluster configuration), and Roll to roll system.

A *continuous roll-to-roll (R2R) process* has been developed by Energy Conversion devices, Inc. of USA. In this approach, a roll of stainless steel substrate is unrolled and fed into the manufacturing process which involves four continuous procedures. It is then rolled back and cut into different sizes suitable to different applications. Stainless steel rolls, 2500 m (mile and a half) long, 36 cm (14 inches) wide, and 125 μm (5 mil) thick move in a continuous manner in four chambers to complete the solar cell fabrication. At present, this process is used by USSC for depositing tandem and triple junction solar cells explained in Section 2.8.4. The process details are given elsewhere (Guha 2005). It is potentially a cheap method, but there are some disadvantages. For instance, if the substrate is made of stainless steel, then the series interconnections cannot be done monolithically; so, the stainless steel merely serves as a carrier for an insulating plastic substrate. The main element of the line is the machine for the a-Si alloy deposition, consisting of a stainless steel web roll-off chamber, six/nine PECVD chambers for six/nine layers of the double/triple junction cell and a roll-up chamber. Stainless steel web, coated with Ag/ZnO, moves continuously through the chambers where the various layers are deposited sequentially (Al back reflector instead of Ag is used in manufacturing to reduce cost). There is an increased risk that cross contamination occurs, since all deposition zones by definition are connected to each other. Since no purging steps can be used as in the single chamber systems, a type of gas gates is used to isolate the adjacent chambers and minimise the adverse effects. The advantages of R2R production are: production is light-weight and flexible; high production yield; can be cut into sizes depending on the required output power.

2.8.6   a-Si/μc-Si tandem solar cell

The micromorph tandem solar cell, composed of an a-Si:H top cell and a μc-Si:H bottom cell, is one of the most promising multijunction candidates for high stabilized efficiency thin film silicon solar cells (Meier et al. 2002a). In fact, the combination of a high a-Si:H band gap (1.7 eV) and a low μc-Si:H band gap (1.2 eV) results in an almost ideal tandem device offering better performance with top cell absorbing photons in the visible and the bottom cell in the red and infra red regions (Meillaud et al. 2006). These merits enabled the launching of the a-Si/μc-Si tandem solar cell production, and the focus of R&D has shifted to a-Si/μc-Si tandem solar cells. At IMT of Neuchatel where the development first started (Meier et al. 1996) stable conversion efficiencies up to 11.7% for 1 cm^2 micromorph cells were demonstrated (Yoshimi et al. 2003). There have been many references on estimating the conversion efficiency of solar cells; and a module conversion efficiency of about 15–16% is expected from a practical viewpoint.

The challenge, however, is to achieve ideal J_{sc} matching between the two subcells because the J_{sc} is limited by the lowest J_{sc} of the subcells in multijunction solar cells. In micromorph tandem cells, the light is weakly absorbed in the top cell (a-Si cell) and the rest reaches the bottom cell (μc-Si cell) where it is eventually absorbed. The light-induced degradation (SW effect) forbids the use of thick a-Si:H solar cells. Therefore, in tandem micromorph cells, the limitation in J_{sc} comes usually from the top cell. One widely accepted and adopted solution in the superstrate configuration (p-i-n) is to introduce a thin intermediate reflector (TCO layer) (Fischer et al. 1996) which enhances the J_{sc} of the top cell without the need for increasing its absorber layer thickness. The role of the TCO interlayer is to reflect part of the incident light towards the top cell by utilizing the difference in refractive index at the junction and to increase the photocurrent fraction from the top cell. The layer is usually thin, between 50 and 150 nm, deposited in situ (Buehlmann et al. 2007; Yamamoto et al. 2005) or ex situ (Domine et al. 2006a). In situ silicon oxide intermediate reflector (SOIR) (Domine et al. 2008), silicon nitride (Vineri et al. 2006) or ex situ zinc oxide intermediate reflector (Domine et al. 2006b) has been reported to be very effective which already implemented (Yamamoto et al. 2004).

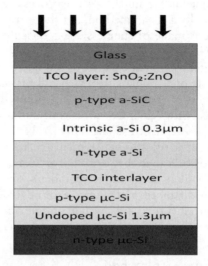

Figure 2.18 Schematic diagram of new type of tandem solar cell with TCO interlayer inserted between top and bottom cells developed by SHARP Corp.

In the substrate configuration (n-i-p), the problem of obtaining high efficiency micromorph tandems with elevated stabilized J_{sc} in the top cell still exists (Soderstrom *et al.* 2010). So far, this problem of current matching is avoided by designing triple junction solar cells which split the Jsc into the three subcells, thus allowing the use of thin a-Si:H top cells (Yang *et al.* 1997). Nonetheless, this strategy demands the implementation of a more complex process, e.g., one more cell and profiling of Ge (Zimmer *et al.* 1998; Guha *et al.* 1998) in the a-SiGe:H middle cells.

Soderstrom *et al.* have recently fabricated tandem micromorph cells on glass and flexible plastic (polyethylene-naphthalate, PEN) substrates and studied the limitation of J_{sc} for the tandem micromorph solar cell in the initial and stabilized state, with and without intermediate layer (Soderstrom *et al.* 2009, 2010). Their results show that an 'asymmetric' intermediate reflector (AIR) is more effective for n-i-p/n-i-p tandem micromorph solar cells in improving Jsc in the top cell. This is because AIR selectively scatters the blue-green light into the top cell whereas the back reflector applied to the substrate scatters the red light into the bottom cell. Such a structure can reach matched J_{sc} in the stabilized state. This research group has reported stabilized efficiencies of 10.1% and 9.8% for the cells fabricated on glass and flexible plastic substrates respectively.

Currently, the highest conversion efficiency has been achieved by a-Si/μc-Si tandem solar cell (Figure 2.18), in which a TCO interlayer is inserted in the tunnel junction between the top and bottom cells. With this structure, the highest conversion efficiency (14.7%) is achieved on small-area cells in the initial stage (Yamamoto *et al.* 2003).

Several excellent research papers appeared on the relevant issues and the reader may refer to them for more details; for example, Bailat *et al.* (2005), Meier *et al.* (2000, 2002, 2003), Mai *et al.* (2005), Yang *et al.* (2008), Soderstrom *et al.* (2008), Huag *et al.* (2009), Yan *et al.* (2007), Green *et al.* (2006) so on.

Table 2.4 Current status of silicon solar cells

	c-Si	μc-Si:H	a-Si:H	a-Si/μc-Si
Lab. record	24.7% (mono) 20.3% (multi)	10.1%	9.5%	11.7%
Module record	22.7% (mono) 15.3% (multi)	8.2%	–	10.4%
Module commercial	12–19%	6–7%	4–7%	–

Source: Green et al 2008 & Roerden et al 2008

Table 2.5 TF a-Si and μc-Si based solar modules presented at Intersolar Europe in June 2011

Producer	Country	Material	Efficiency (%)	Nominal Power (W)	Max. system Voltage (V)
Sharp	Japan	a-Si/μc-Si	9.6	135	1000
Masdar PV	Germany	a.Si/μc-Si	9.4	135	1000
Du Pont Apollo	China	a-Si/μc-Si	9.4	145	1000
Kaneka	Japan	a-Si/μc-Si	9.1	150	1000
GS-Solar	China	a-Si/a-Si/a-SiGe	7.6	60	1000
Schott Solar	Germany	a-Si/a-Si	7.4	107	1000
Moncada Energy	Italy	a-Si single	7.3	420	1000
Sungen	China	a-Si single	6.5	100	1000/600
Xunlight	USA	a-Si/a-SiGe/a-SiGe	6.2	291	1000/600
Senersun	China	a-Si single	5.8	90	1000

Source: Sun & Wind Energy, July 2011, p. 120

2.8.7 Turnkey systems for a-Si solar cell module production

Intense effort has been devoted to achieve the manufacture of Si thin-film solar cell modules using turnkey machines. Some manufacturers have started selling turnkey machines assuring certain level of performance, the dominant being Applied Materials, Inc. and Oerlikon. Recently, Sunfilm AG has released the latest data on their module fabricated using a turnkey machine supplied by Applied Materials, Inc. The dimensions of the substrate used in the module production are $2.2\,m \times 2.6\,m$ ($\sim5.7\,m^2$). a-Si/μc-Si tandem solar cell module was formed on this single-plate substrate achieving $V_{oc} = 285\,V$, $I_{sc} = 2.64\,A$, FF = 62.6%, module conversion efficiency (conversion efficiency per unit exposed module area) = 8.5% and output = 472 W (Stein 2009). The manufacturing cost is expected to markedly decrease with such a large module of $5.7\,m^2$ achieving a conversion efficiency exceeding 10%.

Table 2.4 is a summary of the present laboratory record for stabilized single layer a-Si:H cell efficiency, and the actual commercial modules' stabilized efficiencies (Green et al. 2008; Roedern, et al. 2008). The maximum efficiency for micromorph is 11.7%. Increased efficiency up to 13% was presented in tandem and triple solar cells (Green et al. 2008).

At Intersolar Europe (2011), several manufacturers have presented their new a-Si and μc-Si based thin film modules. The manufacturers of the modules and their features are given in Table 2.5.

2.9 IMPROVING EFFICIENCY OF Si TF SOLAR CELLS

In the Japan's Roadmap for the development of PV systems, PV2030+ (NEDO), the module conversion efficiencies of Si thin-film solar cells are aimed to reach 14% by 2017 and 18% by 2025. To attain these goals, development of triple-junction solar cells becomes crucial because, theoretically, it is very difficult to attain these efficiencies using conventional double-junction solar cells. However, it may become possible if the light induced degradation of a-Si can be completely suppressed. Therefore, technology development for suppressing light induced degradation is very vital in achieving long-term targets.

Figure 2.19 shows the technological developments necessary to realize triple-junction solar cells where both increased conversion efficiency and reduced manufacturing cost must be achieved concurrently. Device technologies to markedly reduce the manufacturing cost are also shown in the figure.

The technological issues, mentioned in detail in the figure, are classified into three groups: (1) material development and performance improvement, (2) the improvement of manufacturing processes, and (3) module related issues.

The research and development effort and the strategies required to achieve these goals is enormous because the issues are very wide ranging.

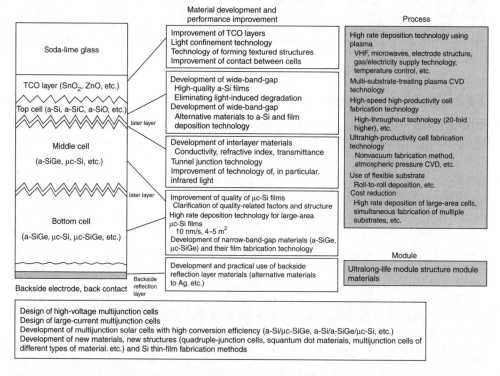

Figure 2.19 Issues in the development of triple-junction solar cells
(*Source*: Konagai 2011, Copyright © 2011 The Japan Society of Applied Physics, reproduced with permission)

Si thin-film triple-junction solar cells

Last few years have seen spectacular R & D effort in this area. The Si thin-film based triple-junction solar cell structures under development include a-Si/μc-Si/μc-Si and a-Si/a-SiGe/μc-SiGe structures as well as a-Si/a-SiGe/μc-Si structure.

United Solar has fabricated and studied the spectrum sensitivity characteristics of a-Si/a-SiGe/μc-Si triple-junction cell. The cell conversion efficiency is 15.4% with $V_{oc} = 2.239\,V$, $I_{sc} = 9.13\,mA/cm^2$, $FF = 0.753$ on a cell area of $0.25\,cm^2$ (Yan et al. 2008). A total photocurrent of $27.9\,mA/cm^2$ is observed indicating the bottom cell (μc-Si cell) has adequately covered the long wavelength region. An improvement in the light absorption is the main issue to be addressed in the realization of triple-junction solar cells that can reach the maximum theoretically possible performance.

United Solar (Xu et al. 2009) has recently reported the fabrication of five different types of a-SiGe:H and μc-Si:H based multi-junction solar cell structures (two doublejunction structures, a-Si:H/a-SiGe:H and a-Si:H/μc-Si:H, and three triple-junction structures, a-Si:H/a-Si:H/ a-SiGe:H, a-Si:H/a-SiGe:H/a-SiGe:H, and a-Si:H/μc-Si:H/μc-Si:H stacks, all deposited on back reflector (Ag/ZnO or Al/ZnO) coated stainless steel substrate) using modified Very High Frequency (MVHF) technology. After optimization, all five structures reached similar initial cell performance, ~12% small active-area ($0.25\,cm^2$) efficiency and 10.6–10.8% large aperture-area ($\geq 400\,cm^2$) efficiency after encapsulation. However, they showed quite different light soaking stability behaviour, which can be attributed to the degradation of component cells. A comparative study between the MVHF deposited solar cells with those deposited by RF was made. Materials studies were also conducted to understand the mechanism responsible for better stability for the MVHF deposited a-SiGe:H solar cells. The best stable efficiency achieved for the large-area encapsulated cells is approaching 10% for both a-SiGe:H and μc-Si:H based multi-junction cells.

Kolodziej and group (Kolodziej et al. 2003), in their strategy of improving the efficiency of thin film Si solar cells, have accomplished a superior method of producing the stable a-Si:H. This group has confirmed that an optimal H-dilution, microstructure, lower thickness and such other operations reduce the light-induced degradation (SWE)

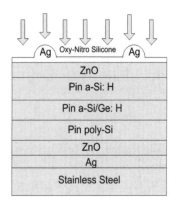

Figure 2.20 Schematic diagram of triple-junction Si TF solar cell (redrawn from Kolodziej et al. 2003)

effect and increase solar cell stability (Kolodziej *et al.* 2000a, 2000b, 2001, 2002a, 2002b). Detailed studies on the tandem structures prepared on glass as well as stainless steel substrates have led the group to fabricate triple junction cell samples on stainless steel of 5×5 cm^2 area. These novel cells consist of three sandwich-layers that include a polycrystalline p-i-n Si structure. The sequence of the layers from the back are: Al/ZnO mirror, polycrystalline p-i-n Si structure prepared at higher temperature of $\sim$450°C, the amorphous p-i-n Si (0.5–1.0%) Ge:H structure, and the p-i-n a-Si:H structure with ZnO electrode and polysilicone protection (Figure 2.20). For the 5×5 cm^2 PV structures, the parameters reported are: $V_{oc} = \sim$2.19 V, $J_{sc} = \sim$6 mA/cm^2, FF $= \sim$0.58 and $\eta = \sim$7.7% at 1 sun luminance.

REFERENCES

Aberle, A.G., & Hezel, R. (1997): Progress in low-temperature surface passivation of silicon solar cells using remote-plasma silicon nitride, *Prog. Photovoltaics*, 5, 29–50.

Aberle, A.G. (2000): Surface passivation of crystalline silicon solar cells: a review, *Prog. Photovoltaics*, 8, 473–487.

Aberle, A.G. (2007): (Ed.), Special Issue 'Recent Advances in Solar Cells', *Advances in OptoElectronics*, vol. 2007, Hindawi Publishing Corp., Egypt.

Aberle, A.G. (2009): Thin Film Solar Cells, *Thin Solid Films*, **517**, 4706–4710.

Agostinelli, G., Delabie, A., Vitanov, P., Alexieva, Z., Dekkers, H.F.W., De Wolf, S., & Beaucarne, G. (2006): Very low surface recombination velocities on p-type silicon wafers passivated with a dielectric with fixed negative charge, *Sol. Energy Mater. Sol. Cells*, **90**, 3438–3443.

Arya, R.R. and Carlson, D.E. (2002): Amorphous silicon module manufacturing at BP Solar, *Prog. Photovolt: Res. and Appl.*, **10**, 69–76.

Bailat, J., Vallat-Sauvain, E., Feitknecht, L., Droz, C. & Shah, A. (2001): Influence of substrate on the microstructure of Microcrystalline layers and cells, Presented at 19th ICAMS 2001, August 2001, Nice, France, Published in *J. Non-Crystalline Solids*, (2002) **299–302**, 1219–1223.

Bailat, J., Terrazzoni-Daudrix, V., Guillet, J., Freitas, F., Niquille, X., Shah, A., Ballif, C., Scharf, T., Morf, R., Hansen, A., Fischer, D., Ziegler, Y., & Closset, A. (2005): Recent development of solar cells on low-cost plastic substrates. *Proc. 20th EU PV Solar Energy Conf*, June 2005, Barcelona, Spain, pp. 1529.

Benick, J., Hoex, B., van des Sanden, M.C.M., Kessels, W.M.M. & Glunz, S.W. (2008): High efficiency *n*-type Si solar cells on Al$_2$O$_3$-passivated boron emitters, *Appl. Phys. Lett.* **92**, 253504–253507.

Beuhlmann, P., Bailat, J., Domine, D., Billet, A., Meillaud, F., Feltrin, A., & Ballif, C. (2007): In situ silicon oxide based intermediate reflector for thin-film silicon micromorph solar cells, *Appl. Phys. Lett*, **91**, 143505.

Bulgarian Academy of Sciences (2010): Buried Contact Technology for Solar Cell, 02 June 2010.

Bultman, J.H., Eikelboom, E.W., Kinderman, R., Tip, A.C., Weeber, A.W., van den Nieuwenhof, M.A., Schoofs, C.F., Schuurmans, F.M., & Tool, C.J.J. (2003): *Proc. 3rd WC on PV Energy Conversion*, May 2003, Okasa, Japan, pp. 979.

Carlson, D.E. and C.R. Wronski (1976): Amorphous Silicon solar cells, *Appl. Phys. Lett.* **28**, 671–673.

Cervantes, D.M. (2008): Silicon heterojunction solar cells obtained by Hot-Wire CVD. Barcelona: *Barcelona University of polytechnica de catalunya, Dissertation thesis*.

Cesar, I., Granneman, E., Vermont, P., Tois, E., Manshanden, P., Geerligs, L.J., Bende, E.E., Burgers, A.R. Mewe, A.A., Komatsu, Y. & Weeber, A.W. (2010): *Proc. 35th IEEE PV Specialists Conf*, June 2010, Honolulu, HI.

Chapin, D.M., Fuller, C.S. & Pearson, G.L. (1954): A new silicon p-n junction photocell for converting solar radiation into electrical power, *J. Appl. Phys.* **25**, 676–677.

Cheek, G.C., Mertens, R.P., van Overstraeten, R., & Frisson, L. (1984): Thin film metallization for Solar cell Applications, *IEEE Trans. Electron Dev.* **31**, 602–609.

Cheng, L.J., & Leung, D.C. (1980): *Proc. 14th IEEE PV Specialists Conf*, January 1980, San Diego, CA, pp. 72.

Conibeer, G., Green, M.A., Cho, E.C., Konig, D., Huang, S., Song, D., Hao, X., Perez-Wurfl, I., Gentle, A., Gao, F., Park, S., Flynn, C., So, Y., Zhang, B., Di, D., Campbell, P., Huang, Y. & Puzzer, T. (2009): *Proc. European Materials Research Society Spring Meet.*, BP2-10.

Cudzinovic, M. and McIntosh, K. (2002): Process simplications to the Pegasus solar cell – Sunpower's high efficiency bifacial silicon solar cell, *Proc. 29th IEEE PV Specialists Conference*, May 2002, New Orleans, La, pp. 70–73.

Cuevas, A., Kerr, M.J. & Schmidt, J. (2003): Passivation of crystalline silicon using silicon nitride, *Proc. 3rdWC on PV Energy Conversion*, May 2003, Osaka, Japan, vol. 1, pp. 913–918.

Deckman, H.W., Wronski, C.R., Witzke, H. & Yablonovitch, E. (1983): Optically enhanced amorphous Silicon solar cells, *Appl. Phys. Lett*, **42**(11), 968–970.

Domine, D., Bailat, J., Steinhauser, J., Shah, A., & Ballif, C. (2006): Micromorph solar cell optimization using a ZnO layer as intermediate reflector, *Proc. 4th WC PV Energy Conversion*, May 2006, Hawaii, HI, pp. 1465–1468.

Domine, D., Buehlmann, P., Bailat, J., Billet, A., Feltrin, A., & Ballif, C. (2008): Optical management in high-efficiency thinfilm Si micromorph solar cells with a silicon oxide based intermediate reflector, *Physica Status Solidi (RRL)*, **2**, 163–165.

Duerickx, F. & Szlufcik, J. (2002): Defect passivation of industrial multi-crystalline solar cells based on PECVD silicon nitride, *Sol. Energy Mater. Sol. Cells*, S72, 231–246.

Egan, R.J., Young, T.L., Evans, R., Schubert, U., Keevers, M., Basore, P.A., Wenham, S.R., & Green, M.A. (2006): Silicon deposition optimization for peak efficiency of csg modules, *Proc. 21st EU PV Solar Energy Conf*, September 2006, Dresden, Germany, 2CV.3.33, pp. 874–876.

Feitknecht, L., Kluth, O., Ziegler, Y., Niquille, X., Torres, P., Meier, J., Wyrsch, N. & Shah, A. (2001): Microcrystalline NIP Solar Cells Deposited at 10 A/sec by VHF-GD, in Solar Energy Materials and Solar Cells, *Sol. Energy Mater. & Sol. Cells*, **66**, 397–403.

Feitknecht, L., Droz, C., Bailat, J., Niquille, X., Guillet, J. & Shah, A. (2003): Towards microcrystalline silicon n-i-p solar cells with 10% conversion efficiency, MRS Proceedings, **762**, Symposium A–Amorphous and nanocrystalline Silicon-based films – 2003: A13.5 (6 pages), Copyright © 2003 Materials Research Society. Reprinted with the permission of Cambridge University Press.

Fischer, H. and Pschunder, W. (1976): Low cost solar cells based on large area unconventional silicon, *Proc. 12th IEEE PV Specialists Conf*, October 1976, Baton Rouge, IEEE Press, Piscataway, pp. 86.

Fischer, D., Dubail, S., Selvan, J.A.A., Vaucher, N.P., Platz, R., Hof, C., Kroll, U., Meier, J., Torres, P., Keppner, H., Wyrsch, N., Goetz, M., Shah, A. & Ufert, K.D. (1996): *Proc. 25th IEEE PV Specialists Conf*, pp. 1053.

Fujiwara, H., Kondo, M. & Matsuda, A. (2003): Interface-layer formation in microcrystalline Si:H growth on ZnO substrates studied by real-time spectroscopic ellipsometry and infrared spectroscopy, *J. Appl. Phys.*, **93**(5), 2400–2409.

Gee, J.M., Hacke, P., Hilali, M., & Jimeno, J.C. (2008): The use of test structures for characterization of back-contact Si solar cells, *Proc. 33rd IEEE PV Specialists Conf*. May 2008, San Diego, CA, pp. 1–5.

Glunz, S.W., Rein, S., Knobloch, J., Wettling, W. & Abe, T. (1999): Comparison of boron- and gallium-doped p-type Czochralski silicon for photovoltaic application, *Prog. Photovoltaics*, 7, 463–469.

Glunz, S.W., Dicker, J., Kray, D. *et al.* (2001): High efficiency cell structures for medium quality silicon, *Proc. 17th EU PV Solar Energy Conf*, Oct. 2001, Munich, Germany, p. 1287.

Goetzberger, A. (1998): *Crystalline Silicon Solar cells*, New York, John Wileys.

Gordon, I., Van Gestel, D., Qiu, Y., Venkatachalam, S., Beaucarne, G., & Poortman, J. (2008): Thin-Film Polycrystalline-Silicon Solar Cells Based on a Seed Layer Approach, *Proc. 23rd EU PV Solar energy Conf*, September 2008, Valencia, Spain, pp. 2053–2056.

Goya, S., Nakano, Y., & Yamashita, N. (2003): *Proc. 3rd WC PV Energy Conversion*, May 2003, Okasa, Japan, 5O-A6-02.

Granek, F., Hermle, M., Reichel, C., Schultz-Wittmann, O., & Glunz, S.W. (2008): High-Efficiency Back-Contact Back-Junction Silicon Solar Cell Research at Fraunhofer ISE, *Proc. 23rd EU PV Solar Energy Conf*, September 2008, Valencia, Spain, pp. 991–995.

Green, M.A. (1982): *Solar cells, Operating principles, Technology, and System Applications*, Englewood Cliffs, N.J, Prentice Hall, Inc.

Green, M.A., Blakers, A.W., Wenham, S.R. & Narayanan, S. (1985): Improvements in silicon solar cell efficiency, *Proc. 18th IEEE PV Specialists Conf*, October 1985, Las Vegas, Nev, pp. 39–42.

Green, M.A. (1995): *Silicon Solar Cells: Advanced Principles and Practices*, Center for PV Devices and Systems, UNSW, Bridge Printery, Sydney.

Green, M.A. (2000): The future of crystalline Si solar cells, *Prog. Photovolt: Res. Appl.* 8(1), 127–139.

Green, M.A. (2001): Crystalline Silicon solar Cells, Ch 4 Green ed MA2.doc (available at p139_chap4Green2001.pdf)

Green, M.A. (2003): *Third Generation PV: Advanced Solar Energy Conversion*, Berlin, Springer.

Green, M.A., Cho, E.C., Huang, Y., Pink, E., Trupke, T. & Lin, A. (2005): *Proc. 20th EU PV Solar Energy Conf*, June 2005, Barcelona, Spain, p. 3.

Green, M.A., Emery, K., King, D.L., Hishikawa, Y. & Warta, W. (2006): Solar efficiency Tables (Version 28), *Prog. in Photovolt*, 14, 455–461.

Green, M.A., Emery, K., Hishikawa, Y., & Warta, W. (2008): Solar cell efficiency Tables (Version 31), *Prog. Photovolt*, 16, 61–67.

Green, M.A., Conibeer, G., Perez-Wurfl, I., Huang, S.J., Konig, D., Song, D., Gentle, A., Hao, X.J., Park, S.W., Gao, F., So, Y.H., & Huang, Y. (2008): Progress with silicon-based tandem cells using group IV quantum dots in a dielectric matrix, *Proc. 23rd EU PV Solar Energy Conf*, September 2008, Valencia, Spain, p. 1.

Green, M.A., Emery K., Hishikawa Y., & Warta W. (2009): Solar cell Efficiency Tables (Version 33), *Prog. Photovolt*, 17, 85–94.

Guha, S., Yang, J., Pawlikiewicz, A., Glatfelter, T., Ross, R., & Ovshinsky, S.R. (1989): Bandgap Profiling for Improving the Efficiency of Amorphous Silicon Alloy Solar Cells, *Appl. Phys. Lett.* 54, 2330–2332.

Guha S., Yang, J., Williamson, D.L., Lubianiker, Y., Cohen, J.D., & Mahan, A.H. (1999): Structural, defect, and device behavior ofhydrogenated amorphous Si near and above the onset ofmicrocrystallinity, *Appl. Phys. Lett.* 74, 1860–1862.

Guha, S., Yang, J. & Banerjee, A. (2000a): Amorphous Silicon Alloy Photovoltaic Research— Present and Future, *Prog. Photovolt: Res. and Appl.* 8, 141–150.

Guha, S. (2000b): Multi junction Solar cells and modules, In: Street, R.A (ed.) *Technology and Applications of a-Si*, Berlin:Springer, pp. 252–305.

Guha, S. (2005): Can your Roof provide your electrical needs? – The growth prospect of BIPV, *31st IEEE PV Specialists Conf*, January 2005, Orlando, Fl, pp. 12–16.

Hacke, P., Gee, J.M., Summer, M.W., Salami, J., & Honsberg, C. (2005): Application of a Boron-source diffusion barrier for the fabrication of back contact silicon solar cells, *Proc. 31st IEEE PV Specialists Conf*, January 2005, Orlando, Fla, pp. 1181–1184.

Hacke, P., Gee, J.M., Hilali, M. *et al.* (2006): Current status of technologies for industrial emitter wrap-through solar cells, *Proc. 21st EU PV Solar Energy Conf*, September 2006, Dresden, Germany, pp. 276.

Hahn, G., Seren, S., Kaes, M., Schoenecker, A., Kalejs, J., Dube, C., Grenko, A., & Belouet, C. (2006): Review on Ribbon Silicon Techniques for Cost Reduction in PV, *Proc. 4th W C on PV Energy Conversion*, Waikoloa, HI, May 2006, pp. 972–975.

Hall, R.N. and Soltys, T.J. (1980): Polka dot solar cell, *Proc. IEEE PV Specialists Conf*, January 1980, San Diego, CA, pp. 550–553.

Hamakawa, Y., Okamoto, H., & Nitta, Y. (1979): A New Type of Amorphous Silicon Photovoltaic Cell Generating More than 2.0 V, *Appl. Phys. Lett.* **35**(2), 187–189.

Hamamoto, S., Ishihara, T., Sato, T., Fujikawa, M., Katsura, T., Morikawa, H., Matsuno, S., Fujioka, H., & Arimoto, S. (2009): Investigation for 19% efficiency at multi-c Si solar cells by industrially probable approach, *24th EU PV Solar Energy Conf*, September 2009, Hamburg, Germany, pp. 1410–1413.

Hartley, O.N., Russel, R., Heasman, K.C., Mason, N.B. & Bruton, T.M. (2002): *Proc. 29th IEEE PV Specialists Conf*, May 2002, New Orleans, La., pp. 118–121.

Haug, F.J., Söderström, T., Python, M., Terrazzoni-Daudrix, V., Niquille, X., & Ballif, C. (2009): Development of micromorph tandem soalr cells on flexible low cost plastic substrates, *Sol. Energy Mater. and Sol. Cells*, **93**, 884–887.

Haverkamp, H., Knauss, H., Rueland, E., Fath, P., Jooss, W., Klenk, M., Marckmann, C., Weber, L., Nussbaumer, H., & Burkhardt, H. (2006): Advancements in the Development of Back contact cell manufacturing Processes, *21st EU PV Solar energy Conf*, September 2006, Dresden, Germany.

Henley, F., Kang, S., Liu, Z., Tian, L., Wang, J., & Cho, Y.L. (2009): Kerf-free 20–150 μm c-Si Wafering for thin PV manufacturing, *Proc. 24th EU PV Solar Energy Conf*, September 2009, Hamburg, Germany, pp. 886–890.

Hoex, B., Schmidt, J., Pohl, P., van de Sanden, M.C.M. & Kessels, W.M.M. (2008): Silicon surface passivation by atomic layer deposited Al_2O_3, *J. Appl. Phys.* **104**, 044903–044914.

Hoornstra, J., & Heurtault, B. (2009): Stencil print applications and progress for crystalline Si solar cells, *Proc. 24th EUPV Solar Energy Conf*, September 2009, Hamburg, Germany, pp. 989–992.

Ichikawa, Y., Yoshida, T., Hama, T., Sakai, H. & Harashima, K. (2001): Production technology for amorphous silicon-based flexible solar cells, *Sol. Energy Mater. Sol. Cells*, **66**, 107–115.

Inoue, S., Sakamoto, T., Komoda, M., Ohwada, H., Fukui, K., & Shirasawa, K. (2008): High efficiency multi-c Silicon B-C solar cells, *Proc. 23rd EU PV Solar Energy Conf*, September 2008, Valencia, Spain, pp. 988–990.

International Standard, IEC 60904-3, Edition 2, 2008, "Photovoltaic devices – Part 3: Measurement principles for terrestrial photovoltaic (PV) solar devices with reference spectral irradiance data", ISBN 2-8318-9705-X, International Electrotechnical Commission, April 2008.

Jayarama Reddy, P. (2010): *Science and Technology of Photovoltaics*, 2nd edition, Leiden, The Netherlands, CRC Press, ISBN 13: 978-0-415-57363-4

Jensen, N., Hausner, R., Bergmann, R., Werner, J., & Rau, U. (2002): Optimalization and Charakterization of Amorphous/Crystalline Silicon Heterojunction Solar Cells, *Prog. Photovolt: Res. and Appl*, **10**, 1–13.

Jooss, W., Fath, P., Bucher, E., Roberts, S., & Bruton, T.M. (2002): Large area Multicrystalline Silicon Buried Contact Solar cells with bulk Passivation and an efficiency of 17.5%,

Proc. 29th IEEE PV Specialists Conference, May 2002, New Orleans, La. Available at 29PVSC_105_9Buried contact-multiSi.pdf.

Kamins, E. (2010): Semiconductor nanowires for Solar cells, at ISETC-2010-Oct20-Ted_Kamins.pdf.

Kanno, H., Ide, D., Tsunomura, Y., Taira, S., Baba, T., Yoshimine, Y., Taguchi, M., Kinoshita, T., & Sakata, H. (2008): Over 22% Efficient Hit Solar Cells, *Proc. 23rd EU PV Solar Energy Conf*, September 2008, Valencia, Spain, pp. 1136–1139.

Kanama, D. and Kawamoto, H. (2008): R & D trends of solar Cell for high efficiency, *Quaterly Review*, No.28, July 2008, available at STTQr 2804 R&Dfor highefficiencysolarcells2008.pdf.

Kayes, B.M., Atwater, H.A., & Lewis, N.S. (2005): Comparison of the device physics principles of planar and radial p-n junction nano-rod solar cells, *J. Appl. Phys*, 97 (11), 114302.

Kluska, S., Granek, F., Hermle, M. & Glunz, S.W. (2008): Loss Analysis of High-Efficiency Back-Contact Back-Junction Silicon Solar Cell, *Proc. 23rd EU PV Solar Energy Conf*, September 2008, Valencia, Spain, pp. 1590–1595.

Kolodziej, A.P. Krewniak, and S. Nowak (2000a): Consideration on wide bandgap amorphous and microcrystalline silicon for solar multijunction cell and x-ray sensor, *Proc. 24th Intl. Conf. IMAPS – Poland, Rytro* 25–29 September,Kraków, pp. 335–342.

Kolodziej, A., Wroñski, C.R., Krewniak, P. & Nowak, S. (2000b): Silicon thin film multijunction solar cells, *Opto-Electron. Rev*, 8, 71–77.

Kolodziej, A., Krewniak, P. & Nowak, S. (2001): Effectiveness of the n$^+$/i and p$^+$/i a-Si:H junctions applied in photovoltaic structures, *Proc. 25th Intl. Conf. and Exhibition IMAPS – Poland*, Polañczyk 26–29 September, Rzeszów, pp. 115–118.

Kolodziej, A., Krewniak, P. & Nowak, S. (2002a): Efficiency optimization ofmicrocrystalline and amorphous silicon solar Cells, *Proc. 26th Int. Conf. of International Microelectronics and Packaging Society*, Warsaw, Poland, 25–27 Sept., pp. 75–78.

Kolodziej, A. & P. Krewniak (2002b): Improvements of silicon thin film solar cells, *Proc. 6th Framework Prog. Workshop*, Ispra, pp. 156–162.

Kolodziej, A., Krewniak, P. & Nowak, S. (2003): Improvement in Si thin film solar cell efficiency, *Opto-Electron. Rev.* 11(4), 281–289.

Konagai, M. (2011): Present Status and future Prospects of Silicon thin film solar cells, *Japanese J. of Appl. Phys.*, 50, 030001 [Online] March 20, 2011.

Kress, A., Tolle, R., Bruton, T., Fath, P., & Bucher, E. (2000): $10 \times 10 \, cm^2$ screen printed back contact cell with a selective emitter, *Proc. 28th IEEE PV Specialists Conf.*, September, 2000, Anchorage, Alaska, pp. 213–216.

Kuwano, Y., Imai, T., Ohnishi, M. & Nakano, S. (1980): *Proc. 14th IEEE PV Specialist Conf*, p. 1402.

Lammert, M.D. and Schwartz, R.J. (1977): The Interdigitated Back contact solar cell: A silicon solar cell for use in Concentrated solarlight, *IEEE Transactions on Electron Devices*, 24, 337–342.

Lange, H. and I. Schwirtlich (1990): Ribbon growth on substrate (RGS) – A new -15-approach to high speed growth of silicon ribbons for photovoltaics, *J. Cryst. Growth*, 104, 108–112.

Lim, B., Hermann, S., Bothe, K., Schmidt, J., & Brende, R. (2008): Permanent Deactivation of the Boron-Oxygen Recombination Center in Silicon Solar Cells, *Proc. 23rd EU PV Solar Energy Conference*, September 2008, Valencia, Spain, pp. 1018–1022.

Lindmayer, J. (1976): Semi-crystalline Silicon Solar Cells, *Proc. 12th IEEE PV Specialists Conference*, Baton Rouge, IEEE Press, Piscataway, p. 82.

Lolgen, P., Leguijt, C., Eikelboom, J.A., Steeman, R.A., Sinke, W.C., Verhoef, L.A., Alkemade, P. & Algra, E. (1993): Aluminum back-surface field doping profiles with surface recombination velocities below 200 cm/s, *Proc. 23rd IEEE PV Specialists Conference*, September 2008, Valencia, Spain, p. 236.

Lossen, J., Mittelstadt, L., Dauwe, S., Lauer, K., & Beneking, C. (2005): Making use of silicon wafers with low lifetimes by adequate POCl3 diffusion, *Proc. 20th EU PV Solar Energy Conference*, June 2005, Barcelona, Spain, p. 1141.

Lu, S., Lingley, Z., Asano, T., Harris, D., Barwicz, T., Guha, S., & Madhukar, A. (2009): Photocurrent induced by nonradiative energy transfer from nanocrystal quantum dots to adjacent silicon nanowire conducting channels: Toward a new solar cell paradigm, *Nano Lett.* **9**, 4548–4552.

Mai, Y., Klein, S., Carius, R., Wolff, J., Lambertz, A., Finger, F., & Geng, X. (2005): Microcrystalline silicon solar cells deposited at high rates, *J. Appl. Phys*, **97**, 114913.

Mason, N., & Jordon, D. (1991): A high efficiency silicon solar cell production technology, *Proc. 10th EU PV Energy Conf*, April 1991, Lisbon, Portugal, pp. 280–283.

Mason, N., Bruton, T., Gledhill, S. *et al.* (2004): The selection and performance of monocrystalline silicon substrates for commercially viable 20% efficient lid-free solar cells, *Proc. 19th EU PV Solar Energy Conf*, June 2004, Paris, France, p. 620.

Mason, N., Artigao, A., Banda, P., Bueno, R., Fernandez, J.M., Morilla, C., & Russel, R. (2004): The Technology and performance of the latest generation Buried contact solar cell manufactured in BP Solar's Tres Canto's facility, *Proc. 19th EU PV Solar Energy Conf*, June 2004, Paris, France, pp. 2653–2655.

Mason, N., Schultz, O., Russel, R., Glunz, S.W. & Warta, W. (2006): 20.1% efficient large area cell on 140 micron thin silicon wafer, *Proc. 21st EU PV Energy Conference*, September 2006, Dresden, Germany, p. 521.

Mataki, H., Padmaperuma, A.B., Kundu, S.N., McGinniss, V.D., Risser, S.M., Nippa, D.W., & Burrows, P.E. (2009): *Proc. 34th IEEE PV Specialists Conf*, June 2009, Philadelphia, PA, p. 600.

Matsui, T., Kondo, M., & Matsuda, A. (2003): Origin of the Improved Performance of High-Deposition-Rate Microcrystalline Silicon Solar Cells by High-Pressure Glow Discharge, *Japan J. Appl. Phys.* **42**, L901–L903.

Maydell, K., Korte, L., Laades, A., Stangl, R., Conrad, E., Lange, F., & Schmidt, M. (2006): Characterization and optimization of the interface quality in amorphous/crystalline silicon heterojunction solar cells, *J. of Non-Crystalline Solids*, **352**, 1958–1961.

McHugo, S.A., Hieslmair, H. & Weber, E.R. (1997): Gettering of metallic impurities in Photovoltaic Silicon, *Appl. Phys.* **A 64**, 127–137.

McIntosh, K., Cudzinovic, M., Smith, D., Mulligan, W. & Swanson, R. (2003): The choice of silicon wafer for the production of low-cost rear-contact solar cells, *Proc. 3rd WC on PV Energy Conversion*, May 2003, Osaka, Japan, vol. 1, pp. 971–974.

Meemongkolkiat, V., Hilali, M., Nakayashiki, K. & Rohatgi, A. (2004): Process andmaterial dependence of Al-BSF in crystalline Si solar cells, *Tech. digest of 14th PV Solar energy Conf*, Bangkok, pp. 401–402.

Meemongkolkiat, V., Nakayashkiki, K., Kim, D.S., Kim, S., Shaik, A. & Rohatgi, A. (2006): Investigation of screen – printing Al paste for local back surface field formation, Proc. 4th WC on PV Energy Conversion, May 2006, Waikoloa, HI, pp. 1338–1341.

Meemongkolkiat, V. (2008): Doctoral thesis, Development of high efficiency mono-Si solar cells through improved optical and electrical confinement, submitted to *Georgia Institute of Technology*, December 2008.

Meier, J., Dubail, S., Flückiger, R., Fischer, D., Keppner, H. & Shah, A. (1994): *Proc. 1st WC on PV Energy Conversion*, December 1994, Honolulu, HI, pp. 409–412.

Meier, J., Torres, P., Platz, R., & Shah, A. (1996): On the way towards high efficiency TF Silicon solar cells by the micromorph concept, MRS Symposium Proc, Spring meeting, San Fransisco, pp. 3–14.

Meier, J., Dubail, S., Cuperus, J., Kroll, U., Platz, R., Torres, P., Anna Selvan, J.A., Pernet, P., Beck, N., Pellaton Vaucher, N., Hof, Ch, Fischer, D., Keppner, H., & Shah, A. (1998): Recent Progress in Micromorph Solar Cells, *J. Non-Cryst. Solids* **227–230**, 1250–1256.

Meier, J., Dubail, S., Golay, S., Kroll, U., Faÿ, S., Vallat-Sauvain, E., Feitknecht, L., Dubail, J. & Shah, A. (2002): Microcrystalline silicon and the impact on micromorph tandem solar cells, *Sol. Energy Mater. & Sol. Cells*, **74**, 457–467.

Meier, J., Spitznagel J., Faÿ, S., Bucher, C., Graf, U., Kroll, U., Dubail, S., Shah, A. (2002a): Enhanced light trapping for micromorph tandem solar cells by LP-CVD. *Proc. 29th IEEE PV Specialists Conf*, May 2002, New Orleans, La, pp. 1118.

Meier, J., Spitznagel, J., Kroll, U., Bucher, C., Fay, S., Moriarty, T., Shah, A. (2003): High-efficiency amorphous and micromorph silicon solar cells, *Proc. 3rd WC PV Energy Conf*, May 2003, Okasa, Japan, pp. 2801–2805.

Meier, J., Sitznagel, J., Kroll, U., Bucher, C., Fay, S., Moriarty, T. & Shah, A. (2004): Potential of amorphous and microcrystalline silicon solar cells, *Thin Solid Films*, **451–452**, 518.

Meillaud, F., Shah, A., Droz, C., Vallat-Sauvain, E., & Miazza, C. (2006): Efficiency limits for single-junction and tandem solar cells, *Sol. Energy Mater. and Sol. Cells*, **90**, 2952–2959.

Mewe, A.A., Lamers, M.W., Bennett, I.J., Koppes, M., Romijn, I.G., & Weeber, A.W. (2009): Reaching 16.4% module efficiency with Back-contacted multi-Si solar cells, *Proc. 24th EU PV Solar Energy Conf*, September 2009, Hamburg, Germany, pp. 946–949.

Miroslav, Mikolesek (2009): Current Status and progress in the New Generation's Si based solar cells, at http://www.posterus.sk/?p=1247 (Accessed on 2nd May 2011).

Miyajima, S., Irikawa, J., Yamada, A. & Konagai, M. (2008): *Proc. 23rd EUPV Solar Energy Conf*, September 2008, Valencia, Spain, p. 1029.

Mulligan, W.P., Rose, D.H., Cudzinovic, M.J. *et al.* (2004): Manufacture of Solar cells with 21% efficiency, *Proc. 19th EU PV Solar Energy Conf*, June 2004, Paris, France, p. 387.

Munzer, K.A., King, R.R., Schlosser, R.E. *et al.* (1995): Manufacturing of back surface field for industrial application, *Proc. 13th EU PV Solar Energy Conf*, Nice, France, October 1995, p. 1398.

Munzer, K.A., Froitzheim, A., Schlosser, R.E., Tolle, R. & Wintsel, M.G. (2006): *Proc. 21st EU PV Solar Energy Conf*, September, 2006, Dresden, Germany, pp. 538.

Munzer, K.A., Winstel, M.G., Krause, A., & Schlosser, R.E. (2008): *Proc. 23rd EU PV Solar Energy Conf*, September 2008, Valencia, Spain, pp. 1875.

Mutitu, J., Shi, S., Barnett, A., Honsberg, C., & Prather, D. (2009): Light Trapping Designs for Thin Silicon Solar Cells Based on Photonic Crystal and Metallic Diffractive Grating Structures, *34th IEEE PV Specialists Conf*, Philadelphia, PA, June 7–12, 2009, Available at file:///E:/34%20IEEE%202009/Abstracts/281.html

Nagel, H., Schmidt, J., Aberle, A.G. & Hezel, R. (1997): Exceptionally high bulk minority-carrier lifetimes in block-cast Multicrystalline, *Proc. 14th EU PV Solar Energy Conf*, June–July 1997, Barcelona, Spain, pp. 762–765.

Nakamura, K., Kohira, M., Abiko, Y., Isaka, T., Funakoshi, Y. & Machida, T. (2008): Development of Back contact Si solar cell and module in pilot production line, *Proc. 23rd EU PV Solar Energy Conf*, 1–5, September 2008, Valencia, Spain, pp. 1006.

Nakajima, K., Fujiwara, K., Usami, N. & Okamoto, S. (2008): IUMRS-ICEM Photovoltaics Symposium, p. 1.

Neuhaus, D.-H., Mehnert, R., Erfurt, G. *et al.* (2005): Loss analysis of solar modules by comparison of IV measurements and prediction from IV curves of individual solar cells, *Proc. 20th EU PV Solar Energy Conference*, June 2005, Barcelona, Spain, pp. 1947–1952.

Neuhaus, D.-H., Kirchner, J., Mehnert, R. *et al.* (2006): Impact of shunted solar cells on the IV characteristics of solar modules, *Proc. 21st EU PV Solar Energy Conference*, September 2006, Dresden, Germany, p. 2556.

Neuhaus, D.-H. and Munzer, A. (2007): Industrial Silicon Wafer Solar cells, *Advances in Optoelectronics*, Article ID 24521, Volume 2007.

Nie, H.B., Xu, S.Y., Wang, S.J., You, L.P., Yang, Z., Ong, C.K., Li, J., & Liew, T.Y.F. (2001): Structural and electrical properties of tantalum nitride thin films fabricated by using reactive RF magnetron sputtering, *Appl. Phy. A: Mater. Sci. Process*, 73, 229–236.

Okamoto, S., Terakawa, A., Maruyama, E., Shinohara, W., Tanaka, M. & Kiyama, S. in '*Amorphous and Heterogeneous Silicon-Based Films – 2001*' (Mater. Res. Soc. Proc. 664, San Francisco, CA, 2001), A11.1.

Ooka, S. (2009): *Proc. 19th Intl. PV Science and Engg. Conf*, November 2009, Jeju, Korea, CSI8-O-4553.

Park, S., Cho, E., Hao, X., Conibeer, G., & Green, M.A. (2008): Study of the Electrical Properties of Si QD Heteroface and p-i-n Junction Devices on Crystalline Silicon Substrate, *Proc. 23rd EU PV Solar Energy Conf*, September 2008, Valencia, Spain, pp. 189–192.

Partain, L.D. (1995): Solar Cells and their Applications, K. Chang (Ed) New York:JohnWileys Photon International, 176, March 2009.

Pearson, G.L. (1985): *Proc. 18th IEEE PV Specialists Conf*, Las Vegas, October 1985; PV Founders award lunch.

PVEducation.org: Buried contact solar cells, [Online] available at http://pveducation.org/pvcdrom/manufacturing/buried-contact-solar-cells.

Ravi, K.V. (1977): The growth of EFG silicon ribbons, *J. Crystal Growth*, 39, 1–16.

Report for the State Committee for Scientific Research Project No. PBZ KBN Kraków, AGH (2003).

Rogol, M. and Conkling, J. (2007): Solar Power in Focus, *Photon Consulting*.

Rohatgi, A. & Rai-Choudhury, P. (1984): Design, fabrication, and analysis of 17–18-percent efficient surface-passivated silicon solar cells, *IEEE Transactions on Electron Devices*, ED-31, 596–601.

Rohatgi, A. & Rai-Choudhury, P. (1986): An approach toward 20-precent-efficient silicon solar cells, *IEEE Transactions on Electron Devices*, ED-33, 1–7.

Roos, O.V. and Anspaugh (1978): *Proc. 13th IEEE PV Specialists Conf*, Washington, D.C, p. 1119.

Rosenblum, M.D., Bathey, B.R., & Kalejs, J.P. (2002): *Proc. 29th IEEE PV Specialists Conf*, May 2002, New Orleans, La. pp. 58–61.

Sachs, E. (1984): Proc. Flat-Plate Solar Array Project Research Forum on the High-Speed Growth and Characterization of Crystals for Solar Cells, p. 279.

Saga, T. (2010): Advances in Crystalline Si solar cell technology for industrial mass production, *NPG Asia Mater.* 2(3), 96–102, doi:10.1038/asiamat.2010.82 published Online 22 July 2010.

Saint-Cast, P., Kania, D., Hofmann, M., Benick, J., Rentsch, J. & Preu, R. (2009): *Appl. Phys. Lett.* 95, 151502.

Sarasin Basic Report (2001): PV-2001 – The market players and Forecasts.

Schaadt, D.M., Feng, B., & Yu, E.T. (2005): Enhanced semiconductor optical absorption via surface plasmon excitation in metal nanoparticles, *Appl. Phys. Lett.* 86, 063106, 3 pages.

Schmidt, J. (2004): Light-induced degradation in crystalline silicon solar cells, *Solid State Phenomena*, 95–96, 187–196

Schmidt, J., Merkle, A., Brendel, R., Hoex, B., van de Sanden, M.C.M. & Kessels, W.M.M. (2008): Surface Passivation of High-efficiency Silicon Solar Cells by Atomic-layer-deposited Al_2O_3, *Prog. Photovoltaics*, 16, 461–466.

Schmidt, J., Veith, B., & Brendel, R. (2009): Effective surface passivation of crystalline silicon using ultrathin Al_2O_3 films and Al_2O_3/SiN_x stacks, *Physica Status Solidi – Rapid Res. Lett.* 3, 287.

Schmidt, J., Werner, F., Veith, B., Zielke, D., Bock, R., Tiba, V., Poodt, P., Roozeboom, F., Li, A., Cuevas, A. & Brendel1, R. (2010): Industrially Relevant Al_2O_3 Deposition techniques for the surface passivation of Si solar cells, *Proc. 25th EU PV Solar energy Conf*, 6–10 September 2010, Valencia, Spain, 2AO.1.6.

Schneider, A., Kopcek, R., Hahn, G., Noel, S. & Fath, P. (2005): Comparison of gettering effects during phosphorous diffusion for one- and double-sided emitters, *Proc. 31st IEEE PV Specialists Conf*, January 2005, Orlando, Fla. pp. 1051–1054.

Schneider, J., & Evans, R. (2006): Industrial Solid phase crystallization of Silicon, *Proc. 21st EU PV Solar Energy Conf*, September 2006, Dresden, Germany, 2CV.4.32, pp. 1032–1035.

Schroter, W., & Kuhnapfel, R. (1990): Model Describing Phosphorus Diffusion. Gettering of Transition Elements in Silicon, *Appl. Phys. Lett.* 56, 2207–2209.

Schroff, R.E.I., & Zeman, R. (1998): *Amorphous and Microcrystalline Solar cells: Modeling, Materials and Device technology*, Kluwer Academic Publs.

Schultz, O., Glunz, S.W., & Willeke, G.P. (2004): Multicrystalline silicon solar cells exceeding 20% efficiency, *Prog. Photovoltaics*, 12, 553.

Schultz, O., Glunz, S.W., Warta, W., Preu, R., Grohe, A., Kober, M., Willeke, G.P., Russel, R., Fernandez, J., Morilla, C., Bueno, R., & Vincueria, T. (2006): High efficiency solar cells with laser grooved buried contact front and laser-fired rear for industrial production, *Proc. 21st EU PV Solar Energy Conf*, September 2006, Dresden, Germany, pp. 826–829.

Schweizer, S., Miclea, P., Henke, B. & Ahrens, B. (2008): Up- and Down-Conversion in Fluorozirconate Based Glass Ceramics for High Efficiency Solar Cells, *Proc. 23rd EU PV Solar Energy Conf*, September 2008, Valencia, Spain, pp. 54–57.

Shah, A., Meier, J., Torres, P., Kroll, U., Fischer, D., Beck, N., Wyrsch, N., & Keppner, H. (1997): Recent progress in Mmicrocrystalline solar cells, *Conf, record. 26th IEEE PV Specialists Conf*, Anaheim, IEEE Press, Piscataway, pp. 569–572.

Shi, Z., Wenham, S. & Ji, J. (2009): Mass production of new high efficiency multi-c Si solar cell technology with selective emitter, *Proc. 24th EU PV Solar Energy Conf*, September 2009, Hamburg, Germany, pp. 1090–1093.

Shi, Z., Wenham, S. & Ji, J. (2009): *Proc. 34th IEEE PV Specialists Conf*, June 2009, Philadelphia, PA.

Sivakov, V., Andrä, G., Gawlik, A., Berger, A., Plentz, J., Falk, F., & Christiansen, S.H. (2009): Silicon nanowire-based solar cells on glass: Synthesis, optical properties, and cell parameters, *Nano Lett.* 9, 1549–1554.

Sinton, R.A., Kwark, Y., Gruenbaum, P., & Swanson, R.M. (1985): *Proc. 18th IEEE PV Specialists Conf*, Las Vegas, Nevada. October 1985, p. 61.

Sinton, R.A., Kwark, Y., Gan, J.Y. & Swanson, R.M. (1986): 27.5% silicon concentrator solar cells, *IEEE Electron Device Lett*, 7, 567–569.

Soderstrom, T., Haug, F.-J., Terrazzoni-Daudrix, V., Niquille, X., & Ballif, C. (2008): N/I Buffer layer for substrate microcrystalline thin film silicon solar cells, *J. Appl. Phys*, 104, 114509-8.

Soderstrom, T., Haug, F.-J., Niquille, X., & Ballif, C. (2008): TCOs for nip thin film Si solar cells, *Prog. Photovolt: Res. Appl.* 17, 165.

Soderstrom T., Haug, F.J., Terrazzoni-Daudrix, V., & Ballif, C. (2008): Optimization of amorphous silicon thin film solar cells for flexible photovoltaics, *J. Appl. Phys*, 103, 114509-8.

Soderstrom, T., Haug, F.J., Niquille, X., Terrazzoni, V., & Ballif, C. (2009): Asymmetric intermediate reflector for tandem micromorph thin film silicon solar cells, *Appl. Phys. Lett*, 94, 063501.

Soderstrom, T., Haug, F.J., Terrazzoni-Daudrix, V. & Ballif, C. (2010): Flexible micromorph tandem a-Si/μc-Si solar cells, *J. Appl. Phys*, 107, 014507 (available at paper_516 micromorph.pdf).

Solar Cells: Chapter 7 – Thin Film Silicon Solar cells, available at CH7_Thin_film_Si_solar_cells.pdf.

Staebler, D.L., & Wronski, C.R. (1977): Reversible Conductivity Changes in discharge-produced Amorphous Si, *Appl. Phys. Lett.* 31, 292–294.

Stangl, R., Froitzheim, A., Elstner, L., & Fuhs, W. (2001): Amorphous/crystalline silicon heterojunction solar cells, a simulation study, *Proc. of 17th EU PV Solar Energy Conf*, October 2001, Munich, Germany, pp. 1387–1390.

Sun, Ke, Karger A., Park N., Madson, K.N., Naughton, P.W., Bright T., Jing Y., & Wang, D. (2011): Compound Semiconductor Nanowire Solar cells, *IEEE Journal of Selected topics in Quantum Electronics*, **17**(4), 1033–1049.

Sun, W.-T., Yu, Y., Pan, H.Y., Gao, X.F., Chen, Q., & Peng, L.M. (2008): CdS Quantum Dots Sensitized TiO_2 Nanotube-Array Photoelectrodes, *J. Amer. Chem. Soc.*, **130**, 1124–1125.

Sun & Wind Energy (2011): Intl. News, 'Q-cells develops multi-crystalline 19.5% solar cell, p. 18, July/2011.

Swanson, R.M. (2008): *Proc. 33rd IEEE PV Specialists Conf*, keynote session, May 2008, San Diego, CA.

Takakura, T., Kidoguchi, S., Yamasaki, I., Okamoto, S., Okamoto, Y., & Taniguchim, K. (2008): Effect of Rapid Thermal Process for CDS Silicon Solar Cell, *Proc. 23rd EU PV Solar Energy Conf*, September 2008, Valencia, Spain, pp. 1472–1474.

Taguchi, M., Tanaka, M., Matsuyama, T. *et al.* (1990): Improvement of the conversion efficiency of polycrystalline silicon thin film solar cell, *Proc. 5th Intl. PV Science & Engineering Conf. (PVSEC '90)*, November 1990, Kyoto, Japan, pp. 689–692.

Taguchi, M., Tsunomura, Y., Inoue, H., Taira, S., Nakashima, T., Baba, T., Sakata, H., & Maruyama, E. (2009): High efficiency HIT Solar cell on thin (<100 μm) Silicon wafer, *Proc. 24th EU PV Solar Energy Conf*, September 2009, Hamburg, Germany, pp. 1690–1693.

Tawada, Y., Okamoto, H. & Hamakawa, Y. (1981): a-Si:C:H/a-Si:H heterojunction solar cells having more than 7.1% conversion efficiency, *Appl. Phys. Lett.* **39**(3), 237–239.

Tawada, Y., Yamagishi, H., & Yamamoto, K. (2003): Mass production of thin film Silicon modules, *Sol. Energy Mater. Sol. cells*, **78**, pp. 647–662.

van Kerschaver, E., & Beaucarne, G. (2005): Back-contact Solar cells: A review, *Prog. Photovolt: Res. Appln.* **14**(2), 107–123.

Veneri, P.D., Mercaldo, L., Usatii, I., Ciani, P., & Privato, C. (2006): Doped Silicon Nitride as Intermediate Reflector in Micromorph Tandem Solar Cell, *Proc. 23rd EU PV Solar Energy Conf*, September 2008, Valencia, Spain, pp. 2343–2346.

Verlinden, P.J., Swanson, R.M. & Crane, R.A. (1994): 7000 high-efficiency cells for a dream, *Prog. in Photovolt*, **2**, 143–152.

Vetterl, O., Finger, F., Carius, R., Hapke, P., Houben, L., Kluth, O., Lambertz, A., Mück, A., Rech, B., & Wagner, H. (2000): Intrinsic Microcrystalline Silicon: A New Material For Photovoltaics, *Sol. Energy Mater. Sol. Cells*, **62**, 97–108.

Washio, H., Katsu, T., Tonomura, Y., Hisamatsu, T., Saga, T., Matsutani, T., Suzuki, A., Yamamoto, Y., & Matsuda, S. (1993): Development of High efficiency thin-film Si space solar cells, *Proc. 23rd IEEE PV Specialists Conf*, May 1993, pp. 1347–1351.

Wenham, S.R. and M.A. Green (1988): US Patent **4**, 726, 850.

Wenham, S.R. and M.A. Green (1988): US Patent **4**, 748, 130.

Wenham, S.R., Robinson, S.J., Dai, X., Zhao, J., Wang, A., Tang, Y.H., Ebong, A., Honsberg, C.B. & Green, M.A. (1994): Rear surface effects in high efficiency silicon solar cells, *Proc. Ist WC on PV Energy Conversion*, Honolulu, HI, pp.1278–1282.

Wenham, S. and M.A. Green (1996): Silicon Solar Cells, *Photovoltaics*. **4**, 3.

Xu, X., Su, T., Beglau, D., Ehlert, S., Pietka, G., Bobela, D., Li, Y., Lord, K., Yue, G., Zhang, J., Yan, B., Worrel, C., Beernink, K., DeMaggio, G., Banerjee, A., Yang, J. & Guha, S. (2009): High efficiency large area multi junction solar cells incorporating a-SiGe:H and nc-Si:H using MVHF technology, *34th IEEE PV Specialists Conf*, June 7–12, 2009, Philadelphia, PA. Available at Mpublic76_090605140610.pdf.

Yan, B., Yue, G., & Guha, S. (2007): Status of nc-Si:H Solar Cells at United Solar and Roadmap for Manufacturing a-Si:H and nc-Si:H Based Solar Panels, in 'Amorphous and Polycrystalline

Thin-Film Silicon Science and Technology 2007', edited by V. Chu, S. Miyazaki, A. Nathan, J. Yang, H-W. Zan, *Mater. Res. Soc. Symp. Proc*, Warrendale, PA, **989**, 335.

Yan, B., Yue, G., Yang, J., & Guha, S. (2008): *Proc. 33rd IEEE PV Specialists Conference*, May 2008, San Diego, CA.

Yang, J., Banerjee, A., & Guha, S. (1997): Triple-junction Amorphous Silicon Alloy Solar Cell with 14.6% Initial and 13.0% Stable Conversion Efficiencies, *Appl. Phys. Lett.* **70**, 2975–2977.

Yang, J., Banerjee, A., Lord, K. & Guha, S. (1998): *Proc. 2nd WC on PV Solar Energy Conversion*, July, 1998, Vienna, Austria, pp. 387–390.

Yang, J., Yan, B., & Guha, S. (2003): Amorphous silicon based photovoltaics – from 'earth' to the 'final frontier', *Sol. Energy Mater. & Sol. Cells*, **78**, 597–612.

Yang, J., Yan, B., & Guha, S. (2005): Amorphous and naocrystalline Si-based MJ solar cells, *Thin Solid Films*, **487**, 162–169.

Yang, X., Yan, B., Yue, G., Jiang, C.-S., Owens, J., Yang, J., & Guha, S. (2008): *Mater. Res. Soc. Symp. Proc*, **1101E**, KK13.2.

Yamamoto, K., Yoshimi, Y., Suzuki, T., Tawada, Y., Okamoto, Y., & Nakajima, J. (1998): Thin film poly-Si solar cell on glass substrate fabricated at low temperature *MRS Spring Meeting*, April, 1998, San Francisco, CA.

Yamamoto, K. (1999): Very thin film crystalline Si solar cells on glass substrate fabricated at low temperatures, *IEEE Trans. Electron Devices*, **46**, 2041–2047.

Yamamoto, K., Toshimi, M., Tawada, Y., Okamoto, Y. & Nakajima, A. (2000): Thin film Si solar cell fabricated at low temperature, *J. Non-Cryst. Solids*, **266–269**, 1082–1087.

Yamamoto, K., Yoshimi, M., Tawada, Y., Fukuda, S., Sawada, T., Meguro, T., Takata, H., Suezaki, T., Koi, Y., Hayashi, K., Suzuki, T., Ichikawa, M., & Nakajima, A. (2002): Large area thin film Silicon module, *Sol. Energy Mater. Sol. Cells*, **74**, 449–455.

Yamamoto, K., Nakajima, A., Yoshimi, M., Sawada, T., Fukuda, S., Suezaki, T., Ichikawa, M., Koi, Y., Goto, M., Takata, H., Sasaki, T., & Tawada, Y. (2003): Novel Hybrid Thin-film Silicon Solar Cell and Module, *Proc. 3rd WC PV Energy Conversion*, May 2003, Okasa, Japan, pp. 2789–2792.

Yamamoto, K., Nakajima, A., Yoshimi, M., Sawada, T., Fukuda, S., Suezaki, T., Ichikawa, M., Koi, Y., Goto, M., Meguro, T., Matsuda, T., Kondo, M., Sasaki, T., & Tawada, Y. (2004): A high efficiency thin film silicon solar cell and module, *Solar energy*, 77, 939–949.

Yamamoto, K., Nakajima, A., Yoshimi, M., Sawada, T., Fukuda, S., Suezaki, T., Ichikawa, M., Koi, Y., Goto, M., & Meguro, T. (2005): *Prog. Photovoltaics*, **13**, 489.

Yi, S.S., Girolami, G., Amano, J., Islam, M.S., Sharma, S., Kamins, T.I., & Kimukin, I. (2006): InP nanobridges epitaxially formed between two vertical Si surfaces by metal-catalyzed chemical vapor deposition, *Appl. Phys. Lett.* **89**, 133121.

Yoshimi, M., Sasaki, T., Sawada, T., Suezaki, T., Meguro, T., Ichikawa, M., Nakajima, A., Yamamoto, K., Matsuda, T., Wadano, K., Santo, K. (2003): High efficiency TF Silicon hybrid solar cell module on 1 m^2-class large area substrate, *3rd WC Photovoltaic Energy Conversion*, Osaka, Japan, pp. 1566–1569.

Zhao, J., Wang, A., Altermatt, P., & Green, M.A. (1995): 24% efficient Silicon solar cells with double layer antireflection coatings and reduced resistance loss, *Appl. Phys. Lett.* **66**, 3628–3636.

Zhao, J., Wang, A., Yun, F., Zhang, G., Roche, D.M., Wenham, S.R., & Green, M.A. (1997): 20,000 PERL silicon cells for the "1996 World Solar Challenge" solar car race. *Prog. Photovoltaics*, **5**, 269–276.

Zhao, J., Wang, A., Green, M.A., & Ferrazza, F. (1998): Novel 19.8% efficient 'honeycomb' textured multicrystalline and 24.4% monocrystalline silicon solar cells, *Appl. Phys. Lett*, **73**, 1991–1993.

Zhao, J., Wang, A., & Green, M.A. (1999): 24.5% efficiency silicon PERT cells on MCZ substrates and 24.7% efficiency PERL cells on FZ substrates, *Prog. Photovoltaics*, **7**, 471–474.

Zhao, J., Wang, A., & Green, M.A. (2001): High-efficiency PERL and PERT silicon solar cells on FZ and MCZ substrates, *Sol. Energy Mater. & Sol. Cells*, **65**, 429–435.

Zimmer, J., Stiebig, H., & Wagner, H. (1998): a-SiGe:H based solar cells with graded absorption layer, *J. Appl. Phys*, **84**, 611(7 pages).

Chapter 3

Polycrystalline CIGS and CdTe thin film solar cells

3.1 INTRODUCTION

Silicon has been the traditional solar cell material, and major portion of the market is taken over by wafer silicon technologies. However, increase in the energy conversion efficiencies has not been very remarkable in recent years despite improvements in manufacturing processes. Research and commercial interests have concurrently started looking at thin-film solar cells based on polycrystalline thin film materials, prominent among them being Copper indium diselenide (CIS), and Copper indium gallium diselenide (CIGS) and Cadmium telluride (CdTe). A high potential for manufacturing cost reduction is expected for these thin-film modules compared to Si wafer based technologies (Woodcock *et al.* 1997). Additionally, thin-film techniques have the advantage of monolithic integration of cells and fewer processing steps and a higher degree of automation.

These compounds are 'direct band gap' semiconductors with fairly high light absorption coefficients ($>10^4$/cm). Therefore, a few microns-thick material is sufficient to absorb the incident solar radiation. This means that the need for long minority carrier diffusion lengths does not exist because majority of the carriers are generated within or very near to the edge of depletion layer. These merits enable to utilize low cost preparation techniques, even on continuous production basis.

Amorphous silicon also exhibits these features. But, the best a-Si based devices have demonstrated around 10% conversion efficiencies only, as shown earlier, requiring large areas for installing the modules. Moreover, the manufacturing costs of tandem and triple junction amorphous devices and micromorph devices are currently high due to the complexity of the structures. Further investigations are underway to improve the efficiencies and to reduce the costs simultaneously of a-Si based PV modules.

3.2 HIGHLIGHTS OF CIGS AND CdTe TECHNOLOGIES

The CIGS thin film belongs to the multinary Cu-chalcopyrite system, where the band gap can be modified by varying the Group III cations among In, Ga, and Al and the anions between Se and S (Rau & Schock 1999; Hengel *et al.* 1996). Wide range of band gaps can be obtained by mixing Ga, and the band gap range of interest for this technology is between 1eV and 1.7eV (Albin *et al.* 1992).

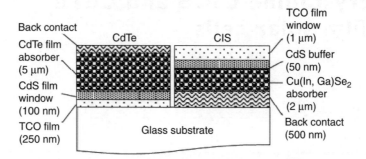

Figure 3.1   Schematic representation of thin-film CIGS and CdTe solar cells
(*Source:* Powalla & Bonnet 2007, Hindawi Publishing Corp., Copyright © 2007 M. Powalla
and D. Bonnet)

Table 3.1   Efficiencies of CIGS and CdTe thin-film solar cells

Cell	Area (cm²)	V_{oc} (V)	J_{sc} (mA/cm²)	FF (%)	η (%)	Comments
CIGSe	0.410	0.697	35.1	79.52	19.5	CIGSe/CdS cell, NREL, *3-stage process*
CIGSe	0.402	0.67	35.1	78.78	18.5	CIGSe/ZnS (O.OH) NREL, Nakada *et al.*
CIGS	0.409	0.83	20.9	69.13	12.0	Cu(In,Ga)S/CdS, Dhere, FSEC
CIAS	–	0.621	36.0	75.50	16.9	Cu(In,Al)S/CdS, IEC, $E_g = 1.15$ eV
CdTe	1.06	0.845	25.9	75.51	16.5	CTO/ZTO/CdS/CdTe, NREL, CSS
CdTe	–	0.840	24.4	65.00	13.3	SnO₂/Ga₂O₃/CdS/CdTe, IEC, VTO
CdTe	0.16	0.815	23.56	73.25	14.0	ZnO/CdS/CdTe/metal, U. of Toledo, *Sputtered*

Source: Noufi *et al.* 2007

The CdTe material in the device mostly exists as a binary with a slight deviation from stoichiometry. Its band gap is about 1.5 eV, a perfect match to the solar spectrum. This band gap may vary slightly as a result of its interaction with its hetero junction partner, CdS (~2.4-eV) during cell processing (McCandless and Sites 2003).

Figure 3.1 shows a schematic of the layer sequence for the thin-film CdTe and CIGS device structures. Individual layer thicknesses are approximate and may vary slightly among laboratories.

The CIGS device is a *substrate configuration* that starts with glass/base electrode, whereas the CdTe device is a *superstrate configuration* that starts with glass/transparent top electrode. The sequence of the growth of the layers in both structures may influence the properties of the front and back junctions, that is, the p-n interface and the back contact and, in turn, the efficiency of the devices.

The best efficiencies so far reported for thin-film CIGS and CdTe solar cells using different techniques are given in Table 3.1.

The CIGS and CdTe modules share common characteristics and device structural elements. Therefore, in principle, the cost per unit area should be similar, and, thus, the conversion efficiency becomes a selective factor for the cost/watt. However, in practice, production processes in terms of throughput and yield can differ significantly and may offset the advantage of higher performance. That is how the cost of producing CdTe modules has an advantage over CIGS. In future years, the costs of semiconductor materials may become more influencing factors.

First Solar has announced (Kaneloss, M., Greentech Media article, July, 26, 2011) that it had produced CdTe solar cell with 17.3% efficiency breaking the earlier world-record of 16.5% set by NREL in 2001. Further, First Solar expects module efficiencies to be at 13.5 to 14.5% by the end of 2014. First Solar modules are cheaper to manufacture and this year, the production cost is an average of US cents 75 per watt. Recently, General Electric, Abound Solar and others have entered the CdTe market. General Electric announced early this year that it produced 12.8% efficiency CdTe cell in the lab and planned to manufacture CdTe modules on mass-scale with record efficiencies in 2013.

Thin-film device research has been in progress for over three decades and has produced encouraging results. However, the long-term potential of the two technologies requires R&D emphasis on science and engineering-based challenges to find solutions to achieve the projected cost-effective module performance and long-term durability. Particularly, transferring knowledge from the laboratory to manufacturing, especially in the area of production processes due to the inherent complexity of the two compound semiconductors, has proven to be much harder than expected. Therefore, much more research focus is needed.

The two technologies are now discussed in detail.

3.3   CIGS THIN FILM SOLAR CELLS

The CIGS device preparation is not simple. It starts with the deposition of molybdenum (Mo) on glass substrate by sputtering. The Mo film properties have to be optimized for adhesion, sheet resistance, and morphology where it allows sodium (Na) from the glass to diffuse through to the CIGS layer. Kuo et al. (2009) have found that the growth parameters, such as argon flow rate, RF power, bi-layer of Mo thin films and substrate temperature, have significant influences on properties of Mo films. The presence of sodium supports the grain growth with a higher degree of <112> texturing for CIGS films, and increases the carrier concentration. All the highest performance devices have used soda lime glass as the substrate material (Bodegard et al. 1999; Hedstrom et al. 1993), which has a good thermal match to CIS thin films (Bodegard et al. 1994). Growth on substrates with no sodium content requires dosing of the CIGS film by laying a 60 to 120 Å NaF layer on the Mo back contact, or introducing NaF during the CIGS deposition. The absence of Na in the device reduces the efficiency by 2% to 3% (absolute). Mo is preferred to other back-contact materials, Pt, Au, Ni, W, Cr, V, Ta, Mn and so on, due to its insignificant diffusion into the absorber layer during cell processing (Scofield et al. 1995; Schmid & Seidel 2005). Then the CIGS absorber is deposited using any of the methods of flux delivery: (a) evaporation of elements simultaneously or in a prescribed sequence, or (b) sputtering of metals followed by selenization with H_2Se, or (c) reactive sputtering of metals with Se vapor, or (d) printing of metals from ink precursors followed by selenization.

The CdS layer is then applied by chemical bath deposition (CBD), followed by sputter deposition of a bilayer consisting of intrinsic and conducting ZnO. The ZnO layer is also applied by using the chemical-vapor deposition process. The industrial processes for both technologies basically adopt combinations of these techniques (Noufi & Zweibel 2006). These methods are further elaborated below.

The first report on chalcopyrite-based solar cell was published in 1974 (Wagner et al. 1974). The cell was prepared from a p-type $CuInSe_2$ (CIS) single crystal onto which a CdS film was evaporated in vacuum with efficiencies up to 12% (Shay et al. 1975). This combination of a p-type chalcopyrite absorber and a wide band-gap n-type window layer is the basic concept even now for the cell designs. Later, polycrystalline thin film of the more general composition Cu (In,Ga)(S,Se)$_2$ has replaced the CIS crystal. Kazmerski et al. (1975, 1976, 1977a, 1977b) produced thin films by dual-source evaporation of $CISe_2$ or CIS_2 and the chalcogen Se or S. The best thin film CdS/CIS solar cells demonstrated efficiencies of 6.6%measured using 100 mW/cm^2 illuminations. Boeing produced all thin film CIS-based cells with record efficiency of 6.6% (Mickelsen & Chen 1980, 1981) which was later improved to >10% (Mickelsen & Chen 1982).

Boeing Company reported the fabrication of CIGS solar cells by co-evaporating elements with (Ga/Ga + In) = 0.23 demonstrating efficiencies of >10% (Chen et al. 1987). Dimmler et al. (1987) at University of Stuttgart produced CIGS cells which reached efficiencies up to 5.8% for $CuGaSe_2$, 9.3% for CIS, and 3% for Cu $In_{0.56}Ga_{0.44}Se_2$ thin films. The increase of bandgap of the absorber layer by introducing Ga has been established by these research groups. Many groups world-wide have developed CIGS solar cells with efficiencies, 15–19%, using different growth processes.

Glass is the most commonly used substrate, but now efforts are being made to develop flexible solar cells on polyimide (Basol et al. 1996; Tiwari et al. 1999; Hanket et al. 2002; Ishizuka et al. 2008; Gecys et al. 2009; Zhang et al. 2008; Caballero et al. 2009; Zachmann et al. 2009) and metal foils (Kessler & Rudmann 2004; Satoh et al. 2000; Herz et al. 2002; Herrmann et al. 2003; Hollars et al. 2005; Otte et al. 2006; Bremand et al. 2007; Shi et al. 2009; Kim et al. 2007; Ishizuka et al. 2008a, 2009; Wuerez et al. 2009; Yagioka & Nakada 2009). Highest efficiencies of 14.1% on polyimide (Bremand et al. 2005) and 17.6% on metal foils (Tuttle et al. 2000) have been reported for CIGS cells. Recently a slight increase in efficiency, 14.7% on polyimide and 17.7% on metal foils has been reported for CIGS solar cells (Ishizuka et al. 2008b).

Several reviews have appeared where investigations related to materials processing and manufacturing processes of CIGS solar cells are discussed (Archer & Hill 2001; Luque & Hegeds 2003; Markvant & Castarier 2003; Romeo et al. 2004; Kemell et al. 2005, Powalla 2006; Miles et al. 2010).

3.3.1 Absorber layer deposition

The phase diagrams of CIS and CIGS thin films are thoroughly investigated (Haalboom et al. 1998; Mikkelsen 1981; Jitsukawa et al. 1998), and stable films of these materials are easily prepared in a wide range of compositions.

The technique that shows potential for commercial module production, in general, is the one relating low cost processes with high yield and reproducibility. To achieve high yield, compositional uniformity over large substrate areas is also important. The device considerations require that the Cu(In,Ga)Se$_2$ layer should be at least 1 µm thick and that the relative compositions of the constituents are determined by the phase diagram. Slightly Cu-deficient compositions of p-type conductivity are found to be well suited for solar cell fabrication (Rau et al. 1998; Nadenau et al. 1999).

Among the deposition methods mentioned earlier, either (i) vacuum co-evaporation of the constituents, Cu, In, Ga, and Se, on to a substrate held at 400°C to 600°C to form the Cu(In,Ga)Se$_2$ film, or (ii) deposition of metallic precursor layers by reactive sputtering or evaporation followed by selenization are generally preferred.

Co-evaporation: The most successful technique for deposition of CIGS absorber layers for highest-efficiency cells is the simultaneous evaporation (Mattox 1998) of the constituent elements, Cu, In, Ga, and Se from multiple sources in single process where Se is present in excess during the deposition. The In/Ga ratio during the deposition process leads to only minor changes in the growth kinetics, but variation of the Cu content strongly affects the film growth.

Four different sequences (Singh & Patra 2007) that have been used to fabricate devices with efficiencies greater than 16% are shown in Figure 3.2.

1   The first process (Figure 3.2a) is the simplest stationary process in which all fluxes as well as substrate temperature is constant throughout the deposition process (Shafarman & Zhu 2000).

2   Figure 3.2b shows the Boeing process (Chen et al. 1987), also called bilayer process, which yields larger grain sizes compared to the constant rate (single stage) process. The larger grain size is attributed to the formation of a Cu$_x$Se phase during the Cu-rich first stage, which improves the mobility of group III atoms during growth (Klenk et al. 1993; Tuttle et al. 1993; Park et al. 2000). Another possibility is the inverted process where first (In,Ga)$_2$Se$_3$ is deposited at a lower temperature (typically ~300°C) followed by the deposition of Cu and Se at an elevated temperature until an overall composition close to stoichiometry is reached (Kessler 1992; Gabor et al. 1993; Zweigart et al. 1994). This process leads to smoother film morphology than bilayer process.

3   In Figure 3.2c, the 'three-stage process' proposed by Kessler (1994) and later modified by Gabor et al. (1994) is presented. This process has yielded the most efficient solar cells. The smoother surface obtained with three-stage process reduces the junction area which is expected to reduce the number of defects at the junction. It also facilitates the uniform conformal deposition of a thin buffer layer and prevents ion damage in CIGS during sputter deposition of ZnO/ZnO:Al.

4   Figure 3.2d shows variations of the Ga/In ratio during deposition which allow the design of graded band-gap structures (Gabor et al. 1996).

A simple schematic of multisource co-evaporation technique is shown in Figure 3.3. With an absorber layer prepared by co-evaporation, the highest efficiency reported is 19.9% (0.419 cm^2 active area) for a CIGS solar cell, and 13% efficiency for a 60 × 120 cm^2 for a CIGS module (Powalla et al. 2004; Repins et al. 2008).

Selenization of pre-deposited precursor layer

This is a two-stage process for optimum CIGS compound formation:

(i)   precursor materials, i.e, Cu, Ga, and In are deposited using low-cost and low-temperature methods that facilitate uniform composition, followed by (ii) thermal

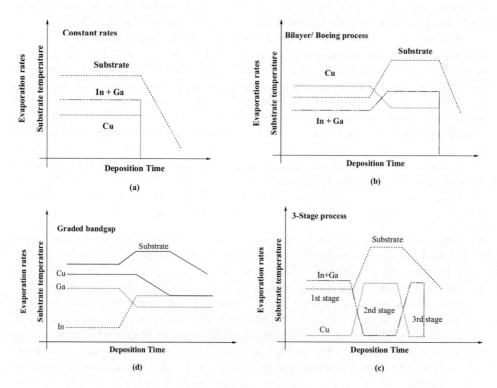

Figure 3.2 Schematic representation of co-evaporation processes (Redrawn from Singh & Patra 2007)

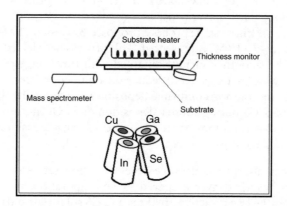

Figure 3.3 Schematic diagram of Multisource evaporation method for CIGS deposition

annealing at elevated temperature in controlled Se or S atmosphere at 400°C to 600°C created by H_2Se/H_2S gas or elemental Se/S. This is called selenization of stacked metal layers. The reaction and anneal step often takes longer time than formation of films by co-evaporation due to diffusion kinetics, but is suitable to batch processing.

Table 3.2 Base-line processes at ZSW for CIGS modules

Process step (Film thickness)	Process technology	Process control
Soda-lime glass/ Mo (0.5–1 μm)	DC magnetron sputtering	Thickness, sheet resistance, adhesion
Mo patterning	Nd-YAG laser patterning of stripes	Electrical insulation
Cu(In,Ga)Se$_2$ (2 μm)	One-step co-evaporation in-line	Atomic absorption spectroscopy, X-ray Fluorescence spectroscopy
CdS (50 nm)	Chemical bath deposition	Visual inspection
i-ZnO (50 nm)	RF magnetron sputtering	Thickness, sheet resistance, transmittance
CIGS/CdS/i-ZnO	Mechanical patterning, scribing	Visual inspection
ZnO:Al (1 μm)	DC magnetron sputtering	Thickness, sheet resistance, transmittance
ZnO patterning	Mechanical scribing	I–V curve, dark and illuminated
Contacts	Bus bars	Visual inspection, resistance
Module packaging	EVA encapsulation	Visual inspection, adhesion

The well established magnetron sputtering is the most common method to prepare metallic precursor layers (Muller *et al.* 2006) which has several advantages over the thermal evaporation method (Kapur & Basol 1990).

The interest in the selenization processes is mainly due to its suitability for large-area film deposition with good control of the composition and film thickness.

3.3.2 State-of-the-art technologies

Wurth Solar has set up base-line processes based on co-evaporation technology developed at the University of Stuttgart. Today, the production is fully integrated and continuous in-line operation with a high level of automation. The production line established at Wurth Solar to fabricate modules of 30 cm × 30 cm is given in Table 3.2. These modules, which have a proven long-time stability and less energy payback time, are installed in several countries, mostly on residential and commercial buildings. Wurth Solar has the distinction of being the first company in Europe to produce CIS modules.

The Japanese group (Nagami *et al.* 2002) has developed a process in which an in-line evaporation system is equipped with a temperature detector to monitor the composition of Ga and In. The CIGS films are deposited by sequential 3-stage process; the Mo, ZnO and ITO films by RF sputtering, and CdS by CBD. The schematic arrangement of the in-line evaporation system is shown in Figure 3.4.

The I-V measurements have given an efficiency of 12.6% and a FF of 0.699 with an aperture of 81.5 sq.cms. The optimization of band-gap grading may further enhance the module performance.

Nanosolar, Inc., Solarion AG, HelioVolt Corporation, Solyndra, Inc., Honda Soltec Co. Ltd, Sulfercell Solartechnik GmbH, Salibro GmbH, Odersun AG, Global Solar energy, Inc. are other prominent manufactures of CIGS thin film modules employing different technologies (Miles *et al.* 2009).

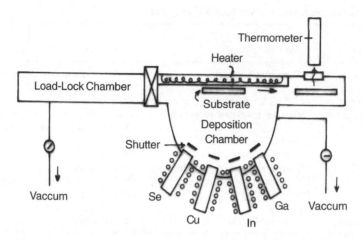

Figure 3.4 Schematic of in-line evaporation system (Jayarama Reddy 2010)

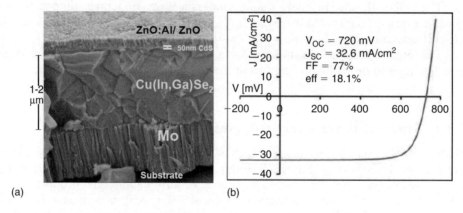

(a)   (b)

Figure 3.5 High efficiency CIGS thin film cell on glass with 18.1% efficiency
(*Source:* Ayodhya Tiwari, EMP, reproduced with permission of the author)

The CIGS thin film solar cell on glass developed at EMPA by Tiwari *et al.* has achieved 18.1% record efficiency. The cell elements – ZnO/ZnO:Al layer is deposited by sputtering, the CdS layer by chemical bath deposition, CIGS absorber layer by co-evaporation and molybdenum contact layer by sputtering. The TEM of cross-sectional view of the cell is shown in Figure 3.5a and the J Vs V curve along with the derived parameters are shown in Figure 3.5b.

3.3.3 Band gap engineering of absorber (CIGS) layer

The current state-of-the-art CIGS solar cells utilize low band gap CIGS material [$E_g = 1.2$ eV for $Ga/(Ga + In) = 0.3$] though the band gap required for optimum solar energy conversion is 1.5 eV. However, the CIGS solar devices fabricated

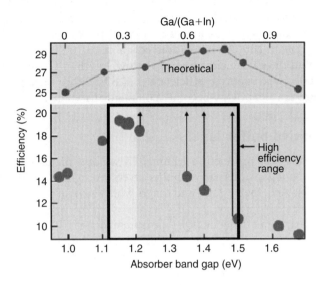

Figure 3.6 Efficiency Vs Absorber bandgap (Noufi *et al.*, NREL, 2007, reproduced with permission of the author)

with $E_g > 1.2\,eV$ and $Ga/(Ga + In) > 0.5$ have demonstrated reduced efficiencies (Ramanathan *et al.* 2002). See Figure 3.6.

Studies at NREL (Noufi *et al.* 2007) have shown that efficiency peaks at $(Ga/Ga + In) = 0.3$ as against theoretical prediction, ~ 0.7–0.8 ($E_g = 1.14\,eV$), and then falls off drastically with higher Ga content. This is due to decreased J_{sc} which is attributed to increased defect density and stronger interfacial recombination, and phase inhomogeneity with the increase of Ga content. This observation requires further understanding of the limiting mechanisms of the efficiency at high Ga content.

3.3.4 Novel absorber layers

By substituting Ga with relatively less quantity of Al, copper indium aluminium selenide, $CuInAlSe_2$ (CIAS) can be produced having same energy gap as that of CIGS. The variation of E_g with Al content is up to 2.7 eV for CAS (copper aluminium selenide) compared to 1.7 eV for CGS. This novel absorber material is suitable to fabricate wider energy band gap solar cells which can be used as top cells in multi junction solar cells with higher performance. Another advantage is that for the equivalent energy band gap, the lattice distortion in CIAS structure is less from the optimum CIS structure compared to CIGS. Efficient solar devices have been reported using CIAS absorber layer, with $\eta > 10\%$ by Woods *et al.* (2005), η up to 16.9% by Marsillac *et al.* (2002) and $\eta > 12\%$ by Minemoto *et al.* (2007).

SnS is another material with direct energy band gap whose electrical properties can be controlled by introducing appropriate dopants. The major advantage is the constituent materials are nontoxic and abundantly available. Moreover, the mass production methods for producing tin and for sulfidizing metals are well established.

These advantages allow SnS highly suitable for absorber layer. SnS-based solar cells are fabricated at laboratory level by several groups using different techniques for the deposition of absorber layer (Noguchi *et al.* 1994; Ristov *et al.* 2001; Sanchez-Juarez *et al.* 2005; Reddy *et al.* 2006; Gunasekaran & Ichimura 2007; Avellaneda *et al.* 2007; Li *et al.* 2009). But, the efficiencies reached are low, less than 1%, except in the case of Reddy *et al.* (2006) who reported efficiencies >1.3%. More focused research efforts are needed before the technology becomes commercially viable.

3.3.5 Alternative buffer layers

Current generation of cell production is limited by window materials, and there is need to develop wider bandgap (alternative) buffer layers alternative to CdS. Materials such as (Cd,Zn)S alloys, ZnS, In_xS_y, ZnSe and others which have suitable features have been investigated.

At NREL, Quantum efficiency (QE) measurements of CIGS solar cells fabricated with alternate buffer layers CdS, ZnS, and CdZnS are performed. The cell using CdZnS buffer layer produced highest efficiency (real breakthrough) of 19.5% (Noufi 2007). Adding Zn to CdS, results in an alloy, $Cd_{1-x}Zn_xS$, with band gap between 2.4 eV and 3.8 eV. Depending on Zn/Cd ratio, this material (buffer) layer yields a better lattice match with the absorber layer and increases the blue response of the solar cell. Further, it should lead to a favourable conduction band offset at the hetero junction interface (Reddy KTR & Jayarama Reddy 1992a).

Studies on $Cd_xZn_{1-x}S/CuGaSe_2$ cells (Reddy KTR & Jayarama Reddy 1992b) have showed an enhanced V_{oc} and I_{sc} leading to a higher conversion efficiency compared to $CdS/CuGaSe_2$ solar cells. Varying the $Zn/(Zn+Cd)$ ratio in the range, 15%–20%, Daveny *et al.* (2006) have developed (CBD-grown)$Cd_xZn_{1-x}S/CIS$ (co-evaporation) cells with efficiencies >13%. Song *et al.* (2006) have reported conversion efficiencies up to 13% for a CIGS solar cell with CBD-grown 40 nm thick $Cd_xZn_{1-x}S$ buffer layer. But, real breakthrough efficiency of 19.5% is achieved, as mentioned above, with the $Cd_{1-x}Zn_xS/CIGS$ cells produced by Bhattacharya *et al.* (2006).

Indium sulfide (In_2S_3) is another promising alternative buffer layer to CdS. Several techniques have been used to prepare these films; but, Atomic layer epitaxy (ALE) grown In_2S_3 films have given record efficiencies with CIGS absorber layers. However, CIGS solar cells with indium sulfide buffer layer prepared by ALE (Naghavi *et al.* 2003), CBD (Velthaus *et al.* 1992, Hariscose *et al.* 1996), Evaporation (Karg *et al.* 1997; Pistor *et al.* 2009), and PVD (Strohm *et al.* 2004) have given varying conversion efficiencies between 11.2% and 16.4%. The initial studies at Stuttgart that used CBD-grown In_2S_3 layers have yielded only 9.5% efficiency. CIEMAT research group has investigated in detail the growth mechanisms of In_2S_3-based buffer layers (Bayon & Herrero 2001). Despite its promise for efficient solar cell development, the current shortage of indium may become an obstacle for large-scale production using this technology.

Zn-based materials, ZnS, ZnSe, ZnSSe and ZnO have also been studied as alternate buffer layers to CdS. ZnS layers grown byCBD (Kessler *et al.* 1993; Ortega Borges *et al.* 1992; Contreras *et al.* 2003), ALE (Platzer-Bjorkman *et al.* 2003), and Vacuum evaporation (Romeo *et al.* 2004); ZnSe layers byCBD (Ennaoui *et al.* 2001, Chaparro

Table 3.3 Efficiencies of chalcopyrite-based solar thin film Modules with different buffer layers

Buffer	Deposition	Area (cm²)	Efficiency	Company	Reference
CdS	CBD	6500	13.0	Wurth Solar	Powalla *et al.* 2004
ZnS	CBD	900	14.3	Showa Shell	Kushiya 2003
ZnS	CBD	3659	13.2	Showa Shell	Kushiya 2003
ZnSe	CBD	20	11.7	HMI/Siemens	Mikami *et al.* 2003
ZnO	ILGAR	20	10.9	HMI/Siemens	Muffler *et al.* 2000
In₂Se₃	ALE	13	13.4	ZSW	Spiering *et al.* 2003
In₂Se₃	ALE	900	12.9	ZSW	Spiering *et al.* 2004
In₂Se₃	CBD	717	9.7	ZSW	Dimmler *et al.* 1998

Adopted from Miles *et al.* 2010

et al. 2002), MOCVD (Munzel *et al.* 2001), ALE (Ohtake *et al.* 1995), and evaporation (Romeo *et al.* 2004); and ZnO layers prepared by CBD (Ennaoui *et al.* 1998), electrodeposition (Gal *et al.* 2001), ALE (Chaisitsak *et al.* 2000), CVD (Olsen *et al.* 1997), and ILGAR (Bar *et al.* 2002) have given considerable conversion efficiencies. Addition of small amount of Mg to ZnO results in ZnMgO whose band gap is higher than ZnO. ALE-grown ZnMgO film has been found to be good buffer layer in CIGS solar cells due to improved blue response, and the best cell has given an efficiency of 18.1% (Negami *et al.* 2002).

In Table 3.3, is given the highest efficiencies achieved using different buffer layers with chalcopyrite-based solar modules. Of them, only ZnS and In₂S₃ buffer layers have been successful in developing $30 \times 30 \, cm^2$ area modules with efficiencies higher than those achieved with CdS layers. The rest of the modules consist of monolithically inter-connected prototype sub-modules of 5×5 or $10 \times 10 \, cm^2$ area.

Showa Shell has produced $900 \, cm^2$ CIGSS modules using thin films synthesized by sulfurization/selenization of metal precursors and chemical bath deposited ZnS which have demonstrated world-record efficiency of 14.3%.

ZSW have fabricated CIGS modules of $900 \, cm^2$ area using In₂S₃ buffer layers deposited by ALE method with evaporated CIGS films which showed efficiencies greater than 12.9%.

3.3.6   Fabrication of CIGSS modules

The Copper Indium Gallium diselenide disulfide, $CIG(S,Se)_2$, usually represented as CIGSS, thin film modules are fabricated by Showa Shell (Japan) by two-stage method. The Cu-In-Ga precursor layers are deposited by magnetron sputtering followed by selenization in H_2Se gas. A sulfurization process has been used to introduce sulfur in the absorber layer. Substrates of $3600 \, cm^2$ size or higher are coated with improved process technologies. An efficiency of 13.4% has been reported for a $Zn(O,S,OH)_x/Cu(In,Ga)(S,Se)_2$ module area of $3600 \, cm^2$ (Kushiya 2003).

Avanics GmbH (Germany) has avoided the use of H_2Se gas which is highly toxic in the fabrication of CIGSS modules. Instead, a Se layer is deposited on top of the precursor layer by evaporation, and rapid thermal processing has been performed in an environment containing both Se and S. Avanics has produced $30 \times 30 \, cm^2$ CIGSS modules with an efficiency of 13.5%, and $60 \times 90\,cm^2$ modules with an efficiency of 13.1% (Probst *et al.* 2004). Avanics technology is environment-friendly and suitable for large-area deposition.

3.3.7 Flexible CIGS solar devices

There are several advantages in using flexible substrates for fabricating thin film cells and modules: (a) possibility of roll-to-roll deposition methods for industrial pro-duction, (b) lower costs of equipment and infrastructure; for example, roll-to-roll equipment used for foils is 10–30 times smaller in size than for in-line glass substrates; high material utilisation and low thermal budget; and no need for robotics, and (c) high rate of deposition. Flexible devices are easy to handle, and cover a wider range of applications including building integrated PV. Stainless steel foils and polymers are generally used as substrates. However, the choice of substrate and the deposi-tion process of CIGS are very important to obtain good performance. Tiwari and his group (EMP, Switzerland) have deposited CIGS films on flexible polymer films. The deposited films exhibit the following characteristics: The layers formed on steel foils have rough conducting surface with kinks and metal impurities; require high temperature (550–600°C) processing; and barrier coatings are needed for mono-lithic interconnection. The highest efficiency achieved is 17.5% with AR coating; whereas the layers deposited on polymer foils have smooth and insulating surface, no metal impurities, no need of barrier coatings for monolithic interconnection, and low temperature (<450°C) processing. The highest efficiency achieved is 14.1% without AR coating.

The defect passivation is essential for flexible foils; hence, controlled sodium is introduced into CIGS by depositing 30 nm NaF at room temperature on CIGS, and annealed in ultra high vacuum at 400°C for about 20 min. There is no change in the microstructure as evidenced by TEM image while SIMS profiles confirm sodium in-diffusion (Figure 3.7) (Tiwari *et al.*, EMP).

Flexible CIGS solar cell on polyimide with world record efficiency of 14.1% is developed at ETH, Zurich (shown in Figure 3.8). This is the highest reported effi-ciency on polymer for any kind of solar cells. The other parameters of the cell are: $V_{oc} = 649\,mV$, $J_{sc} = 31.5\,mA/cm^2$, and FF = 0.691 (Tiwari *et al.* 1999). Their recent result (2009) on the improved device has given an efficiency of 16% with $V_{oc} = 645\,mV$, $J_{sc} = 34.5\,mA/cm^2$, FF = 0.719.

Application of CIGS thin-film modules: Thin-film modules of CIGS are widely used and find themselves as functional design elements in building façades in many countries. Figure 3.9 shows Grain silo 'Schapfenm ühle' located near Ulm in Germany. 1306 CIS modules produced by Wurth Solar are integrated into the façade with an installed capacity of 98 kWp.

3.3.8 Vacuum-free deposition

The usual processes for deposition of thin films of metals or semiconductors require a high vacuum. However, Odersun AG, Germany is manufacturing CIS solar cells using a vacuum-free roll-to-roll process. The process runs as follows: Indium is galvanically applied to a cleaned and polished copper tape of 1 cm wide and 0.1 mm thick; then

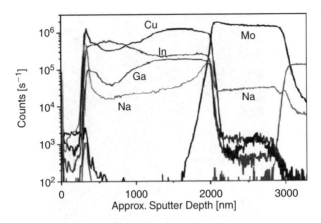

Figure 3.7 SIMS profiles confirming Na diffusion
(*Source:* Ayodhya Tiwari *et al.*, EMP, reproduced with permission of the author)

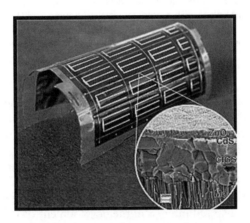

Figure 3.8 Flexible CIGS solar cell on polyimide with record efficiency of 14.1%
(*Source:* Ayodhya Tiwari *et al.*, EMP, reproduced with the permission of the author)

gaseous sulfur is allowed to come in contact with the molten indium layer when the CIS layer is formed. A top layer of copper sulfide is also formed which can be removed by chemical etching. The finished precursor layers are then annealed, and buffer and TCO layers are applied. Only the TCO layer needs to be sputtered in vacuum. The active cell layer is 0.001 mm thin. The cell stripes are processed to variable types of customized solar modules.

3.4 CdTe THIN FILM SOLAR CELLS

CdTe is an excellent direct band gap material with an optimum bandgap (1.45 eV), and a large absorption coefficient requiring only a few microns thick material to absorb

Figure 3.9 Grain silo 'Schapfenmühle' on the outskirts of Ulm, Germany. 1306 CIS modules from the Wurth Solar pilot plant with a nominal installed power of 98 kWp are integrated into the façade
(*Source*: Powalla & Bennet 2007, Hindawi Publishing Corp., Copyright © 2007 M. Powalla & D. Bonnet)

nearly 100% of the solar spectrum with energies >1.45 eV. Theoretically, CdTe solar cells are estimated to achieve efficiencies up to 27% (Loferski 1956).

The typical structure for thin-film CdTe solar cells comprises Glass/SnO_2/CdS/ CdTe/Contacts, shown in Figure 3.1.

Adirovich *et al.* (1969) fabricated the first p-CdTe/n-CdS hetero junction solar cell with an efficiency of 1%. Later, Bonnet and Rabenhorst (1972) produced an all thin film CdTe solar cell with an efficiency >5%. This observed improvement in efficiency and the stability have inspired interest in the development of CdTe/CdS PV devices.

A world-record, total-area efficiency of 16.5% measured under standard AM 1.5 illumination conditions, has been achieved by the research group at NREL (Wu *et al.* 2005) with the device structure, glass/CTO/Zn_2SnO_4/CdS/CdTe/back contact. In this design, the SnO_2 front contact is replaced by the more promising Cd_2SnO_4 contact. The CTO, ZTO, and CdS layers were prepared by RF magnetron sputtering in pure argon at room temperature using hot-pressed oxide targets, and 10 μm thick CdTe layer by close spaced sublimation (CSS). The CdS layer thickness is reduced to improve J_{sc}. The measured absorption and transmission spectra of both SnO_2 and Cd_2SnO_4 layers clearly establish the high quality performance of Cd_2SnO_4 layer as transparent front contact. The measured parameters for this record cell are $V_{oc} = 845$ mV, $J_{sc} = 25.9$ mA/cm^2, FF $= 75.5\%$. It is entirely a dry process, and the process time is relatively reduced while increasing throughput. The CdTe cells produced have good uniformity, reproducibility and device stability.

The CdTe thin films are prepared by a variety of techniques in different laboratories – Thermal evaporation (Moutinho *et al.* 1995; Birkmire and Phillips 1997), Sputtering (Wendt *et al.* 1998), Vapour transport deposition, VTD (Meyers

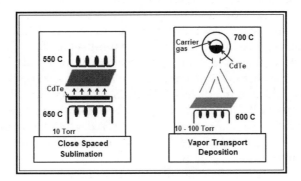

Figure 3.10 Schematic diagrams of CSS and VTD deposition techniques

et al. 1993; Sandwisch 1994), Close-spaced sublimation, CSS (Aramoto *et al.* 1997; Ferekides *et al.* 2000; Ohyama *et al.* 1997; Wu *et al.* 2001; Al-Jassim *et al.* 1998), Electro deposition (Kim *et al.* 1994; Johnson 2000), Spray pyrolysis (Chu & Chu 1995), Screen printing (Ikegami 1988), Metal-organic Chemical vapour deposition (MOCVD) (Irvine *et al.* 2008), and Atomic layer epitaxy (ALE) (Skarp 1991).

Some of these techniques – CSS, spray pyrolysis, and VTD – are more widely used (Figure 3.10). For example, the most efficient cells are produced by CSS technique (Aramoto *et al.* 1997; Ferekides *et al.* 2000; Ohyama *et al.* 1997; Wu *et al.* 2001). CSS is characterized by high deposition rate, $>1\,\mu$m/minute, and produces films with large grain size of several μm (Al-Jassim *et al.* 1998). Spray pyrolysis is a low-cost technique for fabricating large area cells, and the films prepared by this method have large grains and random orientation. VTD is a high rate deposition method developed for commercial production by Solar Cells, Inc. (Meyers *et al.* 1993).

3.4.1 CdTe cell structure elements

The choice of substrate is very important; it has to be transparent, able to withstand process temperatures, and must not contaminate the layers that are deposited subsequently. The general choice is soda-lime glass which is cheap and withstands relatively high processing temperature. Borosilicate glass, though relatively expensive, is used for higher temperature deposition methods. Recently flexible substrates such as metal foils and polymer sheets are used (Mathew *et al.* 2004).

The front contact or TCO is another important layer. Tin oxide (SnO_2), indium tin oxide (In_2O_3:Sn), fluorine-doped tin oxide (SnO_2:F) are mostly used materials. Cd_2SnO_4 which has been used at NREL has shown improved power conversion due to its higher conductivity and transmittance. Generally, to prevent any shunting through thin CdS layer, an intermediate layer between TCO and CdS layers is introduced, and the most promising one is Zn_2SnO_4 (Wu *et al.* 2001).

For buffer layer, CdS is the right wide band gap material with transparency down to about 510 nm. To reduce CdS thickness from the usual $\sim$100 nm has some problem. The diffusion of CdS into CdTe during annealing treatment requires a minimum thickness level for CdS layer to prevent the danger of shunt which may result in a low open circuit voltage (McCandless & Dobson 2004). This problem is minimized by inserting the intermediate layer between TCO and CdS. Chemical bath deposition, a low-temperature technique, has been the most successful technique to prepare CdS films, though subsequent annealing treatment in air or in chlorine atmosphere is needed to decrease the defect density and to increase grain size (Ferekides et al. 1994). This method has yielded most efficient CdTe/CdS and CIGS/CdS solar cells (Wu et al. 2001; Repins et al. 2008). Recently, there has been interest in using wide band gap buffer layers for CdTe cell fabrication, and Irvine et al. (2008) have used $Cd_{0.9}Zn_{0.1}S$, a wider band gap (2.7 eV) material as buffer layer which can cover the lower part of the solar spectrum too.

The absorber CdTe layer is generally 2 to 10 μm thick with a grain size varying between 0.5 and 5 μm. The films have to be heat treated in Cl_2 or O_2 atmosphere to ensure p-type and high conductivity. Post-deposition heat treatment in $CdCl_2$ vapour is very essential for modifying the structural properties of CdTe films. Other methods of annealing are discussed by Miles et al. (2009). This process is very crucial, especially for small-grained evaporated films to promote recrystallization and grain growth (McCandless et al. 1997; Levi et al. 1996); for the passivation of grain boundaries of the polycrystalline CdTe layer (Edwards et al. 1998, 1997); and for CdS/CdTe interfacial mixing, $CdTe_yS_{1-y}/CdS_xTe_{1-x}$ which decreases structural and electrical defects at interface. Al-Allak et al. (1996) have shown remarkable improvement in the performance of CdTe/CdS solar cell following the $CdCl_2$ treatment of CdTe films. Their solar cells with as-deposited CdTe films have shown conversion efficiency of 1.2% and FF, 0.39 whereas the $CdCl_2$- treated films have achieved 10% efficiency and FF of 0.56.

The back contact formation is very ticklish in CdTe solar cell due to high work function of CdTe (~5.7 eV). Creating ohmic contact is a big challenge. To reduce the difference between work functions of CdTe and the back contact metal layer, the carrier concentration in CdTe in the vicinity of the back contact has to be increased. One way is by forming a p^+ layer by chemically etching the top surface of the absorber so that a quasi-ohmic contact with the back contact is formed (Jager & Seipp 1981; Kotina et al. 1998; Sarlund et al. 1996). Alternately, doping of the top surface of the absorber with the required dopant (e.g. Cu doping) and subsequent annealing (Dhere et al. 1997); or direct doping during the growth of the absorber film could be done to increase carrier concentration near the surface. At the end, deposition of an intermediate layer of p-type ZnTe or HgTe or Sb_2Te_3 having a work function lower than CdTe is also used to complement any of above alternatives. Table 3.4 lists the efficiencies of CdTe/CdS solar cells in which CdTe layers are produced by different techniques.

3.4.2 Standard module fabrication

A major advantage of thin-film module processing lies in the monolithic series interconnection of cells to form modules with higher voltages. Thin-film cells are interconnected through simple patterning steps integrated into the processing line. The methods are similar to both CdTe and CIS technologies. The monolithic interconnection for CdTe modules is shown in Figure 3.11. Three scribes between deposition steps accomplish the cell definition, separation, and series interconnection. The TCO properties can also be integrated into the optimization.

Table 3.4 CdTe/CdS solar cells with CdTe layer grown with different methods

CdTe Growth method	Efficiency (%)	Cell area (cm²)	Reference
CSS	16.5*	1.032	Wu et al. 2001
Thermal evaporation	16	0.25	Takamoto et al. 1997
Electrodeposition	14.2	0.02	Woodcock et al. 1991
RF Sputtering	14	–	Gupta & Compaan 2004
ALE	14		Skarp et al. 1991
Spray Pyrolysis	12.7	0.3	Chu & Chu 1995
Screen Printing	12.8	0.78	Ikegami 1998
MOCVD	13.3	0.25	Irvine et al. 2008

*Record efficiency produced at NREL;
Source: Miles et al. 2010

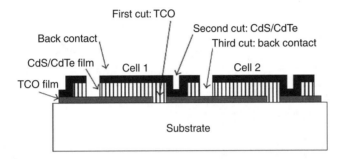

Figure 3.11   Illustration of Monolithic interconnection for CdTe modules
(*Source:* Powalla & Bonnet 2007, Hindawi Publishing Corp., Copyright © 2007 M. Powalla & D. Bonnet)

The 1st and 3rd cuts separate the cells at the front and back contacts respectively. The second cut enables the back contact of cell 1 to connect with the front contact of cell 2. In the case of CIS modules, the interconnection occurs in the same manner but in the reverse sequence. Finally, the contact leads are attached to the first and last cells, and the modules are encapsulated with a lamination foil and glass plate to protect from environmental effects.

For the industrial production of thin-film modules, high throughput in-line deposition systems are required for each processing step and automation to transfer the glass panes from one step to the other. A module processing line used for CdTe module production is shown in Figure 3.12. The circled numbers describe the processing steps starting with cleaning the substrate (step 1) to the characterization of the module and its classification by quality (step 13). Each number in the figure indicates the nature of processing.

3.4.3   Industrial CdTe modules

Several companies have been manufacturing large area modules using different processes. Antec GmbH has developed a low-cost in-line fabrication system. It is a

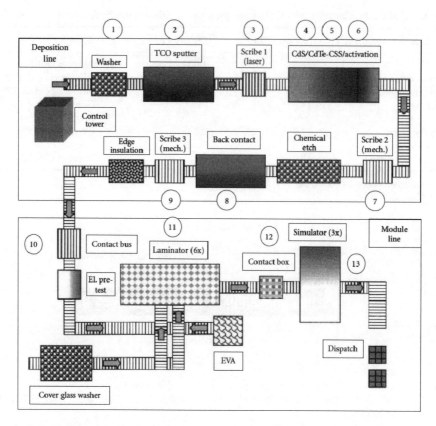

Figure 3.12 Schematic of commercial production line for CdTe thin-film solar modules
(*Source*: Powalla & Bonnet 2007, Hindawi Publishing Corp., Copyright © 2007 M. Powalla &
D. Bennet)

combined 'in-line sputtering and CSS equipment' consisting of four independent processing chambers. The substrates are 10 cm × 10 cm and the structure of the cell
is Ni/ZnTe/p-CdTe/n-CdS/ITO+TO/glass. The system design is such that it avoids
interruption of vacuum and long cooling cycles during the process.

The system developed by Barth *et al.* (2002) is pilot continuous in-line
system performing all fabrication steps in one vacuum zone. The structure is
glass/SnO$_x$:F/CdS/CdTe/carbon/nickel, and cells of 3.6 in × 3.1 in are produced, each
cycle in a process time of two minutes. The continuous process is scalable, uniform
and reproducible. The NREL verified efficiency is 12.4% with FF of 0.71 measured
at 25°C within a variation of 1°C. This process is merited by increased cell reliability,
low production time, and cost viability. More details are given in the earlier book
(Jayarama Reddy 2010).

The US manufacturer, First Solar is the largest producer of CdTe modules with
worldwide capacity of 1.3 GW and manufacturing units in US, Germany and Malaysia.
By 2012, production capacity is to be increased to as much as 2.1 GW by way of the

new plants at Frankfurt, in Malaysia and in France. The characteristics of the 3rd generation CdTe modules are substantially improved (Sun & Wind Energy 8/2010). First Solar achieved considerable progress by being able to increase their power ratings of its 60 cm × 120 cm module from 50 and 55 Wp to 55 to 65 Wp (9.3% total area) respectively within one year (von Roerden 2006). First Solar along with Juwi Solar, Germany has set up a 40 MW CdTe thin film solar field, one of the largest solar fields in the world. Among thin film technologies, CdTe is the leading and cheap technology in the market today. In the case of First Solar's manufacturing cost for thin-film CdTe PV modules, as economies-of-scale have been realized, the manufacturing cost has dropped substantially from US$ 2.94/W (6-MW) in the year 2004 to $1.25/W (90 MW) in 2007 (Ullal & Roedern 2007). The manufacturing cost is expected to come down to $0.70/W (the targeted cost) due to improvements in productivity, module efficiency and yield by 2012, thus making it potentially price competitive with grid-parity electricity. According to recent report (Sun & Wind Energy 8/2010), First Solar's production costs amount to around US$ 0.80/W and are likely to drop to 0.65 to 0.55/W in the medium term.

3.4.4 Flexible CdTe solar cells

The laboratory for Thin Films and Photovoltaics of EMPA, Switzerland has developed highest efficiency flexible CdTe thin-film solar cells on a lightweight polymer (poly-imide) film by using a low temperature (below 450°C) vacuum evaporation process. The cells are subsequently annealed in air. In this device structure, ZnO:Al replaces expensive ITO (indium tin oxide) layer as a transparent electrical contact used in the earlier production that has given conversion efficiency of 11.4%. Substitution of ITO with a bi-layer of ZnO/ZnO:Al also improved process yield and reproducibility of high efficiency solar cells. The efficiency achieved in this design is 12.4% measured under standard AM1.5 illumination conditions. Other parameters of the cell are $V_{oc} = 823$ mV, $J_{sc} = 19.6$ mA/cm^2 and FF = 76.5%. Monolithically interconnected flexible CdTe solar module of 32 cm^2 has demonstrated a record efficiency of 7.5% (Tiwari, EMPA).

3.5 CHALLENGES TO BE ADDRESSED

The main issues that the research groups should address to improve the performance of these polycrystalline thin film solar cells with long range stability and low production costs are briefly mentioned. The success of these efforts will enhance the commercial success of these devices.

1   Lack of adequate science and engineering knowledge base – In order to achieve high throughput and high yield at every process stage, and high degree of reliability and reproducibility leading to higher performance, several issues are to be investigated: (i) measurable material properties that can predict device and module performance, (ii) how materials delivery and film growth are related, (iii) development of control and diagnostics based on material properties and film growth, and (iv) coupling of this knowledge to industrial processes. If the control and diagnostic tools can

respond to rapid processing and facilitate adjusting real-time processes, it will impact throughput and yield, and will make the process reproducible and reliable. Currently, only a few techniques are in practice, based on changes of emissivity from the growing surface, and in-situ monitoring of composition using X-ray fluorescence.

Fundamental research is needed on both reference systems, high-quality absorber films prepared in the laboratory and industrially produced absorber films. New diagnostic techniques such as measuring electron-beam-induced currents in the junction configuration, micro-photoluminescence, and micro-Raman spectroscopy assist to understand the recombination mechanisms (Kniese *et al.* 2007) in real devices and for investigating the function of inhomogeneities (Powalla & Bonnet 2007).

2 Long-Term Stability (Durability) – Both CIGS and CdTe technologies has shown long-term stability. However, performance degradation has also been observed. For example, CdTe and CIGS devices have different sensitivity to water vapour; e.g., oxidation of metal contact, change in properties of ZnO. Therefore, thin film barrier to water vapour, new encapsulants and less aggressive application process are essential. Further, degradation mechanisms at the device level and prototype module level must be better understood. Much work has been done to monitor and investigate performance of CIGS and CdTe modules in the outdoors. To date, the perception level for the causes of performance degradation is inadequate and lacks the coupling of feedback from device- and module-level studies. Recently, Albin *et al.* [2005a, 2005b] at NREL investigated the temperature-dependent degradation of CdTe devices. The findings point out that different mechanism dominates degradation at different temperatures. From 90° to 120°C, the degradation is dominated by Cu-diffusion from the back contact toward the electrical junction, whereas the source of possible degradation from 60° to 90°C is not currently known. Another issue requiring further consideration is the need for encapsulants that can be applied and cured at room temperature and are chemically inert toward the semiconductor layer with which they come in contact.

3 Thinner CIS and CdTe layers – The goal is to achieve <0.5 μm thick layers from the current values of 1.3 to 8 μm. To maintain state-of-the-art performance, modification of deposition parameters is required. Studies to develop (a) models that relate film growth to material delivery, and (b) device structure that maximizes photon absorption have to be pursued.

The major benefit in fabricating thinner cells is less material usage contributing hopefully to reduction in cost. Thinner cells are expected to deliver higher throughput. However, the availability of In and Te is a debatable issue that demands lesser usage. Production at gigawatt level may have problem with the availability of In.

Thinner layers have drawbacks that include micro non-uniformity and its impact on device performance, changes in device physics and structure, and probable lower yield, which require thorough investigations. Studies were undertaken to minimize the thickness of CIGS, CIGSS and CdTe absorber layers with ultimate object of bringing down the cost while maintaining high-efficiency and stability. Table 3.5 shows the cell parameters including efficiency for different thicknesses of the absorber layer (Gupta & Compaan 2005; Ramanathan *et al.* 2007). At

Table 3.5 Thinner cells and their performance

t (μm)	V_{oc} (V)	J_{sc} (mA/cm^2)	FF (%)	Efficiency (%)
1.0 CIGS	0.676	31.96	79.47	17.16 (NREL)
0.75 CIGS	0.652	26.0	74.0	12.5
0.40 CIGS	0.565	21.3	75.7	9.1
0.47 CIGS	0.576	26.8	64.2	9.9 (EPV)
1.3 CIGSS Module	25.26	2.66	69.2	12.8 (Shell Solar)
0.87 CdTe	0.772	22.0	69.7	11.8 (U. of Toledo)

Source: Noufi & Zweibel 2007

U. of Toledo, magnetron sputtering has been used to prepare CdS/CdTe cells with CdTe thickness, 0.5–1.28 μm. The drop in performance currently begins to become significant below 1 μm thickness. Studies guided by device modeling are being pursued to understand the loss mechanism for very thin absorbers (Gloeckler & Sites 2005).

4   Need for Low-cost processes – This issue is more relevant to CIGS technology for two reasons: (a) in the current techniques, throughput is relatively slow, material utilization is poor, and require relatively high vacuum because of complex processes; and (b) indium is highly expensive, about $1000/kg due to limited stock. Hence, deposition by high-rate co-sputtering from cylindrical magnetrons is now being pursued; however, to date, this approach has not demonstrated state-of-the-art performance. A lower-cost process should feature high deposition rates, high material utilization, and simpler equipment capable of processing very large substrates. Examples of such innovative processes are CVD-based, and Nano-technology utilizing nano-components to make CIGS, e.g. printable CIGS (Kapur et al. 2002; Eberspacher et al. 2002).

Fabricating CIS modules on flexible substrates require a lot of research efforts to sort out new issues that arise, like introduction of undesirable impurities, the absence of sodium in the substrate (which improves the absorber film quality), the needed processing temperatures for getting high-quality films, and the necessity of additional layers (insulating films) for monolithic interconnection if formed on conductive substrates.

Both CIS and CdTe technologies are exploring alternative methods like electrodeposition, spraying, and screen printing for material synthesis due to their potential for reducing both materials and production costs. Continuous optimization of several production parameters are needed for improved throughput and yield.

For CdTe solar devices, areas of further research include inter-diffusion at the CdS/CdTe interface where S diffuses into the CdTe film, vapor $CdCl_2$ heat treatment, and the role of Cu 'doping' that is usually used for often proprietary back-contacting procedures, and most importantly whether the maximum open-circuit voltage of CdTe cells could be significantly improved without degradation of FF and J_{sc} (von Roerden, Ullal, Zweibel 2006). Enhanced V_{oc} may be achieved by increasing the net p-type doping of the bulk through extrinsic doping, or, more likely, by better crystal growth conditions that influence more favorable formation of native point defects.

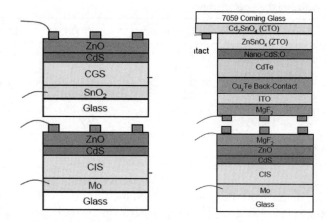

Figure 3.13 Schematic representation of CGS/CIS and CdTe/CIS tandem cells (Noufi *et al.* 2007, reproduced with permission)

3.6   POLYCRYSTALLINE TF MULTI JUNCTION SOLAR CELLS

This research activity, at NREL under a DOE project, was aimed at developing approaches toward improving transparent top cells, an appropriate bottom cell, and interconnects to fabricate a polycrystalline thin film tandem solar cell with an efficiency of 25%. The ultimate goal of this research was to develop a monolithic tandem cell. In this project, the issues related to development of TCOs and mechanical stacks as alternatives to monolithic approach are also investigated. For example, preparing and studying $Cd_{1-x}Mg_xTe$ alloys where largest range of energy gaps is possible with relatively small addition of alloying element, Mg. These alloys are characterized by least deviation from the lattice constant of CdTe, and the energy gap varies linearly with the fraction 'x'.

The research includes several aspects: development of $CuGaSe_2$ top cell (1.7 eV), transparent CdTe top cell (1.5 eV), $CuInSe_2$ bottom cell with 1.0 eV band gap, mechanical-stacked tandems (CGS/CIS and CdTe/CIS structures), and CGS/c-Si monolithic tandem structure (Noufi *et al.* 2007).

(i)   CGS/CIS tandem structure: The top cell is transparent CGS cell (ZnO/CdS/CGS/SnO_2/Glass) and the bottom one is CIS cell (ZnO/CdS/CIS/Mo/Glass). It is mechanically stacked and series-connected by addition. The measured efficiency is 10.7%, and $V_{oc} = 1.32$ V. The NREL official measurement shows efficiency of 9.75 and $V_{oc} = 1.29$ V.

(ii)   CdTe/CIS tandem structure: The top cell is a transparent CdTe solar cell with an efficiency of 13.8% and having transmission, ~40 and 60%. The device structure is: modified borosilicate glass/CTO/ZTO/nano CdS: O/CdTe/ITO/Ni-Al grid. The bottom cell is a record CIS cell with 15% efficiency. The measured tandem parameters as measured by NREL are $V_{oc} = 1.144$ and efficiency $= 15.3\%$. These structures are schematically shown in Figure 3.13.

Table 3.6 Tandem Solar devices and their operating parameters

Organisation	Tandem structure	V_{oc} (V)	J_{sc} mA/cm^2	FF	η (%)	Comment
NREL	Top cell: glass/SnO$_2$/CGS/CdS/ZnO	0.864	15.36	51.25	6.8	4-terminal
	Bottom cell: glass/Mo/CIS//CdS/ZnO	0.456	12.46	69.17	3.9	device
	Mechanical Stack	1.29	–	–	9.7	
NREL	Top cell: glass/CTO/ZTO/nano CdS:O/CdTe/Cu$_x$Te	0.786	25.5	68.9	13.8	4-terminal device
	Bottom cell: glass/Mo/CIS/CdS/ZnO	0.357	6.059	68.01	1.47	
	Mechanical Stack	1.14	–	–	15.3	
Univ. of Delware IEC	Monolithic structure: ZnO/ITO/CdS/ CIGS/ZnO/CdS/CIS/Mo/glass	0.688	10.4	52.8	3.8	
Univ. of Toledo	Monolithic structure: TO:F/CdS/CdTe/ ZnTe:N/ZnO:Al/CdS/HgCdTe	0.960	2	62	1.2	

Source: Noufi 2007

Table 3.7 Module manufacturing cost and profitable price forecast (2006 US$)

Cell technology	2006 Cost/Price ($)	2010 Cost/Price ($)	2015 Cost/Price ($)
Crystalline Si:			
Mono-Si	2.50/3.75	2.00/2.50	1.40/2.20
Multi-Si	2.40/3.55	1.75/2.20	1.20/2.00
Ribbon Si	2.00/3.35	1.60/2.20	1.00/1.70
Conc. Si cells	3.00/5.00	1.50/2.50	1.00/1.70
Amorphous Si-based	1.50/2.50	1.25/1.75	0.80/1.33
CIS	1.50/2.50	1.00/1.75	0.80/1.33
CdTe	1.50/2.50	0.80/1.20	0.65/1.25
Factory profitable Price	2.50/3.75	1.50/2.50	1.20/2.20

Source: Maycock, 33rd IEEE PV Specialists Conf., San Diego, USA 2008

Polycrystalline thin film tandem solar devices studied by different research groups are tabulated (Table 3.6) along with the cell parameters. The structures developed by NREL are mechanical stacks whereas the structures fabricated by IEC, University of Delaware, and University of Toledo are monolithic.

3.7 MANUFACTURING COST OF THIN FILM MODULES

Maycock has projected the manufacturing costs/profitable prices based on the existing costs and presented at 33rd IEEE PV Specialists conference at San Diego. These are given in Table 3.7. Grid parity is expected by 2012–2014 with 16 GW of annual production.

REFERENCES

Adirovich, E.I., Yuabov, Y.M., & Yagudaev, G.R. (1969): Photoelectric effects in film diodes with CdS-CdTe heterojunctions, *Fiz. Tech. Poluprovodnikov*, 3, 81–85

Al-Allak, H.M., Brinkman, A.W., Richter, H. *et al.* (1996): Dependence of CdS/CdTe thin-film solar cell characteristics on the processing conditions, *J. Cryst. Growth*, **159**, 910–915

Albin, D.S., Carapella, J., Tuttle, J.R. & Noufi, R. (1992): The Effect of Copper Vacancies on the Optical Bowing of Chalcopyrite Cu(In,Ga)Se$_2$ Alloys, *Mat. Res. Soc. Procedings*, **228**, 267

Albin, D.S., Berniard, T., Demtsu, S., Noufi, R., & Pankow, J. (2005): Experiments Involving Correlations Between CdTe Solar Cell Fabrication History and Intrinsic Device Stability, *Proc. 31st IEEE PV Specialists Conf*, Jan. 2005, Orlando, Fla., pp. 315

Al-Jassim, M.M., Dhere, R.G., Jones, K.M., *et al.* (1998): The morphology, microstructure, and luminescent properties of CdS/CdTe films, *Proc. 2nd WC PV Solar Energy Conversion*, Vienna, Austria, pp. 1063–1066

Aramoto, T., Kumazawa, S., Higuchi, H., Arita, T., Shibutani, S., Nishio, T., Nakajima, J., Tsuji, M., Hanafusa, A., Hibino, T., Omura, K., Ohyama, H., & Murozono, M. (1997): 16.0% efficient thin film CdS/CdTe solar cells, *Japanese J. Appl. Phys*, **36**, 6304–6305

Archer, M. & Hill, R. (Eds.) (1998): *Clean electricity from Photovoltaics*, Imperial College Press, London

Archer, M.D. & Hill, R., Eds. (2001): Clean Electricity from Photovoltaics, vol. 1, chapter 7 of *Series on Photo conversion of Solar Energy*

Avellaneda, D., Delgado, G., Nair, M.T.S. *et al.* (2007): Structural and chemical transformations in SnS thin-films used in chemically deposited photovoltaic cells, *Thin Solid Films*, **515**, 5771–5776

Bar, M., Fischer, C.H., Muffler, H.J. *et al.* (2002): High-efficiency chalcopyrite solar cells with ILGAR-ZnO WEL-device characteristics subject to the WEL composition, *Proc. 29th IEEE PV Specialists Conf*, May 2002, New Orleans, LA, pp. 636–639

Barth, K.L., Enzenroth, R.A. & Sampath, W.S. (2002). Advances in Continuous, In-Line Processing of Stable CdS/CdTe Devices, *Proc. 29th IEEE PV Specialists Conf*, May 2002, New Orleans, LA, pp. 551–554

Basol, B.M., Kapur, V.K., Leidholm, C.R., Halani, A., & Gledhill, K. (1996): Flexible and light weight copper indium diselenide solar cells on polyimide substrates, *Sol. Energy Mater. and Sol. Cells*, **43**, 93–98

Bayon, R. & Herrero, J. (2001): Reaction mechanism and kinetics for the chemical bath deposition of In(OH)$_x$S$_y$ thin films, *Thin Solid Films*, **387**, 111–114

Birkmire, R.W. & Phillips, J.E. (1997): *Processing and Modelling issues for Thin film Solar cell Devices*, Institute of Energy Conversion, Newark, DE

Birkmire, R.W. & Eser, E. (1997): Polycrystalline Thin film Solar cells: Present Status and Future Potential, *Annual Review of Materials Science*, **27**, 625– 653

Birkmire, R., Eser, E., Fields, S., & Shafarman, W. (2005): Cu(InGa)Se$_2$ solar cells on a flexible polymer web, *Prog.in Photovolt: Res. and Appls*, **13**, 141–148

Bhattacharya, R.N., Contreras, M.A., & Teeter, G. (2004): 18.5% CIGS Device Using Single-Layer, Chemical-Bath-Deposited ZnS(O,OH), *Japanese J. of Applied Physics*, **43**, L1475–L1476

Bodegard, M., Stolt, L., & Hedstrom, J. (1994): The influence of sodium on the grain structure of CuInSe2 films for photovoltaic applications, *Proc. 12th EU PV Solar Energy Conf*, Amsterdam, The Netherlands, pp. 1743

Bodegard, M., Granath, K., Stolt, L., & Rockett, A. (1999): The behavior of Na implanted into Mo thin films during annealing, *Sol. Energy Mater. & Sol. Cells*, **58**, 199–208

Bonnet, D. & Robenhorst, H. (1972): New results on the development of a thin film p-CdTe/n-CdS heterojunction solar cell, *Proc. 9th IEEE PV Specialists Conf*, Silver Springs, MD, pp. 129–132

Bonnet, D. (2001): Cadmium telluride solar cells, In: Archer, M.D., Hill, R. (Eds), Clean electricity from photovoltaics, Imperial College Press, 245–275

Brémaud, D., Rudmann, D., Bilger, G., Zogg, H., & Tiwari, A.N. (2005): Towards the development of flexible CIGS solar cells on polymer films with efficiency exceeding 15%, *Proc. 31st IEEE PV Specialists Conf*, January 2005, Orlando, FL, pp. 223–226

Brémaud, D., Rudmann, D., Kaelin, M., Ernits, K., Bilger, G., Döbeli, M., Zogg, H., & Tiwari, A.N. (2007): Flexible Cu(In,Ga)Se2 on Al foils and the effects of Al during chemical bath deposition, *Thin Solid Films*, 515, 5857–5861

Caballero, R., Kaufmann, C.A., Eisenbarth, T., Cancela, M., Hesse, R., Unold, T., Eicke, A., Klenk, R., & Schock, H.W. (2009): The influence of Na on low temperature growth of CIGS thin film solar cells on polyimide substrates, *Thin Solid Films*, 517, 2187–2190

Chaisitsak, S., Yamada, A., Kongai, M., & Saito, K. (2000): Improvement in performance of ZnO:B/i-ZnO/Cu(InGa)Se$_2$ solar cells by surface treatment for Cu(InGa)Se$_2$, *Japanese J. Appl. Phys*, 39, 1660–1664

Chen, W.S., Stewart, J.M., Stanbery, B.J., Devany, W.E., & Mickelson, R.A. (1987): Development of poly-cryst. CuIn$_{1-x}$Ga$_x$Se$_2$ solar cells, *Proc. 19th IEEE PV Specialists Conf*, New Orleans, LA, pp. 1445

Chu, T.L. & Chu, S.S. (1995): Thin film II-VI Photovoltaics, *Solid State Electronics*, 38, 533–549

Contreras, M.A., Nakada, T., Hongo, M., Pudov, A.O., & Sites, J.R. (2003): ZnO/ZnS(O,OH)/Cu(In,Ga)Se$_2$/Mo solar cell with 18.6% efficiency, *Proc. 3rd WC on PV Energy Conversion*, May 2003, Osaka, Japan, 2LN-C-08

Contreras, M.A., Ramanathan, K., AbuShama, J., Hasoon, F., Young, D.L., Egaas, B., & Noufi, R. (2005): Diode Characteristics in State-of-the-Art ZnO/CdS/Cu(In$_{1-x}$Ga$_x$)Se$_2$ Solar Cells, *Prog. Photovolt: Res. and Appl*, 13, 209

Dhere, R., Rose, D., & Albin, D. (1997): Influence of CdS/CdTe interface properties on the device properties, *Proc. 26th IEEE PV Specialists Conf*, Anaheim, CA, pp. 435–438

Dimmler, B., Dittrich, H., Menner, R., & Schock, H.W. (1987): Performance and optimization of heterojunctions based on Cu(Ga, In)Se$_2$, *Proc. 19th IEEE PV Specialists Conf*, New Orleans, LA, pp. 1454

Dimmler, B., Gross, E., Hariskos, D., Kessler, F., Lotter, E., Powalla, M., Springer, J., Stein, U., Voorwinden, G., Gaeng, M., & Schleicher, S. (1998): *Proc. 2nd WC PV Solar energy Conversion*, Vienna, Austria, pp. 419–422

Dimmler, B. & Wächter, R. (2007): Manufacturing and application of CIS solar modules," *Thin Solid Films*, 515 (15), 5973–5978

Dobson, K., Visoly-Fisher, I., Hodes, G., & Cahen, D. (2000): Stability of CdTe/CdS thin film solar cells, *Sol. Energy Mater. Sol. Cells*, 62, 295–325

Eberspacher, C., Ermer, J.H., & Mitchell, K. (1989): *Process for making thin-film solar cell*, EU Patent Application No.0318315A2.

Eberspacher, C., Pauls, K., & Serra, J. (2002): Non-Vacuum Processing of CIGS Solar Cells, *Proc. 29th IEEE PV Specialists Conf*, May 2002, New Orleans, LA, pp. 684

Edwards, P.R., Halliday, D.P., Durose, K., Richter, H., & Bonnet, D. (1997): The influence of CdCl$_2$ treatment and interdiffusion on grain boundary passivation in CdTe/CdS solar cells, *Proc. 14th EU PV Solar Energy Conf*, June 1997, Barcelona, Spain, pp. 2083–2086

Edwards, P.R., Durose, K., Galloway, S.A., & Bonnet, D. (1998): Front-wall electron beam-induced current studies on thin-film CdS/CdTe solar cells, *Proc. 2nd WC PV Solar Energy Conversion*, Vienna, Austria, pp. 472–476

Ennaoui, A., Weber, M., Scheer, R., Lewerenge, H.J. (1998): Chemical-bath ZnO buffer layer for CuInS$_2$ thin-film solar cells, *Sol. Energy Mater. & Sol. Cells*, 54, 277–286

Ennaoui, A., U. Blieske, & M. CH. Lux-Steiner. (1998): 13.7%-efficient Zn (Se,OH)x/Cu(In,Ga)(S,Se)2 thin-film solar cell, *Prog. Photovolt: Res. and Appl*, 6, no. 6, 447–451

Ferekides, C.S., Marinskiy, D., Viswanathan, V., Tetali, B., Palekis, V., Selvaraj, P., & Morel, D.L. (2000): High efficiency CSS CdTe solar cells, *Thin Solid Films*, **361–362**, 520–526

Gabor, A., Tuttle, J., Contreras, M., Albin, D., Franz, A., Niles, D., & Noufi, R. (1993): High efficiency polycrystalline Cu(In,Ga)Se2-based solar cells, *Proc.12th NREL PV Program Review*, Denver, Colo., 1993, pp. 59–66

Gabor, A.M., Tuttle, J.R., Albin, D.S., Contreras, M.A., Noufi, R., & Hermann, A.M. (1994): Hugh efficiency CuIn$_x$Ga$_{1-x}$Se$_2$ solar cells made from. (In$_x$Ga$_{1-x}$)$_2$Se$_3$ precurser films, *Appl. Phys. Letts*, **65**, pp. 198–200

Gabor, A.M., Tuttle, J.R., Bode, M.H., Franz, A., Tennant, A.L., Contreras, M.A., Noufi, R., Jensen, D.G., & Hermann, A.M. (1996): Bandgap engineering in Cu(In,Ga) Se2 thin films grown from (In,Ga)$_2$Se$_3$ precursors, *Sol. Energy Mater. and Sol. Cells*, **41–42**, 247–260

Gal, D., Hodes, G., Lincot, D., & Schock, H.W. (2000): Electrochemical deposition of Zinc oxide films from non-aqueous solution: a new buffer/window process for thin film solar cells, *Thin Solid Films*, **361–362**, 79–83

Ge¡cys, P., Ra¡ciukaitis, G., Gedvilas, M., & Selskis, A. (2009): Laser structuring of thin-film solar cells on polymers, *EPJ Appl. Physics*, **46**, no. 1, Article ID 12508

Gessert, T.A., Smith, S., Moriarty, T., Young, M., Asher, S., Johnston, S., Duda, A & DeHart, C. (2005): Evolution of CdS/CdTe Device Performance during Cu Diffusion, *Proc. 31st IEEE PV Specialists Conf.*, January 2005, IEEE, Piscataway, N.Y, pp. 291–294

Gloeckler, M. & Sites, J.R. (2005): Potential of Sub-micrometer Thickness Cu(In,Ga)Se$_2$ Solar Cells, *J. Appl. Phys.* **98**, 103703

Green, M.A, Emery, K., Hishikawa, Y., & Warta, W. (2008): Solar Cell Efficiency Tables. (Version 31), *Prog. Photovolt: Res. Appl.* **16**, 61–67

Gunasekaran, M. & Ichimura, M. (2007): Photovoltaic cells based on pulsed electrochemically deposited SnS and photochemically deposited CdS and Cd$_{1-x}$Zn$_x$S, *Sol. Energy Mater. & Sol. Cells*, **91**, 774–778

Gupta, A. & Compaan, A.D. (2004): All sputtered 14% CdS/CdTe thin film solar cell with ZnO:Al transparent conducting oxide, *Appl. Phys. Letts*, **85**, 684–686

Gupta, A. & Compaan, A.D. (2005): High efficiency 1 micron thick sputtered CdTe solar cells, *Conf. recordof the 31st IEEE PV Specialists Conf*, January 2005, (IEEE, Piscataway), N.Y, pp. 235

Haalboom, T., Godecke, T., Ernst, F., Ruhle, M., Herberholz, R., & Schock, H.W. (1998): Phase relations and microstructure in bulk materials and thin films of the ternary system Cu-In-Se, *Proc. 11th International Conf. on Ternary and Multinary Compounds*, Salford, UK, pp. 249–252

Hanket, G.M., Singh, U.P., Eser, U., Shafarman, W.N., & Birkmire, R.W. (2002): Pilot-scale manufacture of Cu(InGa)Se2 films on a flexible polymer substrate, *Proc. 29th IEEE PV Specialists Conf*, New Orleans, LA, pp. 567–569

Hariscose, D., Ruckh, M., & Ruhle, U. (1996): A novel cadmium-free buffer layer for Cu(In,Ga)Se$_2$ based solar cells, *Sol. Energy Mater. and Sol. cells*, **41–42**, 345–353

Hedstrom, J., Ohlsen, H., & Bodegard, M. et al. (1993): ZnO/CdS/Cu(In,Ga)Se$_2$ thin-film solar cells with improved performance, *Proc. 23rd IEEE PV Specialists Conf*, Louisville, KY, pp. 364–371

Hengel, I., Klenk, R., Garcia Villora, E., & LuxSteiner, M.-Ch. (1996): *Proc. 2nd WC PV Solar Energy Conversion*, Vienna, Austria, p. 545

Herz, K. (2002): Dielectric barriers for flexible CIGS solar modules, *Thin Solid Films*, **403–404**, 384–389

Herrmann, D., Kessler, F., Herz, K., Eicke, A., Powalla, M., Niegisch, N., Menning, M., Schulz, A., Schneider, J., & Schumacher, U. (2003): High-performance barrier layers for flexible CIGS thin-film solar cells on metal foils, *Proc. MRS Symp: Compound Semiconductor PV*, San Fransisco, CA, **763**, pp. 287–292

Hollars, D.R., Dorn, R., Paulson, P.D., Titus, J., & Zubeck, R. (2005): Large area Cu(In,Ga)Se2 films and devices on flexible substrates made by sputtering, *Proc. Mater. Res. Society Spring Meeting*, 477–482, F14.34.1

Ikegami, S. (1988): CdS/CdTe solar cells by the screen-printing –sintering technique: Fabrication, photovoltaic properties and applications, *Solar Cells*, **23**, 89–105

Irvine, S.J.C., Barrioz, V., Lamb, D., Jones, E.W., & Rowlands-Jones, R.L. (2008): MOCVD of thin-film photovoltaic solar cells – Next-generation production technology?, *J. Crystal growth*, **310**, 5198–5203

Irvine, S.J.C. ((1997): Metal-Organic Vapour phase epitaxy, In '*Narrow gap II-VI compounds for Opto-electronics and Electromagnetic Applications*, P. Capper (Ed.), Chapman and Hall, London, UK, pp. 71–96

Ishizuka, S., Hommoto, H., Kido, N., Hashimoto, K., Yamada, A., & Niki, S. (2008): Efficiency enhancement of Cu(In,Ga)Se2 solar cells fabricated on flexible polyimide substrates using alkali-silicate glass thin layers, *Appl.Phys. Express*, 1, Article ID 092303, 1–3

Ishizuka, S., Yamada, A., Fons, P., & Niki, S. (2008a): Flexible Cu(In,Ga)Se2 solar cells fabricated using alkali-silicate glass thin layers as an alkali source material, *J. Renewable Sustainable Energy*, 1, Article ID 013102

Ishizuka, S. (2008b): Flexible CIGS photovoltaic cell with energy conversion efficiency of 17.7%—enabling development of a sticker-type high-performance solar cell, *AIST Today*, 8 (10), 20

Ishizuka, S., Yamada, A., Matsubara, K., Fons, P., Sakurai, K. & Niki, S. (2009): Development of high-efficiency flexible Cu(In,Ga)Se$_2$ solar cells: a study of alkali doping effects on CIS, CIGS, and CGS using alkali-silicate glass thin layers, *Current Appl. Phys*, **20**, supplement 1, S154–S156

Jager, H., & Seipp, E. (1981): Transition resistances of ohmic contacts to p-type CdTe and their time-dependant variation, *J. Electronic Materials*, **10**, 605–618

Jayarama Reddy, P. (2010): *Science and Technology of Photovoltaics*, 2nd Edn, CRC Press, Leiden, The Netherlands, ISBN 13:978-0-415-57363-4

Jitsukawa, H., Matsushita, H., & Takizawa, T. (1998): Phase diagrams of the (Cu2Se, CuSe)-CuGaSe2 system and the crystal growth of CuGaSe2 by the solution method, *J. Crystal Growth*, **186**, 587–593

Johnson, D.R. (2000): Microstructure of electrodeposited CdS/CdTe cells, *Thin Solid Films*, **361–362**, 321–326

Kapur, V.K. & Basol, B.M. (1990): Key issues and cost estimates for the fabrication of CuInSe$_2$ (CIS) PV modules by the two-stage process, *Proc. 21st IEEE PV Specialists Conf*, Kissimme, FL, pp. 467

Kapur, V.K., Bansal, A., Le, P., & Asensio, O.I. (2002): Non-Vacuum Processing for CIGS Solar Cells on Rigid and Flexible Substrates, *Proc. 29th IEEE PV Specialists Conf*, May 2002, New Orleans, LA, pp. 688

Karg, F., Aulich, H.a., & Riedl, W. (1997): Persistent photoconductivity in Cu(In,Ga)Se2 heterojunctions, *Proc. 14th EU PV Specialists Conf*, Barcelona, Spain, pp. 2012

Kazmerski, L.L., Ayyagari, M.S., & Sanborn, G.A. (1975): CuInS$_2$ thin-films: Preparation and properties, *J. Appl. Phys*, **46**, 4865–4869

Kazmerski, L.L., White, F.R., & Morgon, G.K. (1976): Thin-film CuInSe$_2$/CdS heterojunction solar cells, *Appl. Phys. Letts*, **29**, 268–270

Kazmerski, L.L. & Sanborn, G. (1977): CuInS$_2$ thin-film homojunction solar cells, *J. Appl. Phys*, **48**, 3178–3180

Kazmerski, L.L. & Juang, Y. (1977): Vacuum deposited CuInTe$_2$ thin-films – Growth, structural and electrical properties, *J. Vacuum Science & Technol*, **14**, 769–776

Kemell, M., Ritala, M. & Leskelä, M. (2005): Thin film deposition methods for CuInSe2 solar cells, *Critical Rev. in Solid State and Materials Sci*, **30**, 1–31

Kessler, J., Velthaus, K.O., Ruekh, M., & Schock, H.W. (1992): *Proc.6th PV Science and Engineering Conf*, New Delhi, pp. 1005–1010

Kessler, J., Ruckh, M., Hariskos, D., Ruhle, U., Menner, R., & Schock, H.W. (1993): Interface engineering between CuInSe$_2$ and ZnO, *Proc. 23rd IEEE PV Specialists Conf*, Louisville, Ky, pp. 447–452

Kessler, J., Schmid, D., Zweigart, S., Dittrich, S., & Schock, H.W. (1994): *Proc. 12th EU PV Solar energy Conf*, Amsterdam, The Netherlands, pp. 648–652

Kessler, F., Herrmann, D., & Powalla, M. (2005): Approaches to flexible CIGS thin-film solar cells, *Thin Solid Films*, **480–481**, 491–498

Kim, M.S., Yun, J.H., Yoon, K.H., & Ahn, B.T. (2007): Fabrication of flexible CIGS solar cell on stainless steel substrate by coevaporation process, *Diffusion and Defect Data B*, **124–126**, pp. 73–76

Klenk, R., Walter, T., Schock, H.-W., & Cahen, D. (1993): A model for the successful growth of polycrystalline films of CuInSe2 by multisource physical vacuum evaporation, *Advanced Materials*, **5**, pp. 114–119

Kim, D., Pozder, S., & Qi, B. (1994): Thin-film CdS/CdTe solar cells fabricated by electrodeposition, *AIP Conference Proceedings*, **306**, pp. 320–328

Kniese, R., Powalla, M., & Rau, U. (2007): Characterization of the CdS/Cu. (In,Ga)Se$_2$ interface by electron beam induced currents, *Thin Solid Films*, **515**, 6163–6167

Kotina, I.M., Tukhkonen, L.M., Patsekina, G.V., Shchukarev, A.V., & Gusinskii, G.M. (1998): Study of CdTe etching process in alcoholic solutions of bromine, *Semiconductor Science & Technology*, **13**, 890

Kuo, S., Jeng, M., Chang, L., Lin, W., Chen, C., Hu, S., Lu, Y & Ching-Wen Wu. (2009): Optimization of Growth Parameters for Improved Adhesion and Electricity of Molybdenum Films Deposited by RF Magnetron Sputtering, *34th IEEE PV Specialists Conf*, June 7–12, 2009, Philadelphia, PA, Available at File:///E:/34%20IEEE%202009/manuscripts/Mpublic289_0623200917.pdf

Kushiya, K. (2003): Progress in large-area Cu(InGa)Se$_2$-based thin-film modules with the efficiency of over 13%, *Proc. 3rd WC PV Energy Conversion*, Osaka, Japan, pp. 319–324

Levi, D.H., Moutinho, H.R., Hasoon, F.S., Keyes, B.M., Ahrenkiel, R.K., Al-Jassim, M., Kazmerski, L.L., & Birkmire, R. (1996): Micro through nanostructure investigations of polycrystalline CdTe: Correlations with processing and electronic structures, *Sol. Energy Mater. and Sol. cells*, **41–42**, 381–393

Li, W., Wei-Ming, S., Juan, Q. *et al.* (2009): The investigations of heterojunction solar cells based on the p-SnS thin-films, *Tech. Digest. 18th PV Science & Engineering Conf*, January 2009, Kolkata, India

Loferski, J.J. (1956): Theoretical considerations governing the choice of the optimum semiconductor for photo-voltaic solar energy conversion, *J. Appl. Phys*, **27**, 777–784

Luque, A. & S. Hegedus, Eds. (2003): *Handbook of PV Science & Engineering*, John Wiley & Sons, New York, NY

Markvant, T. & L. Castarier, Eds. (2003): *Practical Handbook of Photovoltaics: Fundamentals and Applications*, Elsevier, Amsterdam, The Netherlands

Mattox, D. (1998): *Handbook of Physical Vapor Deposition Processing*, Noyes, Park Ridge, NJ, USA

Mathew, X., Enriquez, J.P., Romeo, A., & Tiwari, A.N. (2004): CdTe/CdS solar cells on flexible substrates, *Solar Energy*, **77**, 831–838

Marsillac, S., Paulson, P.D., Haimbodi, M.W., Birkmire, R.W., & Shafarman, W.N. (2002): High-efficiency solar cells based on Cu(InAl)Se$_2$ thin films, *Appl. Phys. Letts*, **81**, 1350–1352

Maycock, P.D. (2008): The future of Photovoltaics, *Proc. 33rd IEEE PV Specialists Conf*, May 2008, San Diego, CA

McCandless, B.E., Moulton, L.V., & Birkmire, R.W. (1997): Recrystallization and sulfur diffusion in $CdCl_2$-treated CdTe/CdS thin-films, *Prog. Photovolt: Res. and Appls*, 5, 249–260

McCandless, B., & Sites, J. (2003): Cadmium telluride solar cells, In *Handbook of PV Science and Engineering*, (Ed.) A. Luque and S. Hegedus, Wiley& Sons, Chichester, UK, pp. 617–657

McCandless, B.E. & Dobson, K.D. (2004): Processing options for CdTe thin film solar cells, *Solar Energy*, 77, 839–856

Meyers, P.V., Zhou, T., Powell, R.C., & Reiter, N. (1993): Elemental vapour deposited polycrystalline CdTe thin film photo-voltaic modules, *Proc. 23rd IEEE PV Specialists Conf*, Louisville, KY, pp. 400–404

Meyers, P.V. (2006): First solar polycrystalline CdTe thin film PV, *Proc. 4th IEEE WC on PV Energy Conversion*, May 2006, Waikoloa, HI, vol. 2, pp. 2024–2027

Meyers, P.E. (1988): Design of a thin-film CdTe solar cell, *Solar Cells*, 23, 59–67

Michelsen, R. & Chen, W. (1980): High photocurrent polycrystalline thin-film $CdS/CuInSe_2$ solar cell, *Appl. Phys. Letts*, 36, 371–373

Mickelsen, R.A. & Chen, W. (1981): Development of a 9.4% efficient thin-film $CuInSe_2$/CdS solar cell, *Proc.15th IEEE PV Specialists Conf*, Orlando, FL, pp. 800

Michelsen, R. & Chen, W. (1982): Polycrystalline thin film $CuInSe_2$ solar cells, *Proc. 16th IEEE PV Specialists Conf*, San Diego, CA, pp. 781–785

Mikami, R., Miyazaki, H., Abe, T., Yamada, A., & Konagai, M. (2003): Chemical bath deposited. (CBD)-ZnO buffer layer for CIGS solar cells, *Proc. 3rd WC on PV Energy Conversion*, May 2003, Osaka, Japan, pp. 519–522

Minimoto, T., Hayashi, T., & Araki, T. (2007): Electronic properties of $Cu(In,Al)Se_2$ solar cells prepared by three-stage evaporation, in ENERGEX 2007, *The 12th International Energy Conf. & Exhibition*, Singapore.

Mikkelsen, J.C. (1981): Ternary phase relations of the chalcopyrite compound $CuGaSe_2$, *J. Electronic Materials*, 10(3), 541–558.

Miles, R.W., Hynes, K.M., & Forbes, I. (2005): Photovoltaic solar cells: An overview of state-of-the-art cell development and environmental issues, *Prog. Crystal Growth and Characterization of Materials*, 51, 1–42

Miles, R.W., Zoppi, G., Ramakrishna Reddy, K.T., & Forbes, I. (2010): Thin Film Solar cells based on the use of Polycrystalline Thin film materials, Zhang, S (ed.), In: *Organic Nanostructured Thin film Devices and Coatings for Clean energy*, London, CRC Press, pp. 1–56

Moutinho, H., Hasoon, F.S., Abulfotuh, F.A., & Kazmerski, L.L. (1995): Investigation of polycrystalline CdTe thin-films deposited by physical vapour deposition, close-spaced sublimation, and sputtering, *J. Vac. Sci. and Tech*, A 13, 2877–2883

Muffler, H., Bar, M., Fischer, C.-H., Gay, R., Karg, F., & Lux-Steiner, M.C. (2000): Sulfide buffer layers for $Cu(InGa)(S,Se)_2$ solar cells prepared by ion layer gas reaction. (ILGAR), *Proc. 28th IEEE PV Specialists Conf*, September 2000, Anchorage, AK, pp. 610–613

Muller, J., Nowoczin, J., & Schmitt, H. (2006): Composition, structure and optical properties of sputtered thin-films of $CuInSe_2$, *Thin Solid Films*, 496, 364–370

Nadenau, V., Hariskos, D., Schock, H.-W., Krejci, M., Haug, F.J., Tiwari, A.N., Zogg, H., & Kostorz, G. (1999): Microstructural study of the $CdS/CuGaSe_2$ interfacial region in $CuGaSe_2$ thin film solar cells, *J.Appl. Physics*, 85, 534–542

Naghavi, N., Spiering, S., Powalla, M., Canava, B., & Lincot, D. (2003): High-efficiency copper indium gallium diselenide solar cells with indium sulfide buffer layer deposited by atomic layer chemical vapour deposition (ALCVD), *Prog. Photovolt: Res. Appls*. 11, 437–443

Negami, T., Aoyagi, T., Satoh, T. *et al.* (2002): Cd free CIGS solar cells fabricated by dry processes, *Proc. 29th IEEE PV Specialists Conf*, New Orleans, LA, pp. 656–659

Negami, T., Satoh, T., Hashimoto, Y., Shimakawa, S., Hayashi, S., Muro, M., Inoue, H., & Kitagawa, M. (2002): Production technology for CIGS thin film solar cells, *Thin Solid Films*, **403–404**, 197–203

Noguchi, H., Setiyadi, A., Tanamura, H., Nagatomo, T., & Otomo, O. (1994): Characterisation of vacuum-evaporated tin sulfide film for solar cell materials, *Sol. Energy Mater. and Sol. Cells*, **35**, 325–331

Noufi, R., & Zweibel, K. (2006): High-Efficiency CdTe and CIGS Thin-Film Solar Cells: Highlights and Challenges, Conf. paper: NREL/CP-520-39894, *4th WC PV Energy Conversion*, May 2006, Waikoloa, HI

Noufi, R., Coutts, T., Bhattacharya, R. *et al.* (2007): Polycrystalline CdTe and CIGS thin-film PV Research, *DOE Solar Energy technologies Program Peer Review*, Denver, CO, April 2007

Ohyama, H., Aramoto, T., Kumazawa, S., Higuchi, H., Arita, T., Shibutani, S., Hishio, T, Hakajima, J., Tsuji, M., Hanafusa, A., Hibina, T., Omura, K., & Murozono, M. (1997): 16.0% efficient thin-film CdS/CdTe solar cells, *Proc. 26th IEEE PV Specialists Conf*, Anaheim, CA. pp. 343–346

Olsen, L.C., Lei, W., & Addis, F.W. (1997): High efficiency CIGS and CIS cells with CVD ZnO buffer layers, *Proc. 26th IEEE PV Specialists Conf*, Anaheim, CA, pp. 363–366

Orgassa, K., Schock, H.W., & Werner, J.H. (2003): Alternative back contact materials for thin-film Cu(In,Ga)Se$_2$ solar cells, *Thin Solid Films*, **431–432**, 387–391

Ortega Borges, R., Lincot, D., & Vedel, J. (1992): *Proc. 11th EU PV Solar Energy Conf*, Montreux, Switzerland, pp. 862

Otte, K., Makhova, L., Braun, A., & Konovalov, I. (2006): Flexible Cu (In,Ga)Se$_2$ thin-film solar cells for space application, *Thin Solid Films*, **511–512**, 613–622

Park, J.S., Dong, Z., Kim, S., & Perepezko, J.H. (2000): CuInSe$_2$ phase formation during Cu$_2$Se/In$_2$Se$_3$ interdiffusion reaction, *J. Appl. Physics*, 87, 3683–3690

Pistor, P., Cabellero, R., Hariscose, D., Izquierdoroca, V., Wachter, R, Schorr, S., & Klenk, R. (2009): Quality and stability of compound indium sulfideas source material for buffer layers in Cu(In,Ga)Se$_2$ solar cells, *Sol. Energy Mater. and Sol. cells*, 93, 148–152

Platzer-Bjorkman, C., Kessler, J., & Stolt, L. (2003): Atomic layer deposition of Zn(O,S) buffer layers for high-efficiency Cu(In,Ga)Se$_2$ solar cells, *Proc. 3rd WC on PV Energy Conversion*, May 2003, Osaka, Japan, pp. 461–464

Powalla, M., Dimmler, B., Schaffler, R., Voorwinden, G., Stein, U., & Mohring, H-D. (2004): CIGS solar modules: Progress in pilot production, new development and applications, *Proc. 19th EU PV Solar Energy Conf*, Paris, France, pp. 1663

Powalla, M. (2006): The R&D potential of CIS thin-film solar modules, *Proc. 21st EU PV Solar Energy Conf*, September 2006, Dresden, Germany, pp. 1789

Powalla, M. & Bonnet, D. (2007): Thin-film solar cells based on the Polycrystalline Compound semiconductors CIS and CdTe, *Advances in OptoElectronics*, Vol. 2007, Article ID 97545, 6 pages, doi:10.1155/2007/97545, Hindawi Publishing Corporation.

Probst, V., Stetter, W., Palm, J., Toelle, R., Visbeck, S., Calwar, H., Niesen, T., Vogt, H., Hernandez, O., Wendl, M., & Karg, F.H. (2003): CIS module Pilot processing: From fundamental investigations to Advanced performance, *Proc. 3rd WC PV Energy Conversion*, May 2003, Osaka, Japan, pp. 329–334

Ramakrishna Reddy, K.T., Koteswara Reddy, N., & Miles, R.W. (2006): Photovoltaic properties of SnS based solar cells, *Sol. Energy Mater. and Sol. cells*, **90**, 3041

Ramanathan, K., Hasoon, & F., Smith. (2002): Properties of Cd and Zn partial electrolyte treated CIGS solar cells, *Proc. 29th IEEE PV Specialists Conf*, New Orleans, LA, pp. 523–526

Rau, U., & H.W. Schock. (2001): Cu(InGa)Se$_2$ solar cells:In: Archer, M.D, Hill, R. (Eds). Clean electricity from Photovoltaics, Imperial College Press, 277–332

Rau, U., & Schock, H.W. (1999): Electronic properties of Cu(In,Ga)Se2 heterojunction solar cells – recent achievements, current understanding, and future challenges, *Appl. Phys.* **A69**, 131–147

Rau, U., Schmitt, M., Parisi, J., Riedl, W., & Karg, F. (1998): Persistent photoconductivity in Cu(In,Ga)Se2 heterojunctions and thin films prepared by sequential deposition, *Appl. Phys. Letts*, 73, 223–225

Reddy, K.T.R & Jayarama Reddy, P. (1992a): Studies of ZnxCd1-xS films and Zn_xCd_{1-x} S/CuGaSe$_2$ heterojunction solar cells, *J. of Physics D: Appl. Phys*, 1345

Reddy, K.T.R., Gopalaswamy, H., & Jayarama Reddy, P. (1992b): Polycrystalline CuGaSe2 thin film solar cells, *Vacuum*, **43**, 811–815

Repins, I., Contreras, M.A., Egaas, B., DeHart, C., Scharf, J., Perkins, C.L., *et al.* (2008): 19.9% efficient ZnO/CdS/CuInGaSe2 solar cell with 81.2% fill factor, *Prog. Photovolt: Res. Appl*, **16**, 235–239

Ristov, M., Sinadinovski, G., Mitreski, M., & Ristova, M. (2001): Photovoltaic cells based on chemically deposited p-type SnS, *Sol. Energy Mater. and Sol. cells*, **69**, 17–24

Roedern, B., Ullal, H.S., & Zweibel, K. (2006): Polycrystalline Thin-Film Photovoltaics: From the Laboratory to Solar Fields, Conf. paper: NREL-520-39838, presented at *IEEE 4th WC on PV Energy Conversion*, Waikoloa, HI, May 7–12.

Roedern, B. & Ullal, H.S. (2008): The Role of Polycrystalline Thin-Film PV Technologies in Competitive PV Module Markets, Conf. paper: NREL/CP-520-42494, May 2008, Presented at *33rd IEEE PV Specialists Conf*, San Diego, CA, USA

Rogol, M. (2008): Refining Benchmarks and Forecasts – Rogol's Monthly Market Commentary, *Photon International*, Jan. 2008, p 84

Romeo, A., Terheggen, M., Abou-Ras, D. *et al.* (2004): Development of thin-film Cu(In,Ga)Se2 and CdTe solar cells, *Prog. Photovolt: Res. and Appls*, **12**(2–3), pp. 93–111

Romeo, A., Abou-Ras, D., Gysel, R. (2004a): Properties if CIGS solar cells developed with evaporated II-VI buffer layers, *Tech. Digest of 14th Intl. PV Science & Engineering Conf*, Bangkok, Thailand, pp. 705–706

Sanchez-Juarez, A., Tiburcio-Silver, A., & Ortiz, A. (2005): Fabrication of SnS$_2$/SnS hetero-junction thin-film diodes by plasma-enhanced chemical vapour deposition, *Thin Solid Films*, **480–481**, 452–456

Sandwisch, D.W. (1994): Development of CdTe module manufacturing, *Proc. 1st WC PV Solar Energy Conversion*, Waikoloa, HI, pp. 836–839

Sarlund, J., Ritala, M., Leskela, M., Siponmaa, E., & Zilliacus, R. (1996): Characterisation of etching procedure in preparation of CdTe solar cells, *Sol. Energy Mater. and Sol. Cells*, **44**, 177–190

Satoh, T., Hashimoto, Y., Shimakawa, S., Hayashi, S., & Negami, T. (2000): "Cigs solar cells on flexible stainless steel substrates, *Proc. Conf. Record of the 28th IEEE PV Specialists Conf*, September 2000, Anchorage, AK, pp. 567.

Scofield, J.H., Duda, A., Albin, D., Ballard, B.L., & Predecki, P.K. (1995): Sputtered molybdenum bilayer back contact for copper indium diselenide-based polycrystalline thin-film solar cells, *Thin Solid Films*, **260**, 26–31

Schmid, U., & Seidel, H. (2005): Effect of substrate properties and thermal annealing on the resistivity of molybdenum thin films, *Thin Solid Films*, **489**, 310–319

Shafarman, W.N., & Zhu, J. (2000): Effect of substrate temperature and deposition profile on evaporated Cu(InGa)Se2 films and devices, *Thin Solid Films*, **361**, 473–477

Shay, J.L., Wagner, S., & Kasper, H.M. (1975): Efficient CuInSe$_2$/CdS solar cells, *Appl. Phys. Letts*, **27**, 89–90

Shi, C.Y., Sun, Y., He, Q., Li, F.Y. & Zhao, J.C. (2009): Cu (In,Ga)Se$_2$ solar cells on stainless-steel substrates covered with ZnO diffusion barriers, *Sol. Energy Mater. and Sol. Cells*, **93**(5), 654–656

Singh, U.P. & Patra, S.P. (2010): Progress in polycrystalline thin-film Cu (In,Ga)Se$_2$ solar cells, *International. J. Photoenergy*, Article ID 468147, 19 p, doi: 10.1155/2010/468147

Skarp, J., Koskinen, Y., Lindfors, S. *et al.* (1991): Development and evaluation of CdS/CdTe thin film PV cells, *Proc. 10th EU PV Solar Energy Conf*, April 1991, Lisbon, Portugal, pp. 567–569

Spiering, S., Hariskos, D., Powalla, M., Naghavi, N., & Lincot, D. (2003): Cd-free Cu(In,Ga)Se2 thin-film solar modules with In$_2$S$_3$ buffer layer by ALCVD, *Thin Solid Films*, **431–432**, 359–363

Spiering, S., Eicke, A., Hariskos, D. *et al.* (2004): Large-area Cd-free CIGS solar modules with In2S3 buffer layer deposited by ALCVD, *Thin Solid Films Proc. Symposium D on Thin Film and nano-structured Materials for Photovoltaics, of the E-MRS 2003 Spring Conference*, Strasbourg, France, **451–452**, 562–566

Strohm, A., Schlotzer, T., Nguyen, Q. *et al.* (2004): New approaches for the fabrication of Cd-free Cu(In,Ga)Se$_2$ heterojunctions, *Proc. 19th EU PV Solar Energy Conf*, Paris, France, pp. 1741–1744

Takamoto, T., Agui, T., Kurita, H. *et al.* (1997): Improved junction formation procedure for low temperature deposited CdS/CdTe solar cells, *Sol. Energy Mater. and Sol. cells*, **49**, 219–225

Tiwari, A.N., Krejci, M., Haug, F.-J., & Zogg, H. (1999): 12.8% efficiency Cu. (In,Ga)Se2 solar cell on a flexible polymer sheet, *Prog. Photovolt: Res. and Appls*, 7, 393–397.

Tiwari, A.N: Solar electricity – Prospects of thin film Solar cells, Laboratory for Thin films and Photovoltaics, EMPA, Switzerland.

Tuttle, J.R, Contreras, M., Tennant, A., Albin, D., & Noufi, R. (1993): High efficiency thin-film Cu. (In, Ga) Se2-based photovoltaic devices: progress towards a universal approach to absorber fabrication, *Proc. 23rd IEEE PV Specialists Conf*, New York, NY, pp. 415–421

Tuttle, J.R., Szalaj, A., & Keane, J. (2000): A 15.2% AM0 1433 W/kg thin-film Cu(In,Ga)Se2 solar cell for space applications, *Proc. 28th IEEE PV Specialists Conf*, Anchorage, AK, pp. 1042–1045

Ullal, H.S., Zweibel, K., & von Roedern, B. (2002): Polycrystalline thin film photovoltaics: Research, development, and technologies, *Proc. of the 29th IEEE PV Specialists Conference*, May 2002, New Orleans, La, pp. 472–477

Ullal, H.S. & von Roedern, B. (2007): Thin Film CIGS and CdTe Photovoltaic Technologies: Commercialization, Critical Issues, and Applications, Conf. paper: NREL/CP-520-42058, Sept. 2007, presented at *22nd EU PV Solar energy Conf*, Milan, Italy, September 2–7, 2007

Velthaus, K.O., Kessler, J., Ruckh, M., Schmid, D., & Schock, H.W. (1992): *Proc. 11th EU PV Solar energy Conf*, Montreux, Switzerland, pp. 842

Wagner, S. (1977): In *'Ternary Compounds'*, G.D Holah. (Ed), Inst. Of Physics, London, UK, Vol.35, pp. 205

Wendt, R., Fischer, A., Grecu, D., & Compaan, A.D. (1998): Improvement of CdTe solar cell performance with discharge control during film deposition by magnetron sputtering, *J. Appl. Phys*, 84, 2920–2925

Woodcock, J.M., Turner, A.K., Ozsan, M.E., *et al.* (1991): Thin-film solar cells based on electro-deposited CdTe, *Conf, record of the 22nd IEEE PV Specialists Conf*, Las Vegas, NV, pp. 842–847

Woodcock, J.M., Schade, H., Maurus, H., Dimmler, B., Springer, J., & Ricaud, A (1997): A study of the upscaling of thin film solar cell manufacture towards 500 MWp per annum, *Proc. 14th EU PV Solar Energy Conf*, H. S. Stephens & Associates, Barcelona, Spain, pp. 857–860

Woods, L.M., Kalla, A., Gonzalez, D. *et al.* (2005): Wide-bandgap CIAS thin-film photovoltaics with transparent back contacts for next generation single- and multi-junction devices, *Materials Sci. & Engineering*, B **116**, 297–302

Wu, X., Keane, J.C., Dhere, R., DeHart C., Duda A., Gessert TA., Asher S., Levi D.H., & Sheldon, P. (2001): 16.5%-efficient CdS/CdTe polycrystalline thin-film solar cell, *Proc. 17th EU PV Solar energy Conf*, Munich, Germany, pp. 995–1000

Wuerz, R., Eicke, A., Frankenfeld, M., Kessler, F., Powalla, M., Rogin, P, & Yazdani-Assl, O. (2009): CIGS thin-film solar cells on steel substrates, *Thin Solid Films*, 517, 2415–2418.

Yagioka, T. & Nakada, T. (2009): Cd-free flexible Cu(In,Ga)Se$_2$ thin film solar cells with ZnS(O,OH) buffer layers on Ti foils, *Appl. Phys. Express*, 2 (7), Article ID 072201, 3 pages

Zachmann, H., Heinker, S., Braun, A., Mudryi, A.V., Gremenok, V.F., Ivaniukovich, A.V., & Yakushev, M.V. (2009): Characterisation of Cu(In,Ga)Se$_2$ based thin film solar cells on polyimide. (2009): *Thin Solid Films*, 517, 2209–2212

Zhang, L., He, Q., Jiang, W.-L, Li, C.-J, & Sun, Y. (2008): Flexible Cu(In, Ga)Se$_2$ thin-film solar cells on polyimide substrate by low-temperature deposition process, (2008): *Chinese Physics Letters*, 25, 734–736.

Zweigart, T., Walter, T., Koble, C., Sun, S.M, Ruhle, U., & Schock, H.W. (1994): Sequential deposition of Cu(In,Ga)(S,Se)$_2$, *Proc. IEEE 1st WC on PV Energy Conversion*, December 1994, Waikoloa, HI, pp. 60–67

Zweibel, K. (2005): The Terawatt Challenge for Thin-Film PV, *NREL Technical Report*, NREL/TP-520-38350 (August, 2005)

Chapter 4

Organic and dye-sensitized solar cells

4.1 INTRODUCTION

Currently, the active materials used for the fabrication of solar cells are mainly inorganic materials – silicon (Si), gallium-arsenide (GaAs), cadmium-telluride (CdTe), and copper-indium-gallium-selenide (CIGS). The power conversion efficiency for these solar cells varies from 8 to 30%. About 85% of the PV market is shared by mono- or multi-crystalline silicon solar cells. GaAs solar cells are reliable and highly efficient, but being expensive, are not generally used for terrestrial applications. Amorphous silicon, CdTe, and CIGS are comparatively recent thin-film technologies.

The current status of PV is that it hardly contributes to the energy market. The high production cost for the silicon wafer solar cells is one of the major issues. Even when the production costs get reduced, large-scale production of the current silicon solar cells could be limited by the scarcity of solar-grade silicon as happened around 2008. To ensure a sustainable technology path for PV, efforts to reduce the costs of the current silicon technology need to be balanced with measures to create and sustain diversity in PV technology. The thin film solar technologies have been progressing well, though a few issues are yet to be addressed for long time durability. Also, the limited availability of indium and tellurium stocks may also slow down large-scale production of thin-film solar cells. Therefore, 'techno diversity', implying development of new solar cell technologies, is crucial (Sanden 2003).

A new approach for the fabrication of solar cells made of entirely new materials, such as organic semiconductors has been studied since 1975; and the strongest case for the third-generation solar cell technologies (Green 2004) is certainly the technologies based on organic materials (especially small molecules and conjugate polymers) due to their promise for ultralow costs and several other merits.

Polymers are basically non-conductors of electricity. Conjugated materials can be described as 'organics' consisting of alternating single and double bonds (Figure 4.1). The discovery by Heeger, MacDiarmid, and Shirakawa that the conjugated polymer, polyacetylene (PA) could be made conductive and the conductivity could be increased by seven orders of magnitude upon oxidation with iodine (Chiang et al. 1977) that triggered research on conjugated polymers. These three scientists were awarded for their pioneering work, the Nobel Prize in Chemistry in 2000 (Shirakawa 2001, 2003; MacDiarmid 2001; Heeger 2001).

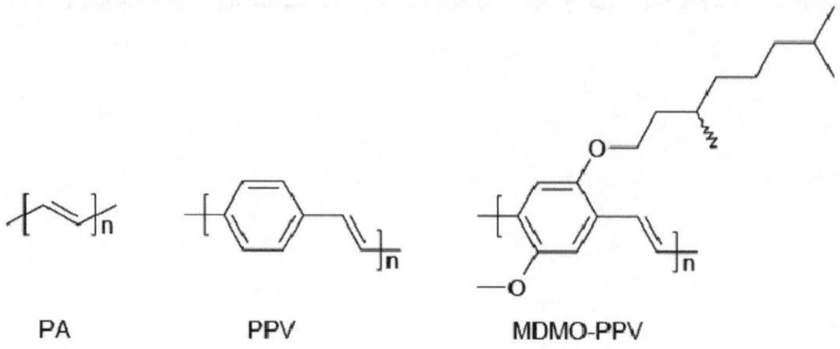

Figure 4.1 Molecular structures of the conjugated polymers: trans-polyacetylene (PA), poly(p-phenylenevinylene) (PPV), and a substituted PPV (MDMO-PPV)

Organic materials used presently in solar cells include conducting polymers, dyes, pigments, and liquid crystals. Among these, the conductive polymers are perhaps the best known for their photo-physical properties, and reviews are published on the subject (e.g., Wallace *et al.* 2000; Brabec & Sariciftci 2001c; Brabec *et al.* 2001b; Peumans *et al.* 2003).

Conjugated polymers and molecules have the immense advantage of simplicity, and chemical tailoring to alter their properties, such as the band gap. Conjugated polymers combine the electronic properties characteristic of traditional semiconductors and conductors with the simplicity of processing and mechanical flexibility of plastics. There is no shortage of the raw materials needed to make organic semiconductors.

Research on molecular thin film devices has started in 1975 (Tang 1975) and the conversion efficiency achieved was 0.001%. Since then, their energy conversion efficiencies have significantly improved to 1% in1986 (Tang 1986) and to 5.5% (Xue *et al.* 2004a, 2004b, 2005; Reyes-Reyes 2005). Several configurations have been studied using 'small molecules' (Xue *et al.* 2004b; Takahasi *et al.* 2000), conjugated polymers (Takahasi *et al.* 2000; Yu *et al.* 1995; Yu & Heeger 1995; Granstrom *et al.* 1998; Jenekhe & Yi 2000; Breeze *et al.* 2004; Brabec *et al.* 2001, 2003; Winder & Sarisiftci 2004; Padinger *et al.* 2003; Wienk *et al.* 2003), small molecules and conjugate polymers (Breeze *et al.* 2002a, 2002b; Nakamura *et al.* 2004), and organic and inorganic materials (Breeze *et al.* 2001) as 'active' layer (a layer in which the majority of the incident photons are absorbed and charges generated).

Molecules with a larger molecular weight (>1000 amu) are referred to as 'polymers' and lighter ones as 'small molecules'. Depending on their molecular structure and chemical composition, polymers can in principle be soluble or insoluble in common solvents. Polymer films can be prepared by several techniques, spin-coating, screen-printing, spray coating, or using recent developments in ink-jet printing, micro-contact printing, and other soft lithography techniques, allowing for large-area, ultra-thin, flexible and low cost devices. Conjugated polymers, and conjugated (small) molecules, however, are mainly thermally evaporated under high vacuum which is much more expensive than solution processing and, therefore, less attractive.

Modules consisting of cells connected in series can be printed in an integrated way. Multijunction solar cells can be made relatively easily since there is no need for crystal lattice matching between layers or high temperature annealing that could cause diffusion between layers, as will be seen in the following pages. In about 10–15 years, it may be possible to manufacture OPV modules with 15% efficiency at a cost of around \$30/m^2. This capability would truly revolutionize the way that the electricity is obtained (Bao et al. 2011).

Compared to existing technologies, there are several factors that attract global commercial interest in these organic semiconductors, such as (a) ability to be deposited at room temperature on a variety of low cost substrate materials (plastic, glass, metal foils), (b) relative ease of processing, (c) materials inherently flexible and readily available, (d) ultra-low cost, and (e) environment-friendly.

4.2 CONFIGURATION & PRINCIPLE OF ORGANIC SOLAR CELL

Photovoltaic cell configurations based on organic materials differ from those based on inorganic semiconductors, because the physical properties of the two groups of materials are significantly different.

Inorganic semiconductors generally have a high dielectric constant and a low exciton binding energy (for GaAs the exciton binding energy is 4 meV). Hence, the thermal energy at room temperature ($k_B T = 0.025$ eV) is sufficient to dissociate the exciton created by absorption of a photon into a positive and negative charge carrier. These electrons and holes are easily transported as a result of the high mobility of the charge carriers (for silicon, hole mobility $= 450$ cm^{2}V^{-1}s^{-1} and electron mobility $= 1400$ cm^{2}V^{-1}s^{-1}) and the internal field of the p-n junction.

Organic semiconducting materials, on the other hand, have a lower dielectric constant and the exciton binding energy is much larger than for inorganic semiconductors, although the exact magnitude remains a matter of debate. The mobilities are also less by several orders of magnitude (Sirringhaus 2005). The highest reported hole mobilities (μ_h) for organic semiconductors reach only about 15 cm^2 V^{-1} s^{-1} for single crystals of small molecules (Sundar et al. 2004) and 0.6 cm^{2}V^{-1}s^{-1} for liquid crystalline polymers (McCulloch et al. 2006). If a photon of sufficient energy is absorbed by the organic semiconductor, an electron is promoted into the lowest unoccupied molecular orbital (LUMO), leaving behind a hole in the highest occupied molecular orbital HOMO). But, this electron-hole pair forms a tightly bound state (called singlet exciton) due to electrostatic interactions. The thermal energy at room temperature is not sufficient to dissociate the exciton and generate free charge carriers even at typical internal fields of ~10^6 to 10^7 V/m (Gregg & Hanna 2003). For example, polydiacetylene, 0.5 eV is needed to split the exciton and, hence, dissociation into free charge carriers does not occur at room temperature. Similarly, in the widely used Poly(2-methoxy-5-(2'-ethyl-hexoxy)-p-phenylene vinylene) (MEH-PPV) (Braun & Heeger 1991) studies have shown that only 10% of the excitons dissociate into free carriers in a pure layer (Miranda, Moses & Heeger 2001), while the remaining excitons decay via radiative or nonradiative recombination. Thus, the energy efficiencies of single-layer polymer solar cells remain low, below 0.1% (Marks et al. 1994; Yu et al. 1994).

To overcome this problem, the hetero junction concept of utilising two different organic materials that differ in electron donating and accepting properties is utilised.

Charges are then created by photo-induced electron transfer between the two components. This photo-induced electron transfer between donor and acceptor boosts the photogeneration of free charge carriers compared to the individual materials.

The photovoltaic properties of crystalline inorganic semiconductor solar cells can be described by energy band models. The situation in molecular or polymeric organic solar cells is however much more complex because of the absence of three dimensional crystal lattice, different intra molecular and intermolecular interactions, local structural disorders, amorphous and crystalline regions, and chemical impurities (Wöhrle & Meissner 1991). The photophysics of organic materials is not yet fully understood (Petritsch 2000b), and no comprehensive theoretical models are available for explaining the organic thin film photovoltaic device characteristics from the basis of molecular properties of the materials (Meissner & Rostalski 2000). Hence, the operation of the organic solar cells is normally explained in the framework and terminology of conventional inorganic p-n junction solar cells to get a qualitative understanding.

In an organic photovoltaic cell, four important processes, namely, absorption of light, charge transfer and separation of the opposite charges, charge transport, and charge collection have to be optimized to obtain high conversion efficiency. For an efficient collection of photons, the absorption spectrum of the PV organic layer should match the solar spectrum and the layer should be sufficiently thick to absorb all incident light. A better match with the solar spectrum is obtained by lowering the band gap of the organic material, but this will ultimately have some bearing on the open-circuit voltage. Increasing the layer thickness is advantageous for light absorption, but burdens the charge transport.

Various structures for organic solar cells have been investigated in recent years. In most organic solar cells, charges are created by photo induced electron transfer. That is, an electron is transferred from an electron donor (D), a p-type material, to an electron acceptor (A), an n-type material, with the help of the additional energy of an absorbed photon (hν). In the photoinduced electron transfer reaction, excitation of the donor (D*) or the acceptor (A*) occurs first, followed by creation of the charge-separated state consisting of the radical cation of the donor (D$^{\bullet+}$) and the radical anion of the acceptor (A$^{\bullet-}$) as shown in the equation (Janssen 2006).

$$D + A + h\nu \rightarrow D^* + A \text{ (or } D + A^*) \rightarrow D^{\bullet+} + A^{\bullet-} \tag{4.1}$$

For an efficient charge generation, the charge-separated state is most important both thermodynamically and kinetically; hence, the energy of the absorbed photon must be used for generation of the charge separated state without loosing via processes like fluorescence or non-radiative decay. Additionally, the charge-separated state needs to be stabilized to enable the photo-generated charges to migrate to one of the electrodes. Therefore, the back electron transfer should be slowed down as much as possible.

The solar radiation falling on donor through a transparent electrode (ITO) results in the photo excited state of the donor, in which an electron is promoted from the highest occupied molecular orbital (HOMO) to the lowest unoccupied molecular orbital (LUMO) of the donor . Subsequently, the excited electron is transferred to the LUMO of the acceptor, resulting in an extra electron on the acceptor (A$^{\bullet-}$) and leaving a hole at the donor (D$^{\bullet+}$). These photo-generated charges are transported and collected at opposite electrodes. A similar charge generation process can occur, when the acceptor is photo excited instead of the donor.

To fabricate a photovoltaic cell, the photoactive material (D + A) is sandwiched between two dissimilar (metallic) electrodes (of which one is transparent), to collect the photo-generated charges. After the charge transfer reaction, the photo-generated charges have to migrate to these electrodes without recombination. It is also to be ensured that the photo-generated charges do not encounter any interface problems at the electrodes.

4.3 TYPES OF ORGANIC SOLAR CELLS

In principle, there are two groups of organic solar cells: Solar cells in which organic molecules are used for absorption of light and charge transport, and cells in which an organic or polymeric material is used for absorption of light only. The first group of solar cells is referred to as 'Organic Photovoltaic cells (OPV)', and the second group as 'dye-sensitized solar cells (DSSC)'. Excellent reviews, for example, by Wallace et al. (2000), Brabec and Saricifitci (2001c), Brabec et al. (2001b), Peumans et al. (2003), and Blom et (2007) for Organic cells, and by Hagfeldt and Gratzel (1995, 2000), Kalyanasundaram and Gratzel (1998), and Gratzel (2006) for Dye-sensitized solar cells have been published.

The remarkable progress in solar cell efficiencies with some organic materials such as dyes in the case of dye-sensitized solar cells and the discovery of efficient charge transfer between certain organic electron donor and acceptor molecules incorporating nanowire components has accelarated research on organic PV materials during the last decade and is very active at the moment.

4.3.1 Single layer organic solar cells

The organic solar cells reported so far can be categorized by their device architecture as having single layer, bilayer, blend, or laminated structure. Initially all-organic solar cells have been prepared by sandwiching a single layer of an organic material between two dissimilar electrodes (work functions being different). Upon short-circuiting the device, electrons move from the low work-function electrode (such as Al) to the high work-function electrode (such as ITO) creating an electric field across the polymer layer. The photovoltaic properties, thus, strongly depend on the nature of the electrodes. Heavily doped conjugated materials resulted in reasonable energy conversion efficiencies of up to 0.3% (Chamberlain 1983). From the conjugated polymers of the first generation, poly (*para*-phenylene vinylene (PPV) appeared to be the most successful candidate for single layer PV device (Marks *et al.* 1994). Antoniadis *et al.* (1993, 1994) reported an IPCE (incident photon to collected electron efficiency) maximum of 5% in ITO/PPV/Al photodiodes and a power conversion efficiency of $\sim$0.1% under low light intensities of $1\,mW \cdot cm^{-2}$. The typical film thickness of the devices varied between 100–600 nm. Riess *et al.* (1994) observed PV effect in PPV Schottky diodes (100–500 nm thick) with power conversion efficiency, 0.1–1.0%, at very low light intensities. V_{oc} varied between 0.7–1.3 V with a low FF, 0.22.

4.3.2 Bi-layer organic solar cells

In 1986, a major breakthrough was achieved by Tang, who introduced a double-layer structure of p- and n-type organic semiconductors (Tang 1986). The preparation of a bi-layer by subliming or by spin-coating a second layer on top of the first resulting in a more or less diffused bi-layer structure is most straight-forward (Jenekhe & Yi 2000; Chen *et al.* 2000; Ingnas *et al.* 2001). In this device, the excitons can be dissociated at the donor acceptor interface due to a relative energy level difference of the donor and acceptor molecules.

The donor and acceptor molecules can be for example conjugated polymers (e.g. Jenekhe & Yi 2000), organic macromolecules (e.g. Meissner & Rostalski 2000), pigments (e.g. Tang 1986) or dyes (e.g. Petritsch *et al.* 2000a). The use of organic bilayers with photo induced electron transfer at the interface has been much investigated over the last decades (Wohrle & Meissner 1991; Chamberlain 1983; Spanggaard & Kerbs 2004).

The bilayer structure is more advantageous than the single layer structure for several reasons: exciton splitting is enhanced by the donor-acceptor interface, the active region is extended to both the donor and the acceptor sides of the junction thereby roughly doubling its width to about 20 nm, and the transport of electrons and holes is separated into different materials reducing the recombination losses. In addition, the band gaps of the two semiconductors can in principle be tuned to match the solar spectrum better (Petritsch 2000b).

Tang *et al.* made a 70 nm thick two-layer device using copper phthalocyanine (CuPc) as the electron donor, and a perylene tetracarboxylic derivative (PTC) as the electron acceptor (Figure 4.2). The photoactive material was placed between two dissimilar electrodes, indium tin oxide (ITO) for collection of the positive charges and silver (Ag) to collect the negative charges. A power conversion efficiency of about 1% was achieved under AM2 illumination (691 W/m^2).

The open circuit voltage of 450 mV and the FF of 65% indicated excellent charge transport (Kietzke 2007). The important aspect in this concept is that the charge generation efficiency is relatively independent of the bias voltage. This was considered the most efficient organic solid state PV cell. For over 20 years, CuPc has been the choice for donor in most small molecule solar cells due to its high stability, high mobility, and easy availability.

The first realization of a conjugated polymer/fullerene diode was achieved shortly after the detection of the ultrafast photo induced electron transfer between MEH-PPV and C60 (Sariciftci *et al.* 1993). The device was comprised of a 100-nm MEH-PPV layer and a 100-nm C60 layer, sandwiched between ITO and Au. On illumination by visible light, an open circuit voltage of 44 mV and a short circuit current density of 2.08 mA/cm^2 were measured. Assuming a FF of 0.48, an energy conversion efficiency of 0.04% was calculated. The most efficient organic bilayer solar cell hitherto reported is, however, a vacuum evaporated pigment based CuPc/C60 device by Peumans & Forrest (2001) with 3.6% efficiency at 150 mW/cm^2 AM1.5 illuminations (Halme 2002).

Several approaches have been investigated for polymer bilayer structures. Alam and Jenekhe (2004) have reported devices based on insoluble PPV as donor and BBL as electron acceptor. BBL was deposited from methanesulfonic acid. Though very high

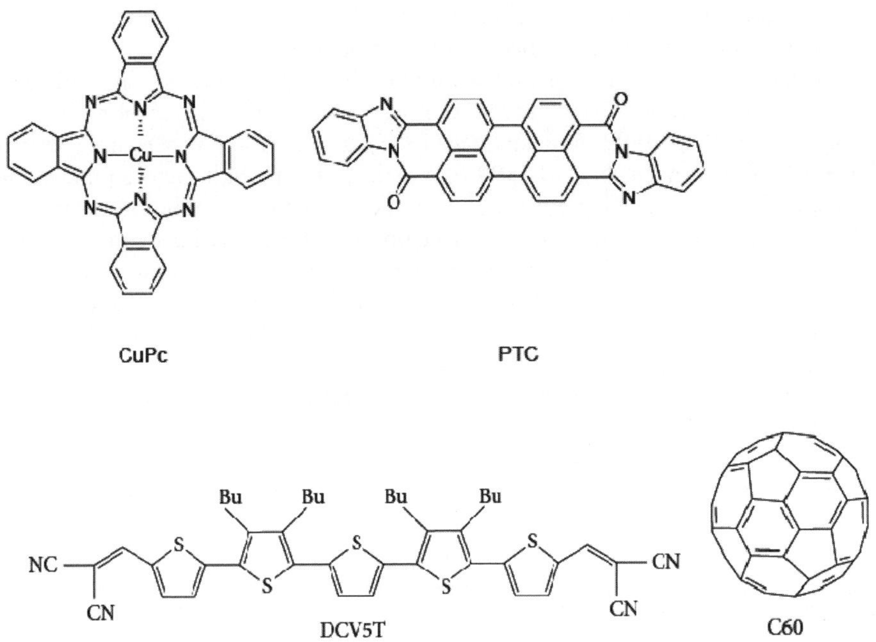

Figure 4.2 Molecular structures of copper phthalocyanine (CuPc), a perylene tetracarboxylic derivative (PTC), C60, and α, α-bis(2,2-dicyanovinyl)-quinquwthiophene (DCV5T)

EQE (upto 62%) was shown, the power conversion efficiency had dropped from 5% at very low light intensities to 1.5% under standard 1 sun.

In the double-layer structure the photoexcitations in the photoactive material have to reach the p-n interface where charge transfer can occur, before the excitation energy of the molecule is lost via intrinsic radiative and non-radiative decay processes to the ground state. Because the exciton diffusion length of the organic material is in general limited to 5–10 nm (Sariciftci et al. 1998), absorption of light within a very thin layer around the interface only contributes to the photovoltaic effect. This limits the performance of double-layer devices, because such thin layer can hardly absorb all the light. A strategy to improve the efficiency of the double-layer cell is related to structural organization of the organic material to increase the exciton diffusion length and, therefore, create more photoactive interfacial area.

4.3.3 Bulk hetero junction solar cells (BHJ cells)

To overcome the limitations of the double layer organic solar cell, the surface area of the donor–acceptor interface needs to be increased. This can be achieved by creating a mixture of donor and acceptor materials with a nanoscale phase separation resulting in a three-dimensional interpenetrating network called 'bulk hetero junction'. This has been achieved in a simple manner (Yu et al. 1995; Yu & Heeger 1995; Halls et al. 1995). By simply mixing the p and n type materials and relying on the inherent tendency of polymer materials to phase separate on a nanometer dimension, junctions throughout the bulk of the material are created that ensure quantitative dissociation of photo generated excitons, irrespective of the thickness.

Polymer-fullerene solar cells were among the first to utilize this bulk-hetero junction principle (Yu *et al.* 1995). The synthesis of 1-(3-methoxycarbonyl)propyl-1-phenyl[6,6]C61 (PCBM) (Hummelan *et al.* 1995), a soluble and processable derivative of fullerene C60, allowed Yu *et al.* (1995) to realize the first bulk-hetero junction solar cell by *blending* it with MEHPPV. Polymer blends (layers containing a mixture of an electron donating polymer and an electron accepting polymer) can be prepared spin-coating from a solution containing both polymers in the same solvent (Yu *et al.* 1995; Brabec *et al.* 2001, 2003). With this soluble derivative, it is possible to prepare more than 80 wt% fullerene-loaded PPV films. However, this attractive solution poses a new challenge. Photo generated charges must be able to migrate to the collecting electrodes through this thoroughly mixed blend. Since the holes are transported by the *p*-type semiconductor and electrons by the *n*-type material, these materials should be preferably mixed into a bicontinuous, interpenetrating network in which inclusions, cul-de-sacs, or barrier layers are avoided. The near-ideal bulk hetero junction cell appears as in Figure 4.3(a) & (b). The BHJ concept is presently the most widely used in organic solar cells. The name 'bulk-hetero junction' solar cell indicates that the interface (hetero junction) between both components is all over the bulk, in contrast to the bilayer-hetero junction. As a result of the intimate mixing, the interface

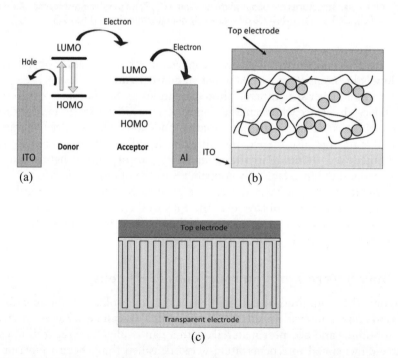

Figure 4.3 (a) The bulk-hetero junction concept. After absorption of light by the photoactive material, charge transfer can easily occur due to the nanoscopic mixing of the donor and acceptor, (b) Blend disordered, (c) ordered blend

where charge transfer can occur has increased enormously. The exciton, created after the absorption of light, has to diffuse towards this charge-transfer interface for charge generation to occur.

Since the diffusion length of the exciton is typically 10 nm or less in organic material, each exciton has to find a donor-acceptor interface within a few nm for efficient charge generation after absorption of light. Otherwise, the exciton will be lost without charge generation. An intimate bi-continuous network of donor and acceptor materials in the nanometer range suppresses exciton loss prior to charge generation. Here, the control of morphology is very vital and is required for providing a large charge-generating interface and suppression of exciton loss, and to ensure percolation pathways for both electron and hole transport to the collecting electrodes.

Small molecule based BHJ solar cells

Various combinations of donor and acceptor materials have been used to build bulk hetero junction solar cells in which the composite active layer is inserted between two electrodes. One of the most promising combinations of materials is a blend of a semiconducting polymer as a donor and a fullerene (C60 derivative) as acceptor. It is well established that at the interface of these materials a sub-picosecond photo induced charge transfer occurs that ensures efficient charge generation (Sariciftci *et al.* 1992; Brabec *et al.* 2001). Surprisingly, the lifetime of the resulting charge-separated state in these blends extends to an order of milli-second (Montanari *et al.* 2002; Offermans *et al.* 2003). This longevity allows the photo-generated charge carriers to diffuse away from the interface (assisted by the internal electric field in the device) to be collected in an external circuit at the electrodes).

Devices based on CuPc: C60 bulk hetero junctions have reached power conversion efficiencies of up to 5% (Xue *et al.* 2004a, 2005; Uchida *et al.* 2004). Xu *et al.* (2004) have achieved highest efficiency for a stacked solar cell comprising two CuPc: C60 bulk hetero junction cells separated via a layer of silver nanoclusters which served a charge recombination layer. However, CuPc as an electron donor can achieve low V_{oc} (<0.6 V) with perylenes or fullerenes as acceptors because large fraction of photon energy is wasted when photo-generated electron on CuPc transfers to C60 or perylene. Mutolo *et al.* (2006) could improve V_{oc} to nearly 1 volt by substituting CuPc by boron subphthalocyanine (SubPc), and the device has demonstrated a power conversion efficiency of 2.1%, providing scope for further improvement. Schulze *et al.* (2006) has used DCV5T as alternative to CuPc (or phtalocyanines), as electron donor and studied the structure, DCV5T: C60. This device has recorded IPCE as high as 52% and conversion efficiency of 3.4%, indicating that DCV5T is a promising electron donor.

Polymer-based BHJ solar cells

In 1995, two research groups (Yu & Heeger 1995; Halls *et al.* 1995) have worked independently on a polymer cell with MEH-PPV (Poly(2-methoxy-5-(2′-ethyl-hexyloxy)-1,4-phenylene-vinylene) as the electron donor and CN-PPV (cyano-para-phenylenevinylene) as the electron acceptor. These studies have shown moderate efficiencies and low external quantum efficiencies (EQE), ∼5 to 6%. The low performance was attributed to the non-optimization of nano-morphology.

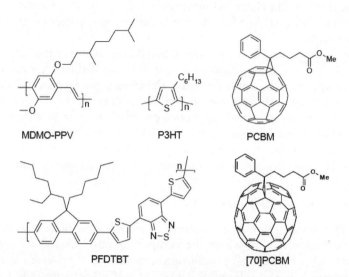

Figure 4.4 Electron Donor and Electron acceptor materials used in polymer-fullerene BHJ solar cells. [Chemical structure of Donors: MDMO-PPV: poly[2-methoxy-5-(3′,7′-dimethyloctyloxy)-p-phenylene vinylene]; P3HT: poly(3-hexylthiophene); PFDTBT: poly[2,7-[9-(2′-ethylhexyl)-9- hexylfluorene]-alt-5,5-(4′,7′-di-2-thienyl-2′,11′,3′-benzothiadiazole)]; Chemical structure of Acceptors: PCBM: 3′-phenyl-3′H-cyclopropa[1,9][5,6]fullerene-C60-Ih-3′-butanoic acid methyl ester; [70]PCBM: 3′-phenyl-3′H-cyclopropa[8,25][5,6]fullerene-C70-D5h(6)-3′-butanoic acid methyl ester]

Breeze *et al.* (2001) has demonstrated EQE of 24% and conversion efficiency of 0.6% using copolymer M3EH-PPV as donor and CN-Ether-PPV as acceptor. In 2004 (Breeze *et al.* 2004), an increased efficiency of 1% with V_{oc} of 1 volt are achieved in this polymer-polymer blend device. Kietzke *et al.* (Kietzke *et al.* 2005; Yin *et al.* 2007) have achieved higher efficiency on this polymer blend by improved processing. V_{oc} of 1.36 V and white light conversion efficiency of 1.7% are obtained. The FF of 35% indicates improved charge transport. Koetse *et al.* (2006) investigated solar cells based on MDMO-PPV as a donor and a novel acceptor polymer PF1CVTP. While high QE of 425 was recorded, the power conversion efficiency decreased at higher light intensities.

Polymer solar cells need to be optimized for doubling FF and EQE in order to reach power conversion efficiencies of 6–7% (Kietzke 2007). The limited performance is attributed to low dissociation efficiency of the photogenerated excitons, and to the amorphous nature of electron accepting polymers. Yin *et al.* (2007) have shown that more crystalline electron acceptor polymers with larger electron mobilities are needed.

A dramatic improvement in device performance with external quantum efficiency reaching 45% under low intensity was realized in bulk-hetero junction solar cells based on MDMO-PPV as a donor and (80 wt% of) PCBM – a solublized form of C60 – as an acceptor (Yu *et al.* 1995b). The structures of these compounds are given in Figure 4.4. The schematic representation of the BHJ cell is shown in Figure 4.5.

In PCBM, the fullerene cage carries a substituent that prevents extensive crystallization upon mixing with the conjugated polymer and enhances the miscibility.

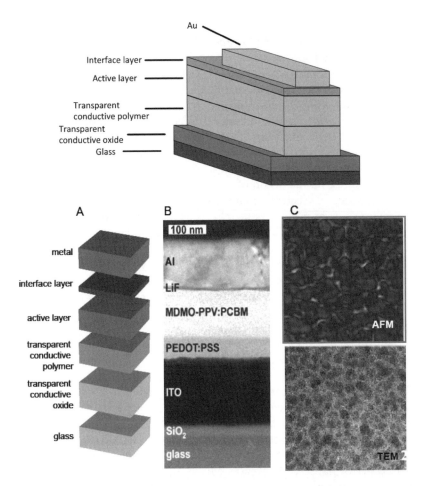

Au

Interface layer

Active layer

Transparent
conductive polymer

Transparent
conductive oxide

Glass

A

metal

interface layer

active layer

transparent
conductive
polymer

transparent
conductive
oxide

glass

B

100 nm

Al

LiF

MDMO-PPV:PCBM

PEDOT:PSS

ITO

SiO₂

glass

C

AFM

TEM

Figure 4.5 (A) Schematic layout of the device architecture of a polymer-fullerene bulkheterojunction solar cell, (B)TEM image of a thin slab of an actual device, showing the individual glass, ITO, PEDOT:PSS, MDMO-PPV/PCBM (1:4 by wt.), LiF, and Al layers. (C) AFM phase and TEM images (1 × 1 μm²) of a MDMO-PPV/PCBM (1:4 by wt.) composite film showing phase separation into a PCBM and polymer-rich phase
(*Source*: R.A.J. Janssen, reproduced with the permission of Prof. Janssen)

In these 2.5%-efficient cells, the photoactive composite layer is sandwiched between two electrodes with different work functions: a transparent front electrode consisting of indium tin oxide covered with a conducting polymer polyethylenedioxythiophene: polystyrenesulfonate (PEDOT:PSS) for hole collection and a metal back electrode consisting of a very thin (~1 nm) layer of LiF covered with Al/Au for electron collection. Except for LiF, all layers are about 100 nm thick.

Although the combination MDMO-PPV and PCBM had been used several times before, Shaheen *et al.* (2001) have improved the performance using a special solvent in the spin-coating of the active layer that improves the nanoscale morphology for charge generation and transport.

Atomic force microscopy (AFM) and TEM studies by van Duren et al. (2004) have resolved the details of the phase separation in these blends. In the best devices, which consist of a 1:4 blend (w/w) of MDMO-PPV/PCBM, nanoscale phase separation occurs in rather pure, nearly crystalline PCBM domains and an almost homogenous 1:1 mixture of MDMO-PPV and PCBM.

There have been numerous studies to get a deeper understanding of these MDMO-PPV/PCBM bulk hetero junctions. Investigations into the morphology, electronic structure, and charge transport have provided detailed understanding of the degree and dimensions of the phase separation in the active layer (van Duren et al. 2004; Martens et al. 2004), on the origin of the V_{oc} (Michaeletchi et al. 2003; Brabec et al. 2001), the influence of electrode materials (Mihaeletchi et al. 2004), and the magnitude of charge carrier mobilities for electrons and holes (Mihaeletchi et al. 2003). These studies revealed that PCBM has high electron mobility compared to many other organic or polymer materials that can be deposited by spin coating. Photo physical studies have provided insights on charge generation, separation, and recombination in these layers, and more recently in working devices. These detailed insights have resulted in quantitative models describing the current voltage characteristics under illumination that serve as a guide for further development.

The electrical current densities are low because of incomplete absorption of the incident light due to a poor match of the absorption spectrum of the active layer with the solar emission spectrum, and low charge carrier mobilities of the organic or polymer semiconductors.

P3HT is a material that can absorb photons at longer wavelengths compared to PPV derivatives (structure shown in Figure 4.4). It is known to have high charge-carrier mobility and lower band gap compared to MDMO-PPV, and has been considered as electron donor for use in solar cells in combination with PCBM. Padinger et al. (2003) and Schilinsky et al. (2002) have shown that P3HT/PCBM blends indeed provide an increased performance compared to MDMO-PPV. These higher efficiencies were obtained through the use of post-production treatment (Padinger et al. 2003). After spin coating of the active layer and deposition of the aluminum top electrode, treating P3HT/PCBM solar cells with a potential higher than the open circuit voltage and a temperature higher than the glass transition temperature led to an improved overall efficiency. This post-production treatment enhances the crystallinity of the P3HT and improves the charge carrier mobility. PV devices of P3HT/PCBM with external quantum efficiencies above 75% and energy conversion efficiencies of up to 3.85% have been reached (Brabeck 2004). Reyes-Reyes et al. (2005) could reach a record power conversion efficiency of 5% due to research contributions of several groups (Reyes-Reyes 2005; Brabec et al. 2001, 2003, 2004; van Duren et al. 2004). The degree of regioregularity, the polydispersities, and the molecular weights of P3HTappear to influence the device efficiency (Hiorns et al. 2006); and with high regioregularity due to improved molecular order, higher efficiencies are observed (Kim et al. 2006) The device physics of PCBM: polymer solar cells is excellently dealt by Blom et al. (2007).

The acceptor material, PCBM which accounts for nearly 75% of the weight of the photoactive layer, has a very low absorption coefficient in the visible region of the spectrum resulting in relatively less contribution to the photocurrent. The low absorption of C60 derivatives is due to their high degree of symmetry that makes many of the low energy transitions forbidden and hence of low intensity. When the C60 moiety is replaced by a less symmetrical fullerene, such as C70, these transitions become allowed and a significant increase in light absorption can be expected. Consequently, when [70]PCBM, which is similar to PCBM but incorporates C70 instead of C60, was used in combination with MDMO-PPV instead of PCBM, the external quantum efficiency increased from $\sim$50% to $\sim$65% while the current density of the solar cell increased by 50% and the energy conversion efficiency to 3.0% (Wienk et al. 2003).

Low-band gap polymers are synthesized to increase the absorption range of P3HT: PCBM cells; but they have not improved the conversion efficiencies due to the low mobilities of low-bandgap polymers. Muhlbacher et al. (2006) has reported a promising low-bandgap donor material, PCPDTBT which in combination with PCBM or $PC_{71}BM$ could extend photo-response to >99 nm. Its FF value of 47% shows the good charge transport characteristic and the overall power conversion efficiency reached is 3%.

Other versions of bulk heterojunction solar cell that has achieved impressive efficiency are the ones where a conjugated polymer is combined with nanocrystalline inorganic material such as CdSe nanocrystals (Greenham et al. 1996), Titania (TiO_2) nanocrystals (Arango et al. 1999), and ZnO nanocrystals (Beek et al. 2004) to make 'blends'. The CdSe nanocrystals can be controlled to give highly elongated molecules resulting in better pathways for electron transport. In polymer-TiO_2 blend, TiO_2 can be patterned into a continuous network for electron transport. TiO_2 has been successfully used to make dye-sensitized solar cells with upto 10% efficiency (O'Regan & Gratzel 1991; Bach et al. 1998; Hagfelt & Gratzel 2000). These are discussed in Chapter 6.

Ordered BHJ Device: In BHJ solar cells, the conjugated polymer and the electron acceptor are randomly interspersed in the films. Such disordered structures create certain problems: In some cases, the two materials phase-separate on too large a length scale resulting in some of the excitons generated when light is absorbed are unable to diffuse to an interface to be dissociated by electron transfer before they decay. In most cases, charge transport is slow enough to enable the charge carriers to reach the electrodes before they recombine with each other unless the films are made so thin that they cannot absorb the entire incident light. Hence, the concept of an 'ordered' cell, though difficult to fabricate, was mooted where well *ordered* conjugate polymer-electron acceptor films would be created as shown in Figure 4.3(c).

In this structure, every exciton formed on the conjugated polymer will be within a diffusion length of an electron acceptor, though studies by Kannan et al. (2003) have shown that even in this case, some light emission still takes place in the polymer. Secondly, after excitons are dissociated by electron transfer, the electrons and holes have straight pathways to the electrodes which minimise the carrier transport time as well as the probability of back electron transfer (Coakley & McGehee 2004). Thirdly, more importantly, ordered structures are much easier to model and understand.

Some approaches have been reported on preparing ordered bulk heterojunctions using conjugated polymers. One of the most attractive approaches is to use a block copolymer which self-assembles to form an array of cylinders oriented perpendicular to the substrate (Hadziiouannou 2002). Another approach is using nonconjugated block copolymers to make ordered films with desired nanostructure (Kim et al. 2004;

Table 4.1 Best in class solar cells: small molecule-based solar cells

Donor	Acceptor	η (%)	V_{oc} (volt)	FF (%)	IPCE	Reference
CuPc	C60	5.7	1.0	59	NA	Xue *et al.* 2004
CuPc	C60	5.0	0.6	60	64%	Xue *et al.* 2005
MeO-TPD-ZnPc(stacked)	C60	3.8	1.0	47	NA	Drechsel *et al.* 2005
CuPc	C60	3.5	0.5	46	NA	Uchida *et al.* 2004
DCV5T	C60	3.4	1.0	49	52%	Schulze *et al.* 2006
CuPc	PTCBI	2.7	0.5	58	NA	Yang *et al.* 2005a,b
SubPc	C60	2.1	1.0	57	NA	Mutolo *et al.* 2006
MeO-TPD, ZnPc	C60	2.1	0.5	37	NA	Dreschel *et al.* 2005
TDCV-TPA	C60	1.9	1.2	28	NA	Cravino *et al.* 2006
Pantacene on PET	C60	1.6	0.3	48	30%	Pandey & Nunzi 2006
SnPc	C60	1.0	0.4	50	21%	Rand *et al.* 2005

Table 4.2 Best in class solar cells: polymer-polymer (blend) solar cells

Donor	Acceptor	η (%)	V_{oc} (volt)	FF (%)	IPCE (%)	Reference
M3EH-PPV	CN-Ether-PPV	1.7	1.4	35	31	Kietzke *et al.* 2005
MDMO-PPV	PFICVTP	1.5	1.4	37	42	Koetse *et al.* 2006
M3EH-PPV	CN-Ether-PPV	1.0	1.0	25	24	Breeze *et al.* 2004

Table 4.3 Best in class solar cells: Polymer-polymer (bilayer) solar cells

Donor	Acceptor	η (%)	V_{oc} (volt)	FF (%)	IPCE (%)	Reference
PPV	BBL	1.5	1.1	50	62	Alam & Jenekhe 2004
MDMO-PPV: PFICVTP	PFICVTP	1.4	1.4	34	52	Koetse *et al.* 2006
M3EH-PPV	CN-Ether-PPV	1.3	1.3	31	29	Kietzke *et al.* 2006
MEH-PPV	BBL	1.1	0.9	47	52	Alam & Zenekhe 2004
M3EH-PPV	CN-PPV-PPE	0.6	1.5	23	23	Kietzke *et al.* 2006

Hamley 2003). This methodology, however, has problem in switching to conjugated blocks. The easier approach appears to be to make a nonporous film or an array of nanowires with inorganic semiconductor such as TiO_2, or ZnO, or CdS and then fill in the pores or the space between the wires with a conjugated polymer (Arango *et al.* 1999; Coakley & McGehee 2003; Ravirajan *et al.* 2004). Several research groups are attracted by this potential area and are actively involved in R & D.

Best Organic Cells: Best Organic solar cells in different combinations so far studied are tabulated in Tables 4.1 to 4.4 (Koetzke 2007). IPCE is 'incident photon to converted electron efficiency' i.e., ratio of number of electrons extracted to number of photons incident. It is a measure of external quantum efficiency.

Table 4.4 Best in class solar cells: blends of polymers and fullerene derivatives

Donor	Acceptor	η (%)	V_{oc} (volt)	FF (%)	IPCE (%)	Reference
P3HT	PCBM	5.0	0.6	68	NA	Ma et al. 2005
P3HT	PCBM	4.9	0.6	54	NA	Reys-Reys et al. 2005
P3HT	PCBM	4.4	0.9	67	63	Li et al. 2005
MDMO-PPV	PC71BM	3.0	0.8	51	66	Wienk et al. 2003
MDMO-PPV on PET	PCBM	3.0	0.8	49	NA	Al-Ibrahim et al. 2004

(Source for Tables 4.1–4.4: Kietzke 2007)

4.4 DYE-SENSITIZED NANOSTRUCTURED SOLAR CELLS

4.4.1 Configuration of the cell

In a dye-sensitized solar cell, an organic dye adsorbed at the surface of an inorganic wideband gap semiconductor is used for absorption of light and injection of the photo excited electron into the conduction band of the semiconductor. By the successful combination of nanostructured electrodes (nanoporous TiO_2) and efficient charge injection dyes, Professor Grätzel and his co-workers developed a solar cell with energy conversion efficiency exceeding 7% in 1991 (O'Regan & Grätzel) and 10% in 1993 (Nazeeruddin et al.). This solar cell is called the dye-sensitized nano-structured solar cell or the Grätzel cell after its inventor.

The research on dye-sensitized solar cells gained considerable attention thereafter. To date, ruthenium dye-sensitized nano-crystalline TiO_2 solar cells reach an energy conversion efficiency of about 10% in solar light (Nazeeruddin et al. 2001).

The schematic of the simplest configuration of the dye-sensitized solar cell (DSSC), recently prepared by Dr. Janne Halme of Aalto University of Science & Technology, Finland, is shown in Figure 4.6. It consists of a transparent conducting glass electrode coated with porous nanocrystalline TiO_2 dye molecules attached to the surface of the nc-TiO_2, an electrolyte containing a reduction-oxidation couple such as I^-/I_3^- and a catalyst coated counter-electrode. As the sunlight falls on the cell, it produces current through an external load connected to the electrodes. The absorption of light in the Cell occurs by dye molecules and the charge separation by electron injection from the dye to the TiO_2 at the semiconductor electrolyte interface.

A single layer of dye molecules however, absorbs only less than one percent of the incoming light (O'Regan & Grätzel 1991). While stacking dye molecules simply on top of each other to obtain a thick dye layer increases the optical thickness of the layer; and the dye molecules in direct contact to the semiconductor electrode surface only can separate charges and contribute to the current generation.

The Grätzel group has developed a porous nanocrystalline TiO_2 electrode structure in order to increase the internal surface area of the electrode so that a large amount of dye could be in contact, at the same time, by the TiO_2 electrode and the electrolyte. The TiO_2 electrode was typically 10 mm thick, with an average particle (as well as pore) size typically in the order of 20 nm, with an internal surface area thousands of times greater than the geometrical (flat plate) area of the electrode (O'Regan & Grätzel

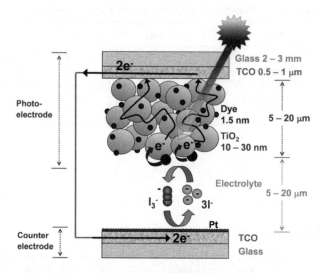

Figure 4.6 A schematic structure of the dye-sensitized solar cell (Credit: Dr. Janne Halme, Aalto University, Personal communication, Nov. 2011)

1991). This porous electrode structure of TiO_2 is, in fact, lets the major part of the solar spectrum available for the dye molecules, since TiO_2, as a large band gap semi-conductor, absorbs only below 400 nm of the solar spectrum. The working principle is explained in the Figure 4.7 (Halme 2002). The incoming photon is absorbed by the dye molecule adsorbed on the surface on the nano crystalline TiO_2 particle and an electron from a molecular ground state S^0 is exited to a higher lying excited state S^* (1). The exited electron is injected to the conduction band of the TiO_2 particle leaving the dye molecule to an oxidized state $S+$ (2). The injected electron percolates through the porous nano crystalline structure to the transparent conducting oxide layer of the glass substrate (negative electrode, anode) and finally through an external load to the counter-electrode (positive electrode, cathode) (3). At the counter-electrode the electron is transferred to tri-iodide in the electrolyte to give iodine (4), and the cycle is closed by reduction of the oxidized dye by the iodine in the electrolyte (5).

The numbers (1), (2), (3), (4), (5) refer to Figure 4.7.

The operating cycle can be summarized in the following chemical reactions (Matthews *et al.* 1996):

Anode: $S + h\upsilon \rightarrow S^*$ Absorption   (4)   (4.2)

$S^* \rightarrow S^- + e^-(TiO_2)$ Electron Injection (5)   (4.3)

$2S^+ + 3I^- \rightarrow 2S + I_3^-$ Regeneration   (6)   (4.4)

Cathode: $I_3^- + 2e^-(Pt) \rightarrow 3I^-$ (7)   (4.5)

Cell: $e^-(Pt) + h\upsilon \rightarrow e^-(TiO_2)$ (8)   (4.6)

Where (6), (7), and (8) are reactions. Due to the positioning of energy level in the system (Figure 4.7), the cell produces voltage between its electrodes and across

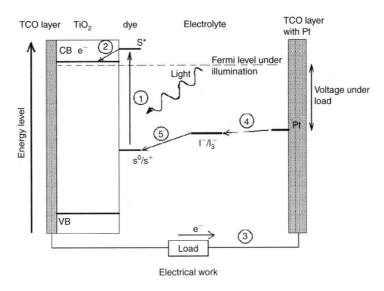

Figure 4.7 Working principle of the dye-sensitized nanostructured solar cell (Credit: Halme 2002, reproduced with permission)

the external load. The maximum theoretical value for the photovoltage at open circuit condition is determined by the potential difference between the conduction band edge of the TiO_2 and the redox potential of the I^-/I_3^- pair in the electrolyte (Cahen *et al.* 2000). The operation of the cell is regenerative in nature because chemical substances are neither consumed nor produced during the working cycle, as visualized in the cell reaction (8) shown above.

The DSCs and the conventional semiconductor p-n junction solar cells differ fundamentally in operation (Halme 2002): (a) While in semiconductor p-n junction solar cells the light absorption and charge transport occurs in the same material, the two functions are separated in the DSCs: photons are absorbed by the dye molecules and transport of charges is carried out in the TiO_2 electrode and electrolyte; (b)The charge separation in the semiconductor p-n junction cells is induced by the electric field across the junction, but no such long-range electric fields exist in the DSC. The charge separation occurs via other kinds of kinetic and energetic reasons at the dye-covered semiconductor-electrolyte interface; (c) In the semiconductor pn-junction cells, the generated opposite charges travel in the same material, while in the DSC, electrons travel in the nanoporous TiO_2 network and holes in the electrolyte. This means, in a DSC, the recombination can occur only at the semiconductor-electrolyte interface; hence, the need for a pure and defect free semiconductor material as in the case of a p-n junction solar cell does not arise.

Due to these fundamental differences in the operation, there is a need for theoretical considerations of the photovoltaic effect in the DSCs. These theoretical aspects are fully dealt in publications by Haggfeldt and Gratzel (1995, 2000), Cahen *et al.* (2000), Pichot and Gregg (2000b), Zaban *et al.* (1997), Ferber *et al.* (1998), Huang *et al.* (1997), Hague *et al.* (1998) and so on.

4.4.2   Performance of DSSCs

(a) Energy conversion efficiency: The Grätzel's group has reported the highest efficiencies (Grätzel 2000; Nazeeruddin *et al.* 1993; O'Regan & Grätzel 1991). Highest
efficiency of $11 \pm 0.5\%$ has been recorded as measured at FhG-ISE in 1996 [The Solar
Cell Efficiency Tables (Green 2001)]. This has been one of the reasons for the rapidly
expanded interest in the DSC research and development.

(b) Long term stability: To become an economically viable and commercially feasible
technology, the DS solar cells should be capable of maintaining non-degrading performance in operation over several years. The long term stability of the dye cells can be
divided into the following major components (Halme 2002).

(i) Inherent photochemical stability of the sensitizer dye adsorbed onto the TiO_2
electrode and in interaction with the surrounding electrolyte; (ii) Chemical and
photochemical stability of the electrolyte; (iii) Stability of the Pt-coating of the counter-
electrode in the electrolyte environment; and (iv) Quality of the barrier properties of
the sealing of the cell against intrusion of oxygen and water from the ambient air, and
against loss of electrolyte from the cell through the sealing. The other possible sources
of degradation in the cell are (Hinsch *et al.* 2001a): direct band-gap excitation of TiO_2
(holes in the TiO_2 valence band act as strong oxidants); catalytic reactions by TiO_2 and
Pt; and changes in the surface structure of the TiO_2. The interaction of these factors
and possibly those not yet known makes the working of the dye cell very intricate.
The chemical composition of the electrolyte has been observed to change over time by
formation of decomposition products of the electrolyte species (Hinsch *et al.* 2001a)
and the dye (Grünwald & Tributsch 1997). These, in turn, may interact with the dye
molecules or other species at the TiO_2 – electrolyte interface adversely affecting the
cell performance in the long term.

Hence, the stability of the dye-sensitized solar cells depends greatly on the designed
chemical composition and the materials of the cell as well as on any other impurities
possibly included during the fabrication of the cells. This is probably the main reason
for a small inconsistency in the stability results reported by different research groups. In
general, the perception has been that the long-term stability is not an intrinsic problem
of the technology, and can be improved by engineering the chemical composition of
the cells (Hinsch *et al.* 2001a). The other serious challenge seems to be the stability at
the elevated temperatures to which the cells can be exposed in operation.

At SHARP, the improvement of cell performance of dyesensitized solar cells (DSCs)
was investigated by means of haze of TiO_2 electrodes and reduction of series-internal
resistance. It was found that the high haze of TiO_2 electrodes effectively improves
the external quantum efficiency of DSCs. The series-internal resistance is successfully
reduced by increasing the surface roughness of platinum counter electrodes and by
decreasing the thickness of electrolyte layer. The highest single cell efficiency of 11.1%
(aperture area: $0.219\,cm^2$) was achieved and confirmed by a public test center (AIST).
Furthermore, large-scaled cell was fabricated using current collecting metal grids, and
the efficiency of 6.8% (aperture area: $101\,cm^2$) was obtained. Moreover, integrated
DSC module was investigated and the efficiency of 6.3% (aperture area: $26.50\,cm^2$)
was also confirmed by the public test center (Han *et al.* 2006).

Even so, the liquid electrolyte represents a major drawback from the technology
point of view. Replacing the liquid electrolyte by a solid hole transporting material is

considered more desirable. The most promising is a solid, wide-band gap hole transporting material resulting in energy conversion efficiencies of 3% (Bach *et al.* 1998). *Solid State DS Solar cells*: Research on the solid state DSC has gained considerable momentum recently as this type was found to be more attractive for flexible solar cells in a roll to roll production (www.konarka.com). The most successful p-type organic conductor utilized has been spiro-OMeTAD whose work function is ∼4.9 eV and hole mobility, 2×10^{-4} cm^2/sec. The conversion yields have remarkably increased over the last few years to over 4% (Schmidt-Mende *et al.* 2005). These cells, however, suffer from fast interfacial recombination of electron-hole reducing the diffusion length of the conduction band electrons to a few microns (Kruger *et al.* 2003) as compared to large value for the electrolyte based DSC. As a result, the film is only 2 μm thick, insufficient to harvest the sunlight by the adsorbed sensitizer, thus reducing the resultant photocurrent. The dye layer, being electrically insulating, blocks this back reaction (Snaith *et al.* 2005). Research efforts are now focused on 'molecular engineering' of the interface to improve the compactness and order of monolayer and prevent the charge carrier recombination (Gratzel 2006).

A new concept for a solid-state Grätzel cell consists of a polymer or organic semiconductor that combines the twin functions of light-absorption and charge (hole) transport in a single material so that both the dye and hole transporting material are replaced (van Hal *et al.* 1999).

The photo induced charge separation at the interface of an organic and inorganic semiconductor has been studied in relation to photovoltaic devices (Huynh *et al.* 2002). When an *organic or polymeric* semiconductor is excited across the optical band gap, the excitation energies and valence band offsets of this molecular semiconductor may allow electron transfer to the conduction band of an inorganic semiconductor, similar to the ruthenium dye. The size of the nano pores in TiO$_2$ is even more important here, because excitations are no longer created at the interface only, but throughout the whole organic material. Essentially all excitons must be able to reach the interface with the TiO$_2$ for efficient charge separation and energy conversion; hence, the distance between the site of excitation and the interface must be within the exciton diffusion length. In most organic materials, the exciton diffusion length is limited to 5–10 nm (Sariciftci ed. 1998) by the fast intrinsic decay processes of the photo excited molecules. Creating nanoporous TiO$_2$ of such dimensions, and filling it completely with an organic semiconductor, is currently one of the challenges in this area. This problem, however, may be overcome by developing oxide films having regular mesoporous channels aligned in a perpendicular direction to the current collector.

However, the appealing point is the recorded V$_{oc}$ with solid-state DSCs reaching one volt due to better match of the hole conductor work function than that of the electrolyte with the redox potential of the sensitizer. If the main issues of recombination and filling of pores can be effectively addressed, the future of solid state DSCs looks very promising (Gratzel 2006).

4.4.3   Dye-sensitized solar modules

Kay & Grätzel (1996) developed a three layer monolithic cell structure. The benefit of the monolithic structure is that all the layers of the cell can be deposited on top of each other on a single TCO coated glass plate, while the opposite glass plate without TCO coating serves merely as a protective barrier and encapsulation. In this structure, the counter-electrode is a porous carbon electrode separated from the TiO_2 electrode by a porous insulating rutile TiO_2 (Kay & Grätzel 1996) or ZrO_2 interlayer (Burnside et al. 2000) with the electrolyte inside the pores. The TCO layer, contacting the carbon electrode of one cell and the TiO_2 electrode of the adjacent cell accomplishes the electrical series connection of the cells. Kay and Grätzel (1996) reported 6.67% efficiency for a single monolithic cell and 5.29% for a small monolithic module. Much larger modules have been developed by industry reaching higher efficiencies of about 7% (Kubo 2005).

The monolithic cell design has been demonstrated to be appropriate for manufacturing DSC modules by screen printing for indoor applications, successfully competing in performance with the commercial amorphous silicon cells currently used for this purpose (Burnside et al. 2000). The monolithic module design has been adopted for low power applications, and has been investigated also by Hinsch et al. (1998).

4.5   LIFETIMES OF POLYMER CELLS

Practical application of bulk-hetero junction polymer-fullerene solar cells requires the stability of the cells. These organic, polymer-based solar cells need to be protected from ambient air (moisture and oxygen) to prevent degradation of the active layer and electrode materials. Even with proper protection there are several other degradation processes to be eliminated to ensure stability. Apart from device integrity, the materials must be photo chemically stable and the nanoscale bicontinuous donor–acceptor in the active layer should preserved. A recent study (Schuller et al. 2004) revealed that MDMO-PPV/PCBM solar cells show an appreciable degradation under increased temperature, although the degradation is not so much associated with the chemical stability. At elevated temperatures, the PCBM molecules can diffuse through the MDMO-PPV matrix and form large crystals, thereby increasing the dimension and extent of phase segregation (Yang et al. 2004). This behavior has been observed for temperatures ~20°C below the glass transition temperature, Tg, of the polymer.

Several strategies that may improve the limited thermal stability of the morphology can be considered. In general, high Tg polymers will increase the stability of as-prepared morphologies. For example, for the combination of poly (3-hexylthiophene) and fullerene derivatives, thermal annealing was considered to enhance the performance (Padinger et al. 2003). And, on cooling to operating temperatures, no further changes may be expected to occur in the morphology. Chemical or radiation induced cross-linking method analogous to that recently employed for polymer LEDs is another attractive method to preserve the as-prepared morphology in these blends (Muller et al. 2003). The use of p-n block copolymers seems a striking option, because the phase separation will be dictated by the covalent bonds between the two blocks (de Boer et al. 2001).

Despite recent encouraging results with regard to stability of these blends, creation of nanoscale bulk hetero junction morphologies that are stable in time and with temperature still remains a challenge. This has to be effectively addressed before polymer photovoltaics can be taken successfully to the field.

4.6   MANUFACTURING STATUS OF DSC AND OPV CELLS

Third-generation thin-film photovoltaic (PV) solar devices are entering the market-place after approximately 20 years of research and development, due to the insight of leading material developers such as Konarka and Plextronics in the organic photo-voltaics (OPV) domain, and Dyesol, EPFL, G24i, Mitsubishi and Peccell in the area of dye-sensitized cells (DSC). Both DSC and OPV technologies lag far behind on the efficiency curve when compared to conventional solar cell technologies (i.e., >20% efficiency). They are, therefore, likely to succeed in markets where their low cost, sub-strate flexibility, and ability to perform in low or variable lighting conditions offer them with a significant competitive advantage. DSC will target larger area BIPV applications while OPV will find its application in lower power consumer applications (Greentech Media Report 2009).

In 2007, there was a major investment in the first DSC manufacturing plant with annual production capacity of 20 MW in UK by G24i (technology licensed by Konarka Technologies), with an additional 25 MW capacity to become operational by the end of 2009. G24i planned for mass production into several markets, including consumer electronics and BIPV. DSCs exhibit the highest efficiencies of any third-generation thin-film solar technology; laboratory cell and tandem cell efficiencies of up to 12% have been reported.

In order to enhance the performance and lifetime of DSCs, several studies in mate-rial development have been undertaken. The dye sensitizers used in commercial cells are fabricated from costly ruthenium-based dyes (from Dyesol and Solaronix), including N-3 and N-719 dyes. Copper-based dyes, however, may become the next generation of inorganic dyes due to their lower cost and increasingly efficient performance. By combining different coloured dyes in a tandem DSC design, the range of light absorp-tion has been extended thereby enhancing cell efficiency and stability. These aspects have been demonstrated by projects undertaken by Panasonic Works, Sony, Kyushu Institute of Technology, and KIST.

Preliminary results are promising with an overall increase in efficiency by up to 50% and stabilities up to 85°C for 12 years, which could even allow for usage in BIPV applications. It seems likely that organic dyes based on carbazoles, indolenes, and porphyrins, will become more important, as they are widely available and are less expensive compared to their inorganic counterparts. The electrolyte system used in DSCs till recently was based on organic liquid solvents, but since these are heat-sensitive and prone to leakage, other more stable choices are being investigated.

Recent research has resulted in providing new insights on the function of polymer solar cells that can be produced speedily even in a roll-to-roll process.

Materials development is now being focused on novel polymers and small molecules, as well as electrode and encapsulation materials. The major drawback is the low efficiency of OPV devices due to their inability to absorb a large enough fraction of the solar spectrum, since the commonly used organic materials are limited to the visible part of the spectrum.

Research has been focussed to surmount this limitation by developing high effi-ciency OPV cells either by stacking and connecting individual cells in series, or using lower-band gap materials with absorption in the infrared thereby optimizing the absorption of incident light. OPV cells are expected to become commercially available from principal developers and suppliers including Heliatek, Konarka Technologies, Mitsubishi, Plextronics, and Solarmer Energy (Greentech Media Report 2009).

4.7   IMPROVING EFFICIENCIES

New combinations of materials that are being developed in various laboratories focus on improving the parameters, V_{oc}, J_{sc}, and the FF that represents the nature of the current-voltage characteristic and determine the energy conversion efficiency of a solar cell.

For ohmic contacts, the V_{oc} of bulk-hetero junction polymer photovoltaic cells is governed by the energy levels of the highest occupied molecular orbital (HOMO) and the lowest unoccupied molecular orbital (LUMO) of donor and acceptor, respectively. In most polymer/fullerene solar cells, the positioning of these band levels of donor and acceptor is such that up to ~0.4–0.8 eV is lost in the electron-transfer reaction. By more careful positioning of these levels, it is possible to raise the open-circuit voltage well above 1 volt. An encouraging result in this respect is PFDTBT that gives $V_{oc} = 1.04$ V in combination with PCBM (Svensson et al. 2003) compared to 0.8–0.9 V for MDMO-PPV and 0.5–0.6 V for P3HT. The tradeoff of increasing the donor-HOMO to acceptor-LUMO energy is that eventually a situation will be reached in which the photo induced electron transfer is held back by a loss of energy gain.

One of the crucial parameters for increasing the photocurrent is the absorption of more photons. This may be achieved by increasing the layer thickness and by shifting the absorption spectrum of the active layer to longer wavelengths. But, an increase of the layer thickness is presently limited by the charge carrier mobility and lifetime. When the mobility is too low or the layer too thick, the transit time of photo generated charges in the device becomes longer than the lifetime, resulting in charge recombination. The use of polymers such as P3HT that are known to have high charge carrier mobilities allows an increase in film thickness from the usual ~100 nm to well above 500 nm, without a loss of current. The absorption of the active layer in state-of-the-art devices currently covers from the UV up to about ~650 nm. In this wavelength range the monochromatic external quantum efficiency can be as high 70% under short-circuit conditions, implying that the vast majority of absorbed photons contribute to the current. The intensity of the solar spectrum, however, maximizes at ~700 nm and extends into the near infrared. Hence, a gain in efficiency can be expected when using low-band gap polymers. The preparation of low-band gap, high mobility, and processable low-band gap polymers is not simple and requires careful design in order to maintain the open-circuit voltage or efficiency of charge separation (Brabec et al. 2002). Because the V_{oc} of bulk hetero junction solar cells is governed by the HOMO of the donor and the LUMO levels of the acceptor, the most promising strategy seems to lower the band gap by adjusting the other two levels, i.e. decrease the LUMO of the donor, or increase the HOMO of the acceptor, or both. Several groups are actively pursuing low-band gap polymers and promising results are emerging (Janssen 2006).

A high FF is advantageous and indicates that fairly strong photocurrents can be extracted close to the open-circuit voltage. In this range, the internal field in the device that assists in charge separation and transport is fairly small. Consequently a high fill factor can be obtained when the charge mobility of both charges is high. Presently the fill factor is limited to about 60% in the best devices, but values up to 70% have been achieved recently (Mozer *et al.* 2004).

Apart from developing improved materials, a further gain in device performance can be expected from the combined optimization of the optical field distribution present in the device. Optical effects, such as interference of light in multilayer cavities, have received only limited attention so far but will likely contribute to a better light-management in these devices.

In essence, R&D efforts need to be focused on the following aspects to push the efficiency from the present achieved level of 5% to 10% which is generally considered as viable for production of PV devices: (i) the development of novel crystalline electron acceptors with high mobilities, and tunable energy levels to replace fullerene derivatives to achieve higher V_{oc}, and (ii) tandem structures for covering broader range of solar spectrum with novel, low-band gap materials.

4.8 NANO-TiO$_2$ DYE/CIGS TANDEM SOLAR CELLS

The fabrication of efficient tandem cells using low-cost thin film technologies remains a challenge (Yang *et al.* 1997; Kim *et al.* 2007; Nakada *et al.* 2006). In a series-connected double-junction device the ideal optical bandgaps are around 1.6–1.7 eV for the top cell and 1.0–1.1 eV for the bottom cell (Bremner *et al.* 2008).

The absorption characteristics of the dye-sensitized solar cell and the CIGS solar cell closely match these requirements. Hence, a wide range of the solar spectrum can be harvested by efficiently converting high energy photons in a top DSC and transmitted low energy photons in an underlying CIGS cell.

Additionally, the main advantage of DSC is its optical transmission and Jsc can be tailored by changing film thickness, pore size, the nature of the dye and the dye loading. Easy preparation of mesoscopic oxide films by screen-printing or doctor-blade methods facilitate these devices well suited for the fabrication of tandem structures for optimum utilization of the spectrum.

A schematic representation of nano-TiO$_2$ dye/CIGS tandem solar cell developed by Tiwari and his group is shown in Figure 4.8. The parameters of this DSC/CIGS tandem are: $V_{oc} = 1.45$ V, $J_{sc} = 14.05$ mA/cm^2, FF $= 0.74$ and $\eta = 15.09\%$. Individually, the DSC top cell has shown $J_{sc} = 13.66$ mA/cm^2, $V_{oc} = 0.798$ V, FF $= 0.75$, and $\eta = 8.2\%$; and for the CIGS bottom cell in the stack, $J_{sc} = 14.3$ mA/cm^2, $V_{oc} = 0.65$ V, and FF $= 0.77$.

The performance of the tandem is superior to that of individual cells. This demonstrates that further gains in efficiencies reaching beyond 20% can be possible from the combination of these two thin film PV technologies.

The J-V characteristic is shown along with the cell structure. This device has given highest efficiency so far recorded. However, the drawbacks of the stacked setup are reflection losses at the stack interface and absorption losses of low energy photons in the conducting glass of the top cell.

A monolithic DSC/CIGS structure to eliminate optical losses from the superfluous layers and interfaces and to reduce material and manufacturing cost is developed (Figure 4.9). This structure has achieved 12.2% conversion efficiency at full sunlight.

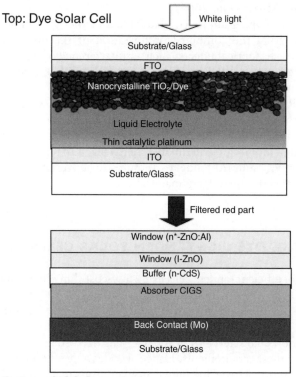

Figure 4.8 Schematic of Nano-TiO$_2$ Dye/CIGS tandem solar cell (Credit: Tiwari, EMPA, reproduced with permission)

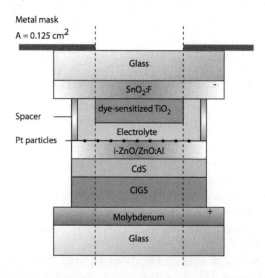

Figure 4.9 Schematic of Monolithic connected DSC/CIGS tandem solar cell (Credit: Tiwari, EMPA, reproduced with permission)

The monolithic device consists of a mesoporous dyesensitized TiO_2 film, which is directly sandwiched with a platinized CIGS solar cell using a spacer, thus avoiding the back glass electrode commonly used in the DSC. The void is filled through a hole in the top electrode with an acetonitrile based electrolyte containing the I^-/I_3^- redoxcouple. The p-type CIGS absorber ($\sim 1\,\mu m$) was grown by sequential coevaporation of elements using a three-stage evaporation process (Gabor *et al.* 1994) on a soda-lime glass substrate, coated with a $1\,\mu m$ thick dc-sputtered layer of molybdenum, and covered with an n-type CdS window layer (50 nm). The detailed fabrication procedure is given by Seyrling *et al.* (2009). The front contact, a 600 nm thick layer of In_2O_3: Sn (ITO), was covered with a transparent layer (<1 nm) of sputtered Pt particles. A 8 μm thick film of 20 nm-sized TiO_2 particles was screen-printed on a SnO_2: F conducting glass electrode (10 Ω/sq.cm) and sensitized by immersing it overnight in a solution of 0.3 mM of C101 dye (Gao *et al.* 2008) and 0.3 mM $3\alpha, 7\alpha$-dihydroxy-5β-cholanic acid. The detailed fabrication procedure for the TiO_2 paste and film has been described by Ito *et al.* (2008).The subcells in the monolithic setup are electrically connected in series. The charges generated in the subcells recombine at the catalytic Pt particles on the electrolyte/ITO interface, that is, the 'holes' from the top cell react with electrons from the bottom cell via $I_3^- + 2e \rightarrow 3I^-$. Thus matching the current densities of the subcells to minimize electronic losses is very crucial. The current density of the DSC can be tuned with choice of the sensitizer, by variation in the optical bandgap, and film thickness, and variation in the optical path length. The current density of the CIGS cell can be tuned varying the bandgap by changing the In/Ga ratio in the absorber (Gao *et al.* 2008). The transparent conductive oxide back electrode is not included in the DSC design to avoid reflection and free charge carrier absorption losses (Coutts *et al.* 2000).

The photovoltaic parameters of a monolithic DSC/CIGS device and its subcells are given in Table 4.5.

The conversion efficiency of the monolithic device (12.2%) slightly exceeds the performance of the CIGS cell (11.6%), justifying the monolithic approach to enhance device efficiency. The open-circuit voltage of the tandem device is close to the sum of the V_{oc}'s of the DSC and CIGS cell, confirming the series connection of the subcells. The J_{sc} of the tandem device is in good agreement with the estimate made using the transmittance spectrum. However, the corrosion of the CIGS cell by the redox mediator (I^-/I_3^- couple) of the dye-sensitized cell and an associated voltage loss (~ 140 mV) limits the performance.

These studies have demonstrated a monolithic DSC/CIGS device with an initial efficiency of 12.2% and further showed that a monolithic DSC/CIGS tandem device has the potential for increased efficiency over a mechanically stacked device due to increased light transmission to the bottom cell.

Table 4.5 Cell parameters

Test device	Efficiency (%)	V_{oc} (V)	J_{sc} (mA/cm²)	FF
DSC (FTO back contact)	8.4	0.74	−15.3	0.74
CIGS (unfiltered)	11.6	0.62	−27.3	0.68
DSC/CIGS monolithic	12.2	1.22	−13.9	0.72

Source: Tiwari et al. 2006

With a suitable protective intermediate layer preventing the degradation mechanism at the electrolyte/CIGS interface, it is expected to make full use of the optical advantages of this setup and to surpass the 15.1% efficiency benchmark given by the stacked device (Liska *et al.* 2006). Substitution of the FTO front electrode with a high mobility TCO, e.g., In_2O_3:Ti which substantially reduces absorption losses in the near infrared (Bowers *et al.* 2011), and careful current-matching may further enhance the device performance.

Seyrling *et al.* (2009), in the context of developing multi junction thin film solar cells, have recently investigated the possibility of optimizing the CIGS composition with respect to different sensitizers for the DSCs to achieve current matching by changing the [Ga]/[In + Ga] ratio. Different combinations of sensitizers and [Ga]/[In + Ga] ratios are investigated to find the best combination to achieve high efficiencies. Factors for performance limitations are identified and alternative remedies are developed to overcome efficiency losses. Limiting factors of the current are determined. Furthermore, to allow for higher temperatures during processing after the CIGS devices are completed, which could be beneficial for monolithically assembled multijunction solar cell production, In_xS_y was investigated as an alternative to the standard CBD-deposited CdS buffer layer. Preliminary work was done on a possible DSC/CdTe/CI(G)S triple junction cell. Transmission of the top and intermediate cell was recorded and the bottom CIGS cell composition varied to potentially achieve current matching in the device. A first stacked triple junction device was constructed.

Semiconductor nanowire – polymer hybrid cells

The nanowire array based polymer hybrid solar cells have been the attraction of several research groups over last two dwcades due to their unique advantages in manufacturing inexpensive large area solar devices compared to conventional solar devices. These are discussed in Chapter 6.

REFERENCES

Alam, M.M. & Jenekhe S.A. (2004): Efficient solar cells from layered nanostructures of donor and acceptor conjugated polymers, *Chemistry of Materials*, **16**, 4647–4656

Al-Ibrahim, M., Roth, H.K., & Sensfuss, S. (2004): Efficient large-area polymer solar cells on flexible substrates, *Appl. Phys, Lett.* 85(9), 1481–1483

Antoniadis, H., Hsieh, B.R., Abkowitz, M.A., Stolka, M., & Jenekhe, S.A. (1993): Photovoltaic Properties of poly (p-phenylene vinylene)/Aluminum Interfaces, *Polym. Preprints*, **34**, 490–491

Antoniadis, H., Hsieh, B.R., Abkowitz, M.A., Jenehke, S.A. & Stolka, M. (1994): Photovoltaic and photoconductive properties of aluminum/ poly(p-phenylene vinylene) interfaces, *Synthetic Materials*, **62**, 265

Arango, A.C., Carter, S.A., & Brock, P.J. (1999): Charge transfer in photovoltaics consisting of polymer and TiO_2 nanoparticles, *Appl. Phys. Lett,* **74**, 1698–1700

Bach, U., Lupo, D., Comte, P., Moser, J.E., Weissörtel, F., Salbeck, J., Speitzer, H., & Grätzel, M. (1998): Solid-state dye-sensitized mesoporous TiO_2 solar cells with high photon-toelectron conversion efficiencies, *Nature, 395,* 550

Bao, Z., Aspuru-Guzik, A., & Toney, M.: 2010 GCEP Report: Organic solar cells, Stanford University, available at 2.2.9_Bao_Public_2011.pdf

Beek, W.J.E., Wienk, M.M., & Janssen, R.A.J. (2004): Efficient hybrid solar cells from zinc oxide nanoparticles and a conjugated polymer, *Adv. Mater*, 16, 1009–1013

Blom, P.W.M., Mihailetchi, V.D., Koster, L.J.A., & Markov, D.A. (2007): Device Physics of Polmer: fullerene bulk hetero-junction solar cells, *Advanced Materials*, 19, 1551–1566

Bowers, J.W., Upadhyaya, H.M., Calnan, S., Hashimoto, R., Nakada, T., & Tiwari, A.N. (2009): Development of nano-TiO$_2$ dye sensitised solar cells on high mobility transparent conducting oxide thin films, *Prog. Photovolt: Res. Appl*, 17, 265–272

Brabec, C.J., Sariciftci, N.S. & Hummelen, J.C. (2001): Plastic solar cells, *Advanced Funtional Materials*, 11, 15–26

Brabec, C.J., Zerza, G., Cerullo, G., De Silvestri, S., Luzzati, S., Hummelen, J.C., & Sariciftci, N.S. (2001a): Tracing photoinduced electron transfer process in conjugated polymer/fullerene bulk heterojunctions in real time, *Chem. Phys. Lett*, 340, 232–236

Brabec, C.J., Cravino, A., Meissner, D., Saraciftci, N.S., Fromherz, T., Rispens, M.T., Sanchez, L., & Hummelan, J.C. (2001b): Origin of the open circuit voltage of plastic solar cell, *Advanced Functional Materials*, 11, 374–380

Brabec, C.J. & Sariciftci, S.N. (2001c): In: Hadziioannou, G. and van Hutten, P.F. (eds.) *Semiconducting Polymers*, Wiley-VHC verlag GmbH, D-69469 Weinheim, Germany

Brabec, C.J., Winder, C., Sariciftci, N.S., Hummelan, J.C., Dhanabalan, A., van Hal, P.A., & Janssen, R.A.J. (2002): A low band gap semiconducting polymer for photovoltaic devices and infra red emitting diodes, *Advanced Functional Materials*, 12, 709–712

Brabec, C.J., Dyakonov, V., Parisi, J., & Sariciftici, N.S. (eds.) (2003): *Organic Photovoltaics: Concepts and Realization*, vol. 60, Springer, New York, NY, USA

Brabec, C.J. (2004): Organic photovoltaics: Technology and Market, *Sol. Energy Mater. and Sol. Cells*, 83, 273–292

Braun, D. & Heeger, A.J. (1991): Visible light emission from semiconducting polymer diodes, *Appl. Phys. Lett*, 58(18), 1982–1984

Breeze, A.J., Schlesinger, Z., Carter, S A., & Brock, P.J. (2001): Charge transport in TiO$_2$/MEH-PPV polymer photovoltaics, *Physical Review B*, 64, Article ID 125205, 9 pages

Breeze, A.J., Salomon, A., Ginley, D.S., Tillmann, H., Horhold, H., & Gregg, B.A. (2002a): Improved efficiencies in polymer-perylene diimide bilayer photovoltaics, In *Organic Photovoltaics III*, vol. 4801 of *Proc. of SPIE*, pp. 34–39, Seattle, USA

Breeze, A.J., Salomon, A., Ginley, D.S., Gregg, B.A., Tillmann, H., & H.-H. Hörhold (2002b): Polymer-perylene diimide heterojunction solar cells, *Appl. Phys. Lett*. 81, 3085–3087

Breeze, A.J., Schlesinger, Z., Carter, S.A., Tillmann, H. & H.-H, Hörhold, (2004): Improving power efficiencies in polymer—polymer blend photovoltaics, *Sol. Energy Mater. and Sol. Cells*, 83, 263–271

Bremner, P., Levy, M.Y., & Honsberg, C.B. (2008): Analysis of tandem solar cell efficiencies under (AM1.5G) spectrum using a rapid flux calculation method, *Prog. Photovoltaics: Res. Appl*. 16, 225–233

Burnside, S.D., Winkel, S., Brooks, K., Shklover, V., Grätzel, M., Hinsch, A., Kinderman, R., Bradbury, C., Hagfeldt, A., & Pettersson, H. (2000): Deposition and characterization of screen-printed porous multi-layer thick film structures from semiconducting and conducting nanomaterials for use in photovoltaic devices, *J. Mater. Sci.: Mater. Electron*, 11, 355–362

Burroughes, J.H., Bradley, D.D.C., Brown, A.R., Marks, R.N., Mackay, K., Friend, R.H., Burn, P.L., & Holmes, A.B. (1990): Light-emitting diodes based on conjugated polymers, *Nature*, 347, 539

Cahen, D., Hodes, G., Grätzel, M., Guillemoles, J.F., & Riess, I.J. (2000): Nature of Photovoltaic Action in Dye-Sensitized Solar Cells, *J. Phys. Chem. B*, 104, 2053–2059

Chamberlain, G.A. (1983): Organic Solar Cells: A Review, *Solar Cells*, 8, 47

Chen, L.C., Godovsky, D., Inganäs, O. *et al.* (2000): Polymer photovoltaic devices from stratified multilayers of donor-acceptor blends, *Advanced Materials*, 12, 1367–1370

Chevalevski, O., Larina, L., & Lim, K.S. (2003): Nanocrystalline tandem photovoltaic cell with twin Dye-sensitized anodes, *Proc. 3rd WC on PV Energy Conversion*, May 2003, Osaka, Japan, pp. 23–26

Chiang, C.K., Fincher Jr., C.R., & Park, Y.W. (1977): Electrical conductivity in doped polyacetylene, *Phys. Rev. Lett*, 30, 1098–1101

Coakley, K.M., McGehee, M.D., Frindell, K.M., & Stucky, G.D. (2003): Infiltrating semiconductor polymers into selfassembeled Mesoporous Titania films for PV applications, *Adv. Funct. Mater*, 13, 301–306

Coakley, K.M., & McGehee, M.D. (2003): Photovoltaic cells made from conjugated polymers infiltrated into mesoporous titania, *Appl. Phys. Lett*, 83, 3380–3382

Coakley, K.M., & McGehee, M.D. (2004): Cojugated Polymer Photovoltaic cells, *Chem. Mater.*, 16, 4533–4542

Coutts, T.J., Young, D.L., & Li, X.N. (2000): Characterization of Transparent Conducting Oxides, *MRS Bulletin*, 25 (8), 58–65

Cravino, A., Leriche, P., Alévêque, O., Roquet, S., & Roncali, J. (2006): Light-emitting organic solar cells based on a 3D conjugated system with internal charge transfer, *Advanced Materials*, 18 (22), 3033–3037

de Boer, B., Stalmach, U., van Hutten, P.F., Melzer, C., Krasnikov, V.V., & Hadziioannou, G. (2001): Supramolecular self-assembly and opto-electronic properties of semiconducting block copolymers, *Polymer*, 42, 9097–9109

Dennler, G., Sariciftci, N.S. & Brabec, C.J. (2006): Conjugated Polymer-based organic solar cells, In: Hadziioannou and Malliaras (eds.) *Semiconducting Polymers – Chemistry, Physics and Engineering*, G.G, WILEY-VCH Verlag GmbH & Co. KGaA, Weinheim, ISBN: 3-527-31271-4

Drechsel, J., Männig, B., Kozlowski, F., Pfeiffer, M., Leo, K., & Hoppe, H. (2005): Efficient organic solar cells based on a double *p-i-n* architecture using doped wide-gap transport layers, *Appl. Phys. Lett.* 86 (24), Article ID 244102, 3pages

Durr, M., Bamedi, A., Yasuda, A., & Nelles, G. (2004): Tandem Dye-Sensitized Solar Cell for Improved Power Conversion Efficiencies, *Appl. Phys. Lett.* 84, p. 3397

Durr, M., Bamedi, A., Schmid, A., Obermaier, M., Yasuda, A., & Nelles, G. (2004): Efficiencies of Co-Sensitized and Tandem-Structure Dye-Sensitized Solar Cells, *Proc. 19th EU PV Solar Energy Conf*, June 2004, Paris, France, pp. 21–25

Ferber, J., Stangl, R., & Luther, J. (1998): An electric model of the dye-sensitized solar cell, *Sol. Energy Mater. Sol. Cells*, 53, 29–54

Friend, R.H., Gymer, R.W., Holmes, A.B., Burroughes, J.H., Marks, R.N., Taliani, C., Bradley, D.C., Dos Santos, D.A., Brédas, J.L., Löglund, M., & Salaneck, W.R. (1999): Electroluminescence in Conjugated Polymers, *Nature*, 397, 121–128

Gabor, A.M., Tuttle, J.R., Albin, D.S., Contreras, M.A., Noufi, R., & Hermann, A.M. (1994): High-efficiency $CuIn_x Ga_{1-x}Se_2$ solar cells made from $(In_xGa_{1-x}) Se_3$ precursor films, *Appl. Phys. Lett.* 65, 198–200

Gao, F., Wang, Y., Shi, D., Zhang, J., Wang, M.K., Jing, X.Y., Humphry-Baker, R., Wang, P., Zakeeruddin, S.M., & Grätzel, M. (2008): Enhance the Optical Absorptivity of Nanocrystalline TiO_2 Film with High Molar Extinction Coefficient Ruthenium Sensitizers for High Performance Dye-Sensitized Solar Cells, *J. Am. Chem. Soc.* 130, 10720–10728

Goetzberger, A. (2000): Photovoltaic materials, Past, Present, Future, *Sol. Energy Mater. & Sol. Cells*, 62, 1–19

Granström, M., Petritsch, K., Arias, A.C., Lux, A., Andersson, M.R. & Friend, R.H. (1998): Laminated fabrication of polymeric photovoltaic diodes, *Nature*, 395, 257–260

Grätzel, M. (2000): Perspectives for Dye-sensitized Nanocrystalline Solar Cells, *Prog. Photovolt: Res. Appl.* 8, 171–185

Gratzel, M. (2005): Mesoscopic solar cells for electricity and hydrogen production, *Chemistry Letters*, **34**, 8–13

Gratzel, M. (2006): The advent of mesoscopic injection solar cells, *Prog. Photovolt: Res. Appl.* **14**, 429–442

Green, M.A. (2001): Solar Cell Efficiency Tables (Version 18), *Prog. Photovolt: Res. Appl.*, **9**, 287–93

Greenham, N.C., Peng, X., & Alivisatos, A.P. (1996): Charge separation and transport in conjugated-polymer/semiconductor-nanocrystal composites studied by photoluminescence quenching and photoconductivity *Phys. Rev. B*, **54**, 17628–17637

Gregg, B.A., & Hanna, M.C. (2003): Comparing organic to inorganic photovoltaic cells: theory, experiment, and simulation, *J. Appl. Phys*, **93**, 3605–3614

Hadziiouannou, G. (2002): MRS Bulletin, **27**, 456

Hagfeldt, A., & Gratzel, M. (1995): Light-Induced Redox Reactions in Nanocrystalline Systems, *Chem. Rev.*, **95**, 49–68

Hagfeldt, A., & Gratzel, M. (2000): Molecular Photovoltaics, *Acc. Chem. Res.* **33**, 5, 269–277

Halls, J.J.M., Walsh, C.A., Greenham, N.C. *et al.* (1995): Efficient photodiodes from interpenetrating polymer networks, *Nature*, **376**, 498–500

Halme, J. (2002): Dye-sensitized Nano-structural and Organic PV: Technical review and Prelimnary tests, Master's thesis, Dept of Engineering Physics and Mathematics, University of Helsinki

Halme, J. (2011): Personal Communication, 21st November 2011

Hamley, I.W., (2003): Nanostructure fabrication using block copolymers, *Nanotechnol.* **14**, R39

Han, L., Fukui, A., Fuke, N., Kide, N., & Yamanaka, R. (2006): High efficiency of Dye-sensitized Solar cell and module, Proc. 4th WC PV Energy Conversion, May 2006, pp. 179–182

Haque, S.A., Tachibana, Y., Klug, D.R., & Durrant, J.R. (1998): Charge Recombination Kinetics in Dye-Sensitized Nanocrystalline Titanium Dioxide Films under Externally Applied Bias, *J. Phys. Chem. B*, **102**, 1745–1749

He, J., Lindstrom, H., Hagfeldt, A., & Lindquist, S.E. (2000): Dye-sensitized nanostructured tandem cell – first demonstrated cell with a dyesensitized photocathode, *Sol. Energy Mater. and Sol. cells*, **62**, 265

Heeger, A.J. (2001): Nobel lecture: semiconducting and metallic polymers, *Angew. Chem. Intl. Edn*, **40**, 2591

Heeger, A.J. (2001): Nobel lecture: Semiconducting and metallic polymers: the fourth generation of polymeric materials, *Reviews of Modern Phys*, **73**, 681–700

Hinsch, A., Kroon, J.M., Ulendorf, I., Meyer, A., Kern, J.M., & Ferber, J. (1998): The performance of dye-sensitized solar cells with a one-facial, monolithic layer built-up prepared by screen printing, *2nd World Conf. and Exhibit. on PV Solar Energy Conversion*, 6–10 July, Vienna, Austria

Hinsch, A., Kroon, J.M., Kern, J.M.R., Uhlendorf, I., Holzbock, J., Meyer, A., & Ferber, J. (2001): Long-term stability of dye sensitized solar cells, *Prog. Photovolt: Res. Appl.*, **9**, 425–438

Hiorns, R.C., de Bettignies, R., Leroy, J. *et al.* (2006): Highmolecular weights, polydispersities, and annealing temperatures in the optimization of bulk-heterojunction photovoltaic cells based on poly(3-hexylthiophene) or poly(3-butylthiophene), *Advanced Functional Materials*, **16**, 2263–2273

Hummelan, J.C., Knight, B.W., LePeq, F., Wudl, F., Yao, J., & Wilkins, C.L. (1995): Preparation and Characterization of Fulleroid and Methanofullerene Derivatives, *J. Org. Chem.* **60**, 532–538

Huang, S.Y., Schlichthoerl, G., Nozik, A.J., Gratzel, M., & Frank, A.J. (1997): Charge Recombination in Dye-Sensitized Nanocrystalline TiO_2 Solar Cells, *J. Phys. Chem. B*, **101**, 2576–2582

Huynh, W.U., Dittmer, J.J., & Alivisatos, A.P. (2002): Hybrid nanorod-polymer solar cells, *Science*, **295**, 2425–2427

Inganas, O., Roman, L.S., Zhang, F., Johansson, D.M., Andersson, R. & Hummelen, J.C. (2001: Recent Progress in thin film Organic photodiodes, *Synthetic Metals*, **121**, 1525–1528

Ito, S., Murakami, T.N., Comte, P., Liska, P., Grätzel, C., Nazeeruddin, M.K., & Grätzel, M. (2008): Fabrication of thin film dye sensitized solar cells with solar to electric power conversion efficiency over 10%, *Thin Solid Films*, **516**, 4613–4619

Jayarama Reddy, P. (2010): *Science and Technology of Solar Photovoltaics*, 2nd edition, CRC Press, Leiden, The Netherlands.

Jenekhe S.A., & Yi, S. (2000): Efficient photovoltaic cells from semiconducting polymer heterojunctions, *Appl. Phys. Lett.*, **77** (17), 2635–2637

Janssen, R. (2006): Introduction to Polymer Solar Cells (3Y280)

Kalyanasundaram, K., & Gratzel, M. (1998): Applications of functionalized transition metal complexes in photonic and optoelectronic devices, *Coord. Chem. Rev.* **177**, 347–414

Kannan, B., Castelino, K., & Majumdar, A. (2003): Design of Nanostructured Heterojunction Polymer Photovoltaic Devices *Nano Lett.* **3**, 1729–1733

Kay, A., & Gratzel, M. (1996): Low cost photovoltaic modules based on dye sensitized nanocrystalline titanium dioxide and carbon powder, *Sol. Energy Mater. Sol. Cells*, **44**, 99

Kietzke, T., Hörhold, H.-H., & Neher, D. (2005): Efficient polymer solar cells based on M3EH-PPV, *Chemistry of Materials*, **17**, 6532–6537

Kietzke, T., Egbe, D.A.M., Hörhold, H.-H., & Neher, D. (2006): Comparative study of M3EH-PPV-based bilayer photovoltaic devices, *Macromolecules*, **39** (12), 4018–4022

Kietzke, T. (2007): Recent advances in Organic solar cells, *Advances in OptoElectronics*, Vol 2007, Article ID 40285, 15p, Hindawi Publishing Corporation.

Kim, Y., Lee, K., Coates, N.E., Moses, D., Nguyen, t.-Q., Dante, M., & Heeger, A.J. (2007): Efficient tandem polymer solar cells fabricated by all-solution processing, *Science*, **317**, 222–225

Kim, Y., Cook, S., Tuladhar, S.M., Choulis, S.A., Nelson, J., Durrant, J.R., Bradley, d.D.C., Giles, M., McCulloch, M., Ha, C.S., & Ree, M. (2006): A strong regioregularity effect in self-organizing conjugated polymer films and high-efficiency polythiophen: fullerene solar cells, *Nature Materials*, **5**, 197–203

Kim, S., Misner, M.J., Xu, T., Kimura, M., & Russel, T.P. (2004): Highly Oriented and Ordered Arrays from Block Copolymers via Solvent Evaporation, *Adv. Mater*, **16**, 226–231

Koetse, M., Sweelssen, M.J., Hoekerd, K.T., Schoo, H.F.M., Veenstra, S.C., Kroon, J.M., Yang, X., & Loos, J. (2006): Efficient polymer: polymer bulk heterojunction solar cells, *Appl. Phys. Lett*, **88**, Article ID 083504, 3pages

Kubo, W., Sakamoto, A., Kitamura, T., Wada, Y., & Yanagida, S. (2004): Dye-sensitized solar cells: improvement of spectral response by tandem structure, *J. Photochemistry and Photobiology A, Chemistry*, **164**, 33

Kubo, T. (Nippon oil) (2005): Design of large area DSC and their Durability, *DSC Symposium, 5th Asian Congress*, Seoul, Korea

Li, G., Shrotriya, V., Huang, J., Yao, Y., Moriarty, T., Emery, K. & Yang, Y. (2005): High-efficiency solution processable polymer photovoltaic cells by self-organization of polymer blends, *Nature Materials*, **4** (11), 864–868

Liska, P., Thampi, K.R., Grätzel, M., Brémaud, D., Rudmann, D., Upadhyaya, H.M., & Tiwari, A.N. (2006): Nanocrystalline dye-sensitized solar cell/copper indium gallium selenide thin film tandem showing greater than 15% conversion efficiency, *Appl. Phys. Lett.* **88**, 203103

Ma, W., Yang, C., Gong, X., Lee, K., & Heeger, A.J. (2005): Thermally stable, efficient polymer solar cells with nanoscale control of the interpenetrating network morphology, *Advanced Functional Materials*, **15** (10), 1617–1622

MacDiarmid, A.G. (2001): Nobel lecture, Synthetic metals: a novel role for organic polymers, *Angew. Chem. Intl. Edition*, **40**, 2581

MacDiarmid, A.G. (2001): Nobel lecture, Synthetic metals: a Novel role for Organic polymers, *Reviews of Modern Phys*, 73, 701–712

Marks, R.N., Halls, J.J.M., Bradley, D.D.C., Friend, R.H. & Holmes, A.B. (1994): The photovoltaic response in poly (pphenylene vinylene) thin-film devices, *J. Phys. Condensed matter*, 6, 1379–1394

Martens, T., D'Haen, J., Munters, T., Beelen, Z., Goris, L., Manca, J., D'Olieslaeger, M., Vanderzande, D., De Schepper, L., & Andriessen, R. (2003): *Synthetic Materials*, 138, 243

Matthews, D., Infelta, P., & Grätzel, M. (1996): Calculation of the photocurrent-potential characteristic for regenerative, sensitized semiconductor electrodes, *Sol. Energy Mater. Sol. Cells*, 44, 119–155

McCulloch, I., Heeney, M., Bailey, C., Genevicius, K., MacDonald, I., Shkunov, M., Sparrowe, D., Tierney, S., Wagner, R., Zhang, W., Chabinyc, M.L., Kline, R.J., McGehee, M.D., & Toney, M.F (2006): Liquid-crystalline semiconducting polymers with high charge-carrier mobility, *Nature Materials*, 5, 328–333

Meissner, D., & Rostalski, J. (2000): Highly efficient molecular organic solar cells, *Proc. of the 16th EU PV solar Energy Conf*, May, 2000, Glasgow, UK, Plenary talk PA 1.2, 6 pages

Mihailetchi, V.D., Blom, P.W.M., Hummelen, J.C., & Rispens, M.T. (2003): Cathode dependence of the open-circuit voltage of polymer:fullerene bulk heterojunction solar cells, *J. Appl. Phys.* 94, 6849–6854

Mihailetchi, V.D., Koster, L.J.A., & Blom, P.W.M. (2004): Effect of metal electrodes on the performance of polymer: fullerene bulk heterojunction solar cells, *Appl. Phys. Lett.* 85, 970

Mihailetchi, V.D., van Duren, J.K.J., Blom, P.W.M., Hummelen, J.C., Janssen, R.A.J., Kroon, J.M.M., Rispens, T., Verhees, W.J.H., & Wienk, M.M. (2003): Electron transport in a methallofullerene, *Advanced Functional Materials*, 13, 43–46

Miranda, P.B., Moses, D. & Heeger, A.J. (2001): Ultrafast photogeneration of charged polarons in conjugated polymers, *Phys. Rev. B*, 64 (8)), Article ID 081201, 4 pages

Montanari, I., Nogueira, A.F., Nelson, J., Durrant, J.R., Winder, C., Loi, M.A., Sariciftci, N.S., & Brabec, C.J. (2002): Transient optical studies of charge recombination dynamics in a polymer/fullerene composite at room temperature, *Appl. Phys. Lett.*, 81, 3001–3003

Mozer, A.J., Denk, P., Scharber, M.C., Neugebauer, H., Sariciftci, N.S., Wagner, P., Lutsen, L., & Vanderzande, D. (2004): Novel Regiospecific MDMO-PPV Copolymer with Improved Charge Transport for Bulk Heterojunction Solar Cells, *J. Phys. Chem. B*, 108, 5235–5242

Muhlbacher, D., Scharber, M., Morana, M. *et al.* (2006): High photovoltaic performance of a low-bandgap polymer, *Advanced Materials*, 18, 2884–2889

Müller, C.D., Falcou, A., Reckefuss, N., Rojahn, M., Wiederhirn, V., Rudati, P., Frohne, H., Nuyken, O., Becker, H. & Meerholz, K. (2003): Multi-colour organic light-emitting displays by solution processing, *Nature*, 421, 829–833

Mutolo, K.L., Mayo, E.I., Rand, B.P., Forrest, S.R. & Thompson, M.E. (2006): Enhanced open-circuit voltage in subphthalocyanine/C60 organic photovoltaic cells, *J. Amer. Chem. Society*, 128, 8108–8109

Nakada, T., Kijima, S., Kuromiya, Y., Arai, R., Ishii, Y., Kawamura, N., Ishizaki, H. & Yamada, N. (2006): Chalcopyrite Thin-Film Tandem Solar Cells with 1.5 V Open-Circuit Voltage, *Conference Record of 4th WC on PV EnergyConversion*, Waikola, HI, pp. 400–403.

Nakamura, J.-I., Yokoe, C., Murata, K. & Takahashi, K. (2004): Efficient organic solar cells by penetration of conjugated polymers into perylene pigments, *J. of Appl. Phys.* 96, 6878–6883

Nazeeruddin, M., Kay, A., Rodicio, I., Humphry-Baker, R., Mueller, E., Liska, P., Vlachopoulos, N., & Graetzel, M. (1993): Conversion of Light to Electricity by cis-X2Bis(2,2'-bipyridyl-4,4'-dicarboxylate)ruthenium(II) Charge-Transfer Sensitizers (X = Cl-, Br-, I-, CN- and SCN-) on Nanocrystalline TiO$_2$ Electrodes, *J. Am. Chem. Soc.*, 115, 6382–6390

Nazeeruddin, M.K., Réchy, P., Renouard, T., Zakeeruddin, S.M., Humphry-Baker, R., Comte, P., Liska, P., Vevey, L., Costa, E., Shklover, V., Spicca, L., Deacon, G.B.,

Bignozzi, C.A. & Grätzel, M. (2001): Engineering of efficient panchromatic sensitizers for nanocrystalline TiO₂-based solar cells, *J. Am. Chem. Soc., 123*, 1613–1624

Offermans, T., Meskers, S.C.J., & Janssen, R.A.J. (2003): Charge recombination in a poly(para-phenylene vinylene)-fullerene derivative composite film studied by transient, nonresonant, hole-burning spectroscopy, *J. Chem. Phys., 119*, 10924–1029

O'Regan, B. and Grätzel, M. (1991): Low cost high-efficiency solar cells based on the sensitization of colloidal TiO₂, *Nature, 353*, 737–740

Padinger, F., Rittberger, R.S., & Sariciftci, N.S. (2003): Effects of postproduction treatment on plastic solar cells, *Adv. Funct. Mater, 13*, 85–88

Pandey, A.K., and Nunzi, J.M. (2006): Efficient flexible and thermally stable pentacene/C60 small molecule based organic solar cells, *Appl. Phys. Lett.* 89 (21), Article ID 213506, 3 pages

Petritsch, K. *et al.* (2000a): Dye based donor/acceptor solar cells, *Sol. Energy Mater. & Sol. Cells, 61*, 63–72

Petritsch, K. (2000b): Organic Solar Cell Architectures, PhD Thesis, Technisch-Naturwissenschaftliche Fakultät der Technischen Universität Graz (Austria)

Peumans, P., Yakimov, A.V. & Forrest, S.R. (2003): Small molecular weight organic thin-film photodetectors and solar cells, *J. Appl. Phys, 93*, 3893–3723

Pichot, F. and Gregg, B.A. (2000): The Photovoltage-Determining Mechanism in Dye-Sensitized Solar Cells, *J. Phys. Chem. B, 104*, 6–10

Rand, B.P., Xue, J., Yang, F., & Forrest, S.R. (2005): Organic solar cells with sensitivity extending into the near infrared, *Appl. Phys. Lett.* 87 (23), Article ID 233508, 3 pages

Ravirajan, P., Haque, S.A., Durrant, J.R., Poplavskyy, D., Bradley, D.D.C., & Nelson, J. (2004): Hybrid nanocrystalline TiO₂ solar cells with a fluorene-thiophene copolymer as a sensitizer and hole conductor, *J. Appl. Phys, 95*, 1473–1480

Reiss, W., Karg, S., Dyakonov, V., Meier, M., & Schworer, M. (1994): *J. Lumin.* **60/61**, 906

Reyes-Reyes, M., Kim, K. & Carroll, D.L. (2005): High-efficiency photovoltaic devices based on annealed poly (3-hexylthiophene) and 1-(3-methoxycarbonyl)-propyl-1- phenyl-(6,6)C61 blends, *Appl. Phys. Lett.* 87, Article ID 083506

Sariciftci, N.S. (Ed.) (1998): *Primary photoexcitations in conjugated polymers: Molecular Exciton versus Semiconductor Band Model,* World Scientific Publishers, Singapore

Sariciftci, N.S., Smilowitz, L., Heeger, A.J., & Wudl, F. (1992): Photoinduced electron transfer from a conducting polymer to buckminsterfullerene, *Science, 258*, 1474

Sariciftci, N.S., Baun, D., Zhang, C., Srdanov, V.I., Heeger, A.J., Stucky, G., & Wudl, F. (1993): Semiconducting polymer-buckminsterfullerene heterojunctions: Diodes, photodiodes, and photovoltaic cells, *Appl. Phys. Lett.* **62**, 585–587

Sandén, B.A. (2003): *Proc. 18th EU PV Energy Conference*, Bologna, Spain.

Schilinsky, P., Waldauf, C., & Brabec, C.J. (2002): Recombination and Loss Analysis in Polythiophene Based Bulk Heterojunction Photodetectors, *Appl. Phys. Lett., 81*, 3885–3887

Schmidt-Mende, L., Bach, U., Humphry-Baker, R., Horiuchi, T., Miura, H., Ito, S., Uchida, S., & Gratzel, M. (2005): Organic dye for highly efficient solid-state dye-sensitized solar cells, *Advanced Materials, 17*, 813–815

Schuller, S., Schilinsky, P., Hauch, J., & Brabec, C.J. (2004): Accelerated Lifetime Testing and Stability Aspects of Polymer Solar Cells, *Appl. Phys. A, 79*, 37–40

Schulze, K., Uhrich, C., Schüppel, R., Leo, K., Pfeiffer, M., Brier, E., Reinold, E., & Bäuerle, P. (2006): Efficient vacuum deposited organic solar cells based on a new low-bandgap oligothiophene and fullerene C60, *Advanced Materials, 18*, 2872–2875

Seyrling, S., Calnan, S., Bücheler, S., Hüpkes, J., Wenger, S., Brémaud, D., Zogg, H. & Tiwari, A.N. (2009): CuIn(1-x)Ga(x)Se2 photovoltaic devices for tandem solar cell application, *Thin Solid Films, 517*, 2411–2414

Seyrling, S., Buecheler, S., Chirila, A., Perrenoud, J., Verma, R., Wenger, S., Graetzel, M., & Tiwari, A.N. (2009): Development of multijunction thin film solar cells, *34th IEEE Photovoltaic Specialists Conf,* June 2009, Philadelphia, PA, Available at file:///E:/34%20IEEE%202009/Abstracts/301.html

Shaheen, S.E., Brabec, C.J., Sariciftci, N.S., Padinger, F., Fromherz, T. & Hummelen, J.C. (2001): 2.5% efficient Organic plastic solar cells, *Appl. Phys. Lett,* **78**, 841–843

Shaheen, S.E., Radspinner, R., Peyghambarian, N., & Jabbour, G.E. (2001): Fabrication of bulk heterojunction plastic solar cells, *Appl. Phys. Lett.,* **79**, 2996 (8 pages)

Shaheen, S.E., & Ginley, D.S. (2004): Photovoltaics for the Next Generation: Organic-Based Solar Cells, Schwarz, Contescu, and Putyera, (Eds.); In *Dekker Encycl.Nanosci. Nanotechnol,* Marcel Dekker Inc.: New York, pp. 2879–2895

Shirakawa, H. (2001): Nobel lecture: The discovery of polyacetylene film, the dawning of an era of Conducting polymers, *Reviews of Modern Physics,* **73**, 713–718

Shirakawa, H. (2001): Nobel lecture: The discovery of polyacetylene film, the dawning of an era of Conducting polymers, *Angew. Chem. Intl. Ed.,* **40**, 2574

Shirakawa, H. *et al.* (2003): Twenty five years of conducting Polymers, *Chem. Commun,* p. 1

Sirringhaus, H. (2005): Device physics of solution-processed Organic field-effect transistors, *Advanced Materials,* **17**, 2411–2425

Spannggard, H., & Kerbs (2004): A brief history of the development of organic and polymeric photovoltaics, *Sol. Energy Mater. Sol. Cells,* **83**, 125–146

Sundar, V.C., Zaumseil, J., Podzorov, V., Menard, E., Willett, R.L., Someya, T., Gershenson, M.E., & Rogers, J.A. (2004): Elastomeric transistor stamps: reversible probing of charge transport in Organic crystals, *Science,* **303**, 1644–1646

Svensson, M., Zhang, F., Veenstra, S.C., Verhees, W.J.H., Hummelen, J.C., Kroon, J.M., Inganäs, O., & Andersson, M.R. (2003): Highperformance polymer solar cells of an alternating polyfluorene copolymer and a fullerene derivative, *Advanced Materials,* **15**, 988

Takahashi, K., Kuraya, N., Yamaguchi, T., Komura, T. & Murata, K. (2000): Three-layer organic solar cell with high-power conversion efficiency of 3.5%, *Sol. Energy Mater. and Sol. Cells,* **61**, 403–416

Tang, C.W., & Albrecht, A.C. (1975): Photovoltaic effects of metal – chlorophyll-a – metal sandwich cells, *J. Chem. Phys,* **62**, 2139–2149

Tang, C.W. (1979): US Patent 4164431, August 1979

Tang, C.W. (1986): Two-layer Organic Photovoltaic cell, *Appl. Phys. Lett.* **48**, 183–185

Tiwari, A.N. (2006): Solar electricity: Prospects of Thin film Solar cells, Laboratory for Thin films & Photovoltaics, EMPA, Switzerland

Uchida, S., Xue, J., Rand, B.P., & Forrest, S.R. (2004): Organic small molecule solar cells with a homogeneously mixed copper phthalocyanine: C60 active layer, *Appl. Phys. Lett,* **84**, 4218–4220

van Duren, J.K.J., Yang, X., Loos, J., Bulle-Liewma, C.W.T., Sieval, A.B., Hummelan, J.C., & Janssen, R.A.J. (2004): Relating the morphology of poly(p-vinylene)/ methanofullerene blends to solar-cell performance, *Advanced Functional Materials,* **14**, 425–434

van Hal, P.A., Christiaans, M.P.T., Wienk, M.M., Kroon, J.M., & Janssen, R.A.J. (1999): *J. Phys. Chem. B,* **103**, 4352–4359

Wallace, G.G., Dastoor, P.C., Officer, D.L., & Too, C.O. (2000): Conjugated Polymers: New materials for Photovoltaics, *Chemical Innovation,* **30**, 1, 14–22

Wenger, S., Seyrling, S., Tiwari, A.N., & Grätzel, M. (2009): Fabrication and performance of a monolithic dye-sensitized TiO_2/Cu (In, Ga) Se_2 thin film tandem solar cell, *Appl. Phys. Lett.* **94**, 173508

Wienk, M.M., Kroon, J.M., Verhees, W.J.H. *et al.* (2003): Efficient methano [70]fullerene/ MDMO-PPV bulk heterojunction photovoltaic cells, *Angewandte Chemie International Edition,* **42**, 3371–3375

Winder, C., and Sariciftci, N.S. (2004): Low bandgap polymers for photon harvesting in bulk heterojunction solar cells, *J. Materials Chemistry,* **14**, 1077–1086

Wohrle, D., & Meissner, D. (1991): Organic Solar Cells, *Adv. Mater.* **3**, 129

Xue, J., Uchida, S., Rand, B.P. & Forrest, S.R. (2004a): 4.2% efficient Organic photovoltaic cells with low series resistances, *Appl. Phys. Lett.* **84**, 3013–3015

Xue, J., Uchida, S., Rand, B.P. & Forrest, S.R. (2004b): Asymmetric tandem Organic photovoltaic cells with hybrid planar-mixed molecular heterojunctions, *Appl. Phys. Lett.* **85**, 5757–5759

Xue, J., Rand, B.P., Uchida, S., & Forrest, S.R. (2005): Mixed donor-acceptor molecular heterojunctions for photovoltaic applications. II. Device performance, *J. Appl. Phys.* **98**, Article ID 124903, 9 pages

Yang, F., Shtein, M., & Forrest, S.R. (2005a): Morphology control and material mixing by high-temperature organic vaporphase deposition and its application to thin-film solar cells, *J. Appl. Phys.* **98** (1), Article ID 014906, 10 pages

Yang, F., Shtein, M., & Forrest, S.R. (2005b): Controlled growth of a molecular bulk heterojunction photovoltaic cell, *Nature Materials*, **4** (1), 37–41

Yang, J., Banerjee, A., & Guha, S. (1997): Triple-junction amorphous silicon alloy solar cell with 14.6% initial and 13.0% stable conversion efficiencies, *Appl. Phys. Lett.* **70**, 2975–2977

Yang, X., Van Duren, J.K.J., Janssen, R.A.J., Michels, M.A.J., & Loos, J. (2004): Morphology and Thermal Stability of the Active Layer in Poly(p-phenylenevinylene)/Methanofullerene Plastic Photovoltaic Devices, *Macromolecules*, **37**, 2151–2158

Yin, C., Kietzke, T., Neher, D. & Hörhold, H.-H. (2007): Photovoltaic properties and exciplex emission of polyphenylenevinylene-based blend solar cells, *Appl. Phys. Lett*, **90**, Article ID 092117, 3 pages

Yu, G., Zhang, C., & Heeger, A.J. (1994): Dual-function semiconducting polymer devices: light-emitting and photodetecting diodes, *Appl. Phys. Lett*, **64**, 1540-1542

Yu, G., Gao, J., Hummelen, J.C., Wudl, F. & Heeger, A.J. (1995): Polymer Photovoltaic cells: Enhanced efficiencies via a network of internal donor-acceptor heterojunctions, *Science*, **270**, 1789–1791

Yu, G., & Heeger, J. (1995): Charge separation and Photovoltaic conversion in Polymer composites with internal donor/acceptor heterojunction, *J. Appl. Phys.*, **78**, 4510–4515

Yu, G., & Heeger, J. (1997): High efficiency photonic devices made with semiconducting polymers, *Synthetic Metals*, **85**, 1183-6

Zaban, A., Meier, A., & Gregg, B.A. (1997): Electric Potential Distribution and Short-Range Screening in Nanoporous TiO_2 Electrodes, *J. Phys. Chem. B*, **101**, 7985–7990

Chapter 5

High-efficiency solar devices

5.1 INTRODUCTION

This chapter is devoted to III-V multi junction solar devices and Concentrator Photo-voltaic systems (CPV). These devices/systems are developed with an aim to improve the conversion efficiencies of the solar cells while reducing costs on the material and fabrication processes. The enormous potential of these systems have been amply demonstrated. The research and development effort in both the areas is still underway in several laboratories to optimize the production processes for mass production of the devices.

Silicon thin-film multijunction solar cells and tandem structures based on a-silicon and μc-silicon and their current status, the research efforts on polycrystalline CIGS and CdTe thin-film based tandem solar devices, and DSC/CIGS tandem structures are discussed in earlier chapters. In this chapter, multi-junction solar cells based on III-V compounds are dealt.

5.2 III-V MULTI JUNCTION SOLAR CELLS

5.2.1 Introduction

The concept of multi-junction using III-V semiconductors with a variety of bandgaps and lattice constants had been well known since the 1950s, but was not pursued for about two decades. The renewed interest was created by several technological break-throughs; for example, the realisation of monolithic two terminal AlGaAs/GaAs dual junction cells based on a tunnel junction interconnect between the two photovoltaic junctions and subcells, respectively (Bedair *et al.* 1979), and the development of commercial MOVPE (metal-organic vapour phase epitaxy) reactors during the 1980s and so on. While LPE (liquid phase epitaxy) was used in the beginning, MOVPE technology had become the preferred growth method due to its ability for preparing multi-layer structures of different materials with high quality, large area and high throughput. Another breakthrough was the use of GaInP as top cell material (Olsen *et al.* 1985) instead of AlGaAs, which suffered from low diffusion length due to the affinity of aluminium for oxygen. These technological breakthroughs led to the development of the GaInP/GaAs/Ge triple-junction cell, which became commercially available in the 1990s by US manufacturers.

Multi junction III-V solar cells (MJ cells) are third generation PV devices with advantages of unparalleled high efficiency, ability to be integrated into very lightweight panels, and reliability in the space environment. As a result, these cells have become the standard for space power generation. Remarkable technology explosion in the area of photovoltaic cells designed for space purposes (Meusel *et al.* 2005; King *et al.* 2002; Sharps *et al.* 2004) happened since 2000. Commercially-available multi junction solar cells with 30% conversion efficiency under the AM0 space spectrum have become a reality.

Since radiation exposure, thermal cycling, vibration, atomic oxygen effect, contamination from volatile materials, and electrostatic discharge are the factors that impact operation in the space environment, these aspects must be the elements of qualification process for space solar cells and panels; and a better comprehension of these issues now has resulted in new standards of reliability. Hectic research activity by several groups has facilitated to develop very-thin, flexible and extremely lightweight space solar cells and panels, capable of being folded or rolled into a smaller stowage volume for launch. Earlier, these features have been normally associated with thin-film polycrystalline or amorphous Si PV technology only.

Due to high cost of production, these cells are not directly used currently for terrestrial applications, for e.g., power generation. Several research groups have studied the possibility of employing III-V solar cells in conjunction with light concentration for terrestrial PV applications (Swanson 2000; Bosi & Pelosi 2007), and are presently used in certain designs of CPV. But III-V semiconductors based on arsenides and phosphides find use as the materials of choice for commercial optoelectronic devices (Mawst *et al.* 1996; Hung 1988).

Using optical devices such as mirrors, Fresnel lenses, dichroic films, and light guides, the possibility to collect solar light and concentrate its energy on a single small area solar cell has been demonstrated. This concept reduces the total cell area an amount equal to the concentration ratio. The cost of the PV system therefore decreases because a relatively inexpensive optical concentrator replaces the expensive semiconductor material. Moreover, using light concentration boosts the cell conversion efficiency. Further, concentration could facilitate more careful use of land for installations as well as boosting the energy production of large solar power plants. Concentrator cells could also challenge conventional rooftop Si panels, particularly since improved kWp/m^2 generation would make a concentration system installation possible in large buildings. Moreover, a better integration of PV elements with architectural design could be achieved.

In fact, concentrator photovoltaics is presently drawing more and more attention encouraged by the increasing terrestrial market as well as by the prospect of using III-V multi-junction concentrator cells, which have now become commercially available with average efficiencies of about 35% (AM1.5d low AOD, C×500) and best cells recently crossing the efficiency threshold of 40% (King *et al.*).

5.2.2   Basic principles of multi-junction solar cells

The basic aspects of single junction and multi junction III-V solar cells are available in the literature. Tandem devices are naturally the most advanced multi-junction technology and there are many examples of combinations of first and second generation devices to produce third generation devices. The highest performing devices are, however, made of expensive III-V semiconductors that can only be thought of for concentrator or space applications (Karam *et al.* 1999). More cost effective terrestrial multi junction devices combine the polycrystalline-silicon and amorphous thin-film silicon technologies (Yoshima *et al.* 2003; Yang *et al.* 1994). These devices are relatively inexpensive for their efficiency but do not realise the full potential of multi-junction efficiency improvements.

Technology based on the use of multiple-junctions is the only proven third generation technology (Green 2006; Yoshima *et al.* 2003; Yang *et al.* 1994). Single junction devices perform optimally at the wavelength equal to the bandgap, inherently becoming inefficient at all other wavelengths across the solar spectrum. Multi-junction devices stack different solar cells with multiple bandgaps tuned to utilise the entire spectrum. Light is first incident upon a wide band-gap device that can produce a relatively high voltage and thereby make better use of high energy photons, then lower energy photons pass through narrow band-gap sub-devices that can absorb the transmitted IR-photons (Burnett 2002). Maximum efficiencies of 55.9%, 63.8%, and 68.8% are predicted for 2 (tandem), 3- and 4-junction devices (Green 2006). However costs go up as fabrication becomes increasingly complex with the increasing number of interfaces and cells.

Multi-junction solar cells therefore use a combination of semiconductor materials to more efficiently capture a larger range or entire photon energies (Yamaguchi 2005; Strobl 2006; Dimroth & Kurtz 2007; Burnett 2002; Marti & Luque 2004). Depending on the particular technology, present-day multi-junction solar cells are capable of generating approximately twice as much power under the same conditions as traditional silicon solar cells. Availability of materials with most favorable band gaps that simultaneously allow high efficiency through low defect densities is the basic limitation for the fabrication of multi-junction solar cells. III-V semiconducting compounds are recognized as good candidates for fabricating such multi junction cells. Their band gaps cover a wide spectral range, and most of these materials have direct electronic structure, implying a high absorption coefficient. They can be grown with extremely high crystalline and optoelectronic quality by high-volume growth techniques despite their complex structures (Yamaguchi 2001; Roman 2004; Dimroth *et al.* 2001; King 2005).

Multi-junction solar cells have been studied for over five decades (Wolf 1960). The first multi-junction device demonstrated in early 1980 converted 16% of the solar energy into electricity (Luque and Hegedus 2003). Then, by the end of 2000, a triple junction InGaP/GaAs/Ge device had achieved 30% efficiency (Cotal *et al.* 2000). III-V Multi-junction solar cells have highest theoretical efficiency conversion, 86.8% (Dimroth 2005) as compared to other photovoltaic technologies (Schokley & Queisser 1961; Green *et al.* 2007; Honsberg *et al.* 2001). The present-day record efficiency of 40.7% was achieved with a multi-junction solar cell fabricated by Boeing Spectrolab Inc. in December 2006 (Spectrolab website).

Band gap engineering

Material systems based on GaAs, InP or GaN can use different compositions of aluminium or indium alloys to modify band-gaps while maintaining lattice constants

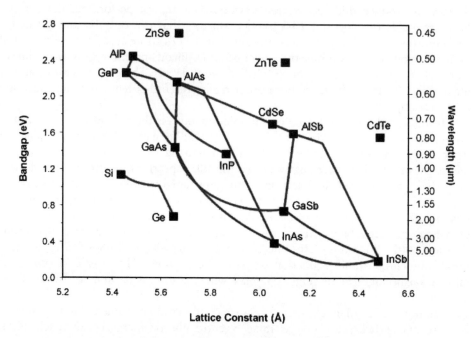

Figure 5.1 Ternary and quaternary III-V compounds – lattice constants and bandgaps

(Figure 5.1). These can produce monolithic multi-junction devices in a single growth run (with tunnel junctions between each device).

Triple-junction solar cells currently produced are made of GaInP (1.9 eV), GaAs (1.4 eV), and Ge (0.7 eV). In early multi junction designs, lattice matching was considered as a secondary concern. However, work at NREL showed that lattice mismatching as low as ±0.01% causes significant degradation of PV quality. Therefore, with great effort, GaInP, GaAs, and Ge which have a desirable complement of bandgap energies and matching lattice constants for the fabrication of triple-junction cell were selected at NREL.

Advanced multi-junction solar cell designs anticipate use of AlGaInP (2.2 eV), AlGaAs (1.6 eV), GaInP (1.7 eV), GaInAs (1.2 eV), GaInNAs (1.0–1.1 eV) (Dimroth *et al.* 2001). For example, Spectrolab's record-breaking cell used $Ga_{0.5}In_{0.5}$ P (or $GaInP_2$) with band gap energy of 1.85 eV. Less gallium and more indium would result in a lower band gap compound material, up to the resulting InP with band gap energy of 1.3 eV and the lattice constant of 5.88 Å. However, such an adjustment in band gaps should be made in conjunction with lattice-constant constraints (Burnett 2002).

Electric current matching

The series connection of monolithically-grown multi-junction solar cells makes matching of currents a desirable characteristic (Smestad 2002). The output current of the multi junction solar cell is limited to the lowest of the currents generated by any of the individual cells. Therefore, the currents through each of the subcells are controlled

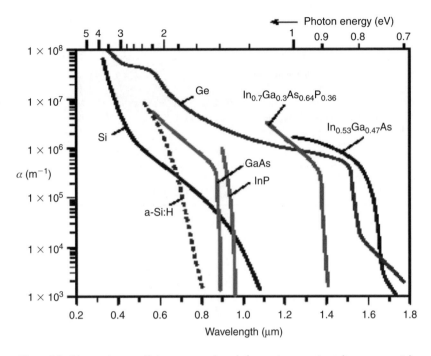

Figure 5.2 Absorption coefficient vs wavelength for various semiconductor materials
(*Source*: Kasap 2001, Copyright © 2002, Prentice Hall, New York)

to have the same value. The current is proportional to the number of incident photons with energy beyond the semiconductor's band gap, and the absorption coefficient of the material. So, a thin layer is sufficient if the photons that exceed the band gap are in abundance. At the same time, if the absorption coefficient is low, the layer must be made thicker, since on average a photon must travel through more of the material before being absorbed. After materials are selected with desired band gaps and lattice constants, the thickness of each layer must be determined based on the material's absorption coefficient and the number of incident photons with a given energy, so that each layer will generate the same photocurrent. The absorption coefficient for various semiconductors as a function of photon wavelength is shown on Figure 5.2.

The design of GaInP/GaAs/Ge solar cell implies a relatively thick Ge layer because of its lower absorptivity, while other layers are of different thickness; terrestrial and space versions of the cell vary to account for the differing solar spectra, particularly UV and near-IR radiation in these two different environments (Burnett 2002).

5.2.3 Fabrication of Triple-junction solar cells

Multi-junction solar cells are fabricated by either of the two methods: (i) by mechanical stacking of independently grown layers or (ii) by growing each semiconductor layer monolithically on top of the other as one single piece by metal organic chemical vapour

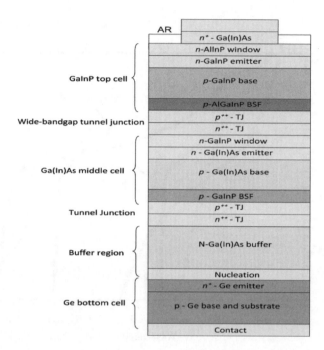

Figure 5.3 Schematic structure of a triple-junction photovoltaic cell (Redrawn from King *et al.* 2007a)

deposition (MOCVD) or molecular beam epitaxy (MBE) (Luque & Hegedus 2003; Poortmans & Arkhipov 2006). The mechanically stacking, though pragmatic, is less-desirable because the method suffers from the bulkiness, additional expense, and heat-sinking (Burnett 2002). MOCVD is preferred to MBE because it not only ensures high crystal quality all through the device, but makes it easy for scaling up to commercial production capacities.

The transport of electrons between layers in monolithic multi-junction solar cells is addressed using a tunnel junction, which is a stack of highly-doped layers, producing an effective potential barrier for both minority-carriers. The strong doping is necessary in order to have a thin depletion region, promoting tunneling across the junction and minimizing optical losses. While S, Se, Te, Sn, Si, C, Ge are used as n-type dopants, Zn, Be, Mg, Cd, Si, C, Ge are used as p-type dopants; and the last three elements act as n-type or p-type depending on whether they replace a Ga or an As atom in the crystalline structure (Roman 2005). The AR coating is a broadband two-layer dielectric stack, such as TiO_2/Al_2O_3, or Ta_2O_5/SiO_2 or ZnS/MgF_2, whose spectral reflectivity characteristics are designed to reduce device's typically large reflectance ($\sim$30%) in the related spectral region to <1%. Currently, the most efficient multi-junction photovoltaic cells are fabricated using GaInP, GaAs, and Ge layers on Ge substrate. The schematic of such a triple-junction solar cell, usually containing about 20 layers, is shown in Figure 5.3. The first layer of this record-breaking cell is composed of GaInP (1.85 eV), converting short wavelength portions of spectrum, blue and UV photons, the second GaAs-layer (1.42 eV) captures near-infrared light photons, and the third layer is made of Ge, effectively absorbs the lower photon energies of the IR radiation that are above 0.67 eV (Dimroth *et al.* 2001; King 2005).

The photons with wavelengths below around 650 nm that pass through GaInP layer into GaAs layer are not being efficiently captured. A larger portion of the spectrum is absorbed by the Ge layer, since the difference between the band gap of the top two layers is 0.4 eV while the difference between the bottom two layers is 0.7 eV. Layers in this configuration are lattice matched with one another (Sherif *et al.* 2006).

In order to match currents among the layers, two variations of the layers' thicknesses exist: one is having a thicker top cell for terrestrial applications, and another is for the absorption of the larger amounts of high-energy photons in space. The configuration of the record-efficient triple-junction device fabricated by Spectrolab is $Ga_{0.44}In_{0.56}P/Ga_{0.92}In_{0.08}As/Ge$. The performance of the cell is measured at NREL under standard spectrum. Under 236 suns (23.6 W/cm^2) intensity, with an aperture area of 0.2691 cm^2, at $25 \pm 1°C$, the cell parameters are: $V_{oc} = 3.089$ V, $J_{sc} = 3.377$ A/cm^2, FF = 88.24%, $V_{mp} = 2.749$ V, and $\eta = 39.0 \pm 2.3\%$ (Sheriff *et al.* 2006).

5.2.4 Future design considerations

(a) Design optimization of the existing layers

The efficiency of the present-day triple-junction solar cell can be improved by design optimization of each subcell.

In the past, triple-junction cell efficiency has been improved by using disordered GaInP which has higher bandgap, 1.88 eV instead of ordered GaInP with 1.78 eV bandgap as top cell material (King 2005; Takamoto *et al.* 2003). Besides, the top GaInP layer could be made thicker to increase its current production, so that the multi-junction cell would generate a higher matched current, and thereby more power (Luque & Hegedus 2003).

As an alternative to top GaInP layer, $Al_{0.37}Ga_{0.63}As$ or AlGaInP with the bandgap of 1.98 eV which has similar lattice constant and bandgap energy can be used. This approach is not favourably considered in the past as these materials are highly sensitive to oxygen and water contamination, but recent results demonstrate their potential (Takamoto *et al.* 2003; King *et al.* 2002; Fetzer *et al.* 2002). The efficiency can be increased by replacing the second layer, GaAs with a 1.25 eV-bandgap material. This second layer could collect a larger current, while reducing the number of photons transmitted to the Ge layer (King 2005; Honsberg 2005).

(b) Increasing the number of junctions

Another possible improvement to the design is to develop devices with more junctions. Four junction solar cells have been suggested using a material with a bandgap of 1.0 eV (Luque & Hegedus 2003). GaInNAs is considered in this context as it is the most studied and can be grown lattice-matched to Ge.

Progress has been achieved in reducing the background doping and defect concentration by special growth and annealing conditions, but there is a problem of low current generation (Strobl 2006; Meusel 2004). Thus, present-day four-junction solar cells do not lead to higher efficiencies than triple-junction devices.

Five- and six-junction cell designs, contrary to triple-junction cells, divide the solar spectrum into narrower wavelength ranges that allows all the subcells to be better current matched to the low current- producing subcell (Luque & Hegedus 2003; King 2005; Dimroth 2005). Besides, the finer division of the incident spectrum reduces thermalization losses from the photo-generated electron-hole pairs created by photons far above the band gap energy; also, the smaller current density in these cells lowers resistive losses (King *et al.* 2002).

The theoretical efficiency limits for multi junction devices based on thermo-dynamic basics are 37, 50 and 56% for 1, 2 and 3 band gaps respectively (Luque & Hegedus 2003; Honsberg 2005). The improvement in efficiency on going from one to two or three band gaps is considerable, but the benefits are found to reduce as more junctions are added. So, the reality of the solar cell with more than four or five junctions is doubtful. However, theoretical studies show that efficiencies of up to 86.8% can be achieved using an infinite number of band gaps.

(c) Incorporating semiconductor quantum dots

In recent years it has been proposed and experimentally verified that the use of nanostructures, such as quantum wells, quantum wires, superlattices, nanorods, or nanotubes, offer the potential for high photovoltaic efficiency by tailoring the properties of existing materials, and for reducing of cost using self-assemblance of nanostructures (Myong 2007; Barnham 2005; Morf 2002; Bailey *et al.* 2003; Ekins-Daukes 2002; Gur 2005; Nosova *et al.* 2005; Das *et al.* 2001; Honsberg 2006; King *et al.* 2004).

Semiconductor quantum dots (QD) are currently the most fascinating subject mainly due to their size-dependent electronic structures, and hence tunable optoelectronic properties (Nozik 2002; Raffaelle 2006; Suwaaree 2006). The application of III-V compounds and other combinations as material systems for the QDs are discussed in Chapter 6.

5.2.5   Metamorphic (lattice-mismatched) solar cells

To achieve higher efficiencies, a more radical departure from conventional three-junction cell design which has a higher theoretical efficiency ceiling is required. One such approach is the use of metamorphic or lattice-mismatched materials, to tune the bandgaps of the individual subcells of a multi junction cell to the solar spectrum for maximum conversion efficiency (King *et al.* 2005, 2000; Dimroth *et al.* 2000; Wanlass *et al.* 2005).

The essential distinguishing feature of III-V multi junction cells is the very wide range of subcell and device structure band gaps that can be grown with high crystal quality, and correspondingly high minority-carrier recombination lifetimes which is true for lattice-matched multi junction cells. But, when metamorphic semiconductors are used, the flexibility in band gap selection becomes entirely different, because all subcells need not have the same crystal lattice constant. Due to this advantage, there has been growing interest in the study of metamorphic solar cell materials (King *et al.*

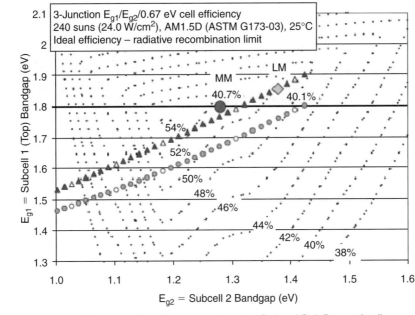

Figure 5.4 Ideal iso-efficiency contours for 3-junction terrestrial concentrator cells with variable top and middle subcell bandgaps, based on radiative recombination and the terrestrial solar spectrum at 240 suns. Subcell 1 and 2 bandgap pairs of GaInP and GaInAs at the same lattice constant are shown for both disordered and ordered GaInP. The measured efficiencies and bandgap combinations for these MM and LM cells are also plotted, showing the theoretical advantage of the metamorphic design
(*Source:* King et al. 2007c, Hindawi Publishing Corp., Copyright © 2007 R.R. King et al.)

2007a, 2007b, 2006a, 2006b, 2005, 2000; Dimroth *et al.* 2000; Takamoto *et al.* 2003; Wanlass *et al.* 2005; Bett *et al.* 2005; Geisz *et al.* 2007; Sherif *et al.* 2005).

The benefits of flexibility in subcell band gap selection are clear from the theoretical efficiency contours for 3-junction terrestrial concentrator cells shown as a function of top (subcell 1) band gap Eg_1 and middle (subcell 2) band gap Eg_2 (King *et al.* 2007c) in Figure 5.4. These contours are based on the diode characteristics of subcells limited only by the primary mechanism of radiative recombination, and on the shape of the terrestrial solar spectrum. The cell model is discussed in greater detail by King *et al.* (2006b). Efficiencies up to 54% seem to be possible in principle at this concentration for 3-junction cells in the radiative recombination limit, increasing to over 58% for 4-junction terrestrial concentrator cells (King *et al.* 2006b).

In 3-junction GaInP/GaInAs/Ge metamorphic solar cells, the GaInP and GaInAs subcells can be grown on a metamorphic buffer such that these two subcells are lattice-matched to each other, but are both lattice-mismatched to the Ge substrate and subcell. Metamorphic (MM) cells appear to bring the cell design closer to the region of Eg1 and Eg2 space that has the highest theoretical efficiencies. The lower band gaps of MM

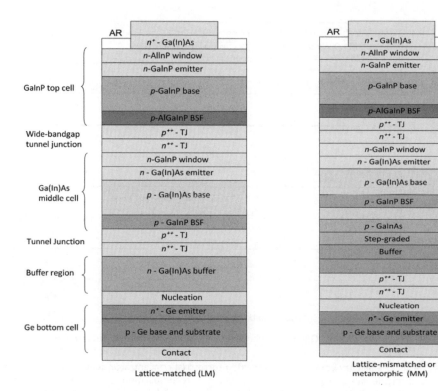

Figure 5.5   Schematic cross-sectional diagrams of lattice-matched (LM) and metamorphic (MM) GaInP/GaInAs/Ge 3-junction cell configurations, corresponding to the LM 40.1% and MM 40.7%-efficient concentrator cells (Redrawn from King *et al.* 2007c)

subcells can use a larger part of the solar spectrum that is generally wasted as excess photo-generated current in the Ge bottom cell in most lattice-matched three junction cells. In the past, recombination at dislocations in MM materials has often affected this promise of higher theoretical efficiency. However, for the recent metamorphic 40.7%-efficient and lattice matched 40.1%-efficient cells, the studies have established that the density and behaviour of dislocations have been adequately controlled to demonstrate the efficiency advantage of the MM design both theoretically and *experimentally*.

The analysis was extended taking into account the shadowing and specific resistance associated with the metal grid pattern used on the 40.7% record cell. Fill factor calculated for the 3-junction cell with the band gap combination of the MM 40.7% cell is 87.5% with series resistance included, essentially identical to that measured experimentally for the 40.7% cell at 240 suns. If additional real-life effects are included, the calculated contours show a good estimate of the efficiencies that can be achieved in practical, state-of-the- art, three-junction cells as a function of bandgap.

Schematic diagrams of LM and MM cells are presented in Figure 5.5, showing the step-graded metamorphic buffer used in the MM case to transition from the lattice constant of the substrate to that of the upper subcells.

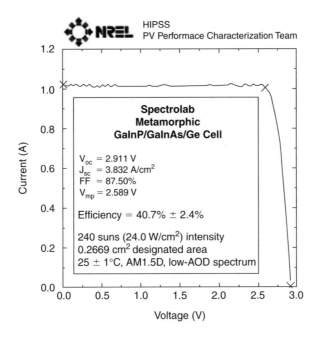

Figure 5.6 Illuminated I-V curve for the record 40.7% metamorphic three-junction cell, independently verified at NREL
(*Source*: King *et al.* 2007c, Hindawi Publishing Corp., Copyright © 2007 R.R. King *et al.*)

High efficiency multi junction cells

Metamorphic semiconductor materials have made possible the band gap engineering of subcells in 3-junction solar cells resulting in higher measured efficiencies than even for the best lattice-matched cells. Experiments on step-graded buffers, used to transition from the substrate to the subcell lattice constant, have been used to control the usual problem of dislocations in the active cell regions due to the lattice mismatch. The band gap-voltage offset, $(Eg/q) - V_{oc}$, is a key indicator of the quality and suppression of SRH recombination in semiconductors of variable band gap; lower offset value is preferred, since it is a measure of the separation between electron and hole quasi-Fermi levels and the conduction and valence band edges (King *et al.* 2005, 2006a, 2006b). Metamorphic GaInAs single-junction cells with 8% In are fabricated and tested with a band gap-voltage offset of 0.42 V at one sun, essentially the same as GaAs control cells, indicating the long minority-carrier lifetimes that can be achieved in metamorphic materials.

Extensive experimental studies were carried out on GaInP/GaInAs/Ge terrestrial concentrator cells, using a variety of metamorphic and lattice-matched 3-junction cell configurations, wide-band-gap tunnel junctions and other high efficiency semiconductor device structures, current matching conditions, cell sizes, grid patterns, and fabrication processes, resulting in new understanding of the limiting mechanisms of terrestrial multi junction cells, and new heights in performance.

Figure 5.6 shows the measured illuminated I-V curve for the record efficiency 40.7% metamorphic GaInP/GaInAs/Ge three-junction cell at 240 suns (King *et al.*

2007c), under the standard spectrum for concentrator solar cells. This is the first solar cell to reach the highest solar conversion efficiency yet achieved for any type of photovoltaic device.

A lattice-matched three-junction cell has also achieved efficiency of 40.1% measured at 135 suns (AM1.5D, low-AOD, 13.5 W/cm^2, 25°C). These measurements have been *independently* verified at NREL. King *et al.* (2007c) have studied the I-V characteristics of both the record MM and LM devices and compared. The higher current and lower voltage of the metamorphic design is evident from the data. The efficiency, V_{oc}, and fill factor are analysed as a function of incident intensity, or concentration ratio, for the record 40.7% MM and 40.1% LM cells, as well as for an additional MM cell with good performance at high intensities. Interestingly, the efficiencies of both the record MM and LM cells track very closely at the same concentration, with the possibility of extending measurements to a higher concentration for the MM cell. Fill factors for both types of cell are quite high at about 88% in the 100–200 sun range. V_{oc} increases at rates of ~210 mV for MM cell and ~190 mV for LM cell per decade in the 100–200 suns range. Thus the MM subcells increase in voltage rather more rapidly with excess carrier concentration (than in the LM case), as one would expect if defects in the MM materials become less active at mediating recombination at higher injection levels.

Of all the PV technologies, the III-V multi junction concentrator cells not only offer the highest efficiency but also the highest rate of increase. These high III-V cell efficiencies have led to concentrator PV *module* efficiencies of over 30%, more than double the typical ~15% efficiencies of wafer silicon modules. This high efficiency can balance the PV system economics (Sherif *et al.* 2005), as it reduces all area-related costs of the module. With cost/Wp advantage, these high efficiency multi junction concentrator cells could create an explosive market growth for concentrator PV at multi-GW/year production levels.

5.2.6   Four-junction (terrestrial) solar cells

A 4-junction terrestrial concentrator solar cell was developed by King *et al.* (2006b) with the structure, (Al)GaInP/AlGa(In)As/Ga(In)As/Ge. The elements in the brackets indicate optional elements in the subcell composition. This type of cell divides the photon flux available in the terrestrial solar spectrum above the band gap of the GaInAs among the three subcells, as against 2 subcells as in the case of 3-junction cell. As a result, the current density of a four-junction cell is roughly 2/3rd of a corresponding 3-junction cell and the resistive power loss is around $(2/3)^2 = 4/9$ or less than half of a 3-junction cell. This is very crucial for concentrator cells.

Ideal efficiencies of over 58%, and practical cell efficiencies of 47% are possible for 4-junction terrestrial concentrator cells with a band gap combination of 1.90/1.43/1.04/0.67 eV. These band gaps are reachable with metamorphic materials and the use of transparent graded buffer layers, for example, in inverted metamorphic cell designs (Wanlass *et al.* 2005; King *et al* 2005, 2006b; Geisz *et al.* 2007).

Figure 5.7 shows a lattice-matched 4-junction cell, with all the subcells at the lattice constant of the Ge substrate; lattice mismatched versions of the four-junction cell are also possible, giving greater flexibility in bandgap selection (King *et al.* 2007c). The practical 4J cell efficiencies are about five absolute efficiency points over those

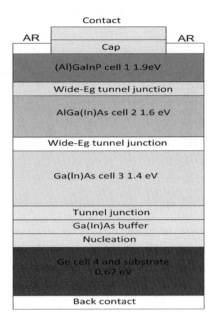

Figure 5.7 Schematic representation of LM 4-junction cell with all the subcells at the lattice constant of Ge (Redrawn from King et al. 2007c)

for 3-junction cells. Further, these cells benefit from reduced resistive power losses as described above, and from more efficient use of the terrestrial solar spectrum for this band gap combination.

Four-junction cells designed for the terrestrial applications as well as for the high current densities of concentrator operation are grown by metal-organic vapor-phase epitaxy (MOVPE), processed into devices, and studied. The external quantum efficiency of one such 4-junction (Al)GAInP/AlGa(In)As/Ga(In)As/Ge terrestrial concentrator solar cell is studied and the band gaps of each subcell are determined: GaInP subcell 1 (top cell) = 1.86 eV, AlGaInAs subcell 2 = 1.62 eV, GaInAs subcell 3 = 1.38 eV, and Ge subcell 4 = 0.70 eV.

By convoluting with the terrestrial AM1.5D (ASTM G173-03) spectrum, the current densities of the subcells are determined to be 9.24, 9.24, 9.58, and 21.8 mA/cm^2 for subcells 1, 2, 3, and 4 respectively. The data shows the subcells are very close to being current matched.

Illuminated light I-V curves for the 4-junction (Al)GaInP/AlGa(In)As/Ga(In)As/Ge cell are measured at 256 suns, and also for a similar solar cell with only upper 3 active junctions (inactive Ge subcell)(King et al. 2006b). The open circuit voltage of the 4-junction cell is 4.364 V compared to 3.960 V for the cell with an inactive subcell 4, indicating the Ge bottom cell accounts for about 400 mV of the Voc at this concentration. Preliminary measured efficiency for this (non-optimized) 4J cell is 35.7% at 256 suns. Independently confirmed measurements of the 40.1% lattice-matched and 40.7% metamorphic 3- junction cells are shown in Table 5.1 along with 4J cell for comparison.

Table 5.1 Comparison of 3J (LM & MM) and 4J Conc. Cells

	3J Concentrator Cells		4J Concentrator Cells	
	Lattice-matched	Metamorphic	4J Cell	Inactive Ge
V_{oc} (V)	3.089	2.922	4.364	3.960
J_{sc}/inten. (A/W)	0.1431	0.1575	0.0923	0.0923
V_{mp} (V)	2.749	2.565	3.949	3.572
FF	88.2%	85.5	88.6	88.2
Area (cm^2)	0.269	0.378	0.256	0.256
Conc. (suns)	236	179	256	254
Efficiency, η	39.0%	39.3%	35.7	32.3

5.2.7   Five- and six-junction solar cells

Another approach to improve the efficiency is to divide the solar spectrum more thinly using a greater number of junctions, such as 5- and 6-junction cells (King et al. 2004, 2005a).

Monolithic series-inter-connected 5- and 6-junction solar cells are developed by Spectrolab, Inc. (Reference: U.S. Pat.No. 6,316,715, Spectrolab, Inc., filed 3/15/00 and issued 11/13/01, and King et al. 2007d). The solar spectrum is divided by the bandgaps in AlGaInP (2.0eV)/GaInP (1.8 eV)/AlGaInAs (1.6 eV)/GaInAs (1.41 eV)/GaInNAs (1.1 eV)/Ge(0.67 eV) six-junction cell. The values given in the brackets are the bandgap energies of the concerned materials.

In the Quantum efficiency measurents as a function of photon energy, the subcell response cutoff showing the bandgap of each subcell has been clearly observed. Subcell 1 (the top subcell) is current matched to the other subcells in part by making it very thin, since its bandgap difference from subcell 2 is not sufficient to allow subcell 2 to current match the other subcells otherwise. As a result, the GaInP subcell 2 appears to have significant response at high photon energies (>2 eV). The measurements show that GaInNAs (subcell 5) is the lowest performing subcell. However, even with this relatively low quantum efficiency, the $\sim$1.1-eV GaInNAs subcell is capable of generating greater than 8 mA/cm^2, allowing it to be current matched to the other subcells in the series-interconnected, six-junction cell stack. The thermal treatment experienced by GaInNAs subcell during the growth of the upper four subcells on top influences the subcell behaviour.

Fully integrated six-junction solar cells with active GaInNAs subcells are grown and tested (King et al. 2005a; 2004). The measured light I-V curves of three 6-junction AlGaInP/GaInP/ AlGaInAs/GaInAs/GaInNAs/Ge solar cells with three different anneal conditions for the GaInNAs subcell, have given same value of over 5.3 V for Voc. The short-circuit current is 7.3 mA/cm^2 and active-area efficiency is 23.6%, for these early prototype 6-junction cells with heavily metalized grid pattern (King et al. 2006a). GaInNAs subcells and fully-integrated 6-junction cells with an active $\sim$1.1 eV GaInNAs subcell are grown on Ge wafers of 100 mm dia. at Spectrolab with good uniformity in a production scale MOVPE reactor (King et al. 2006a).

5.2.8   Prospects for multi junction solar cells

Multi-junction cells are established ones, but require technological advances aimed at cost reduction. These cost reductions can be possible either by producing much cheaper devices or by using high concentration systems. A key issue with concentrator systems is the need for mechanical tracking with associated higher production and maintenance costs which counterbalance the efficiency gains.

At the device level key requirements are in material growth, contacting, bonding and insulation. It is presumed that nanotechnology and the principles of self-alignment and self organisation may help solve some of these main issues over the next two decades, and large volumes of high-efficiency multi-junction devices will be well positioned for terrestrial applications.

The most promising outlook for multi junction approaches may be based on polymer semiconductors, or combinations of polymer semiconductors with crystalline semiconductors. The intrinsic ability to tune polymer band gaps and the ultra low-cost, low temperature, versatile deposition capability make polymers a natural choice, although the contacting issues common to multi junction cells and the long-term stability problems that affect polymer solar cells remain problematic and need solutions. By 2050, it is possible to imagine that self-organised semiconducting polymer structures, in multi junction arrangements could be the mainstream 50% efficient photovoltaic technology.

5.3 HIGH CONCENTRATION PV TECHNOLOGY (HCPV)

5.3.1 Introduction

Concentrator PV technology (CPV) is another concept that has emerged from the proposal to improve the efficiency of solar devices while driving down the costs.

A typical basic concentrator unit consists of a lens to focus the light, a cell assembly, a housing element, a secondary concentrator to reflect off-centre light rays onto the cell, a mechanism to dissipate excess heat produced by concentrated sunlight, and various contacts and adhesives. These basic units may be combined in any configuration to produce the desired sized module. The primary reason for using concentration is to decrease the area (size) of solar cell becuase solar cell is the most expensive component of a PV system, on a per-area basis. A concentrator uses relatively inexpensive materials (plastic lenses, metal housings, etc.) to capture a large area of solar energy and focus it onto a small area, where the solar cell resides. One measure of the effectiveness of this approach is the concentration ratio (how much concentration the cell is receiving). A tracker is also used to allow the optical and solar cell combination to follow the sun. Thus, the development of a CPV system involves optical, electronic, and mechanical components, each of which offering its own options.

5.3.2 Classification of CPV

Fraunhofer ISE distinguishes two classes of CPV: (1) low concentrator PV (LCPV) which operates at a concentration of up to 50 suns (1 sun corresponds to irradiation of 1000 W/m^2) with silicon solar cells and a single-axis or sometimes twin-axis tracking systems; (2) high concentrator PV (HCPV) with concentration factor over 300 suns, using high efficiency III-V multi junction solar cells and a high precision twin-axis tracking system. NREL report, however, classifies differently: (a) high-concentration (multi junction) cells – >400×; (b) medium-concentration (Si- or other) cells – ~3× to 100×; (c) enhanced concentration (Si-modules) – <3×.

Fresnel lenses made of plastic are used as concentrators, though glass Fresnel lenses are used by a few manufacturers. Mirror systems such as flat Cassegrain design consisting of a concave primary and a convex secondary mirror are also used.

Since the solar cells are exposed to high temperatures by the concentrators, heat dissipation arrangement is essential. A highly conducting plate is fixed below the cell (passive cooling); an active cooling system is also used.

In order to see the focused beams always strike the cells as ideally as possible, very precise trackers are arranged to track the sun.

Due to the interplay of the three individual systems (optical, electronic, and mechanical), there are several challenges in making the CPV system to work efficiently. In general, HCPV makes use of the multi-junction solar cells and LCPV allows Si cells; the HCPV is expected to reach module efficiencies of 30% (S&W Energy 7/2011).

5.3.3 Merits of CPV

Besides increasing the power and reducing the size or number of cells used, concentrators have the additional advantage that cell efficiency increases under concentrated light. The increase in efficiency depends largely on the cell design and the cell material used. Another advantage of the concentrator is that it can use small individual cells because it is harder to produce large-area, high-efficiency cells than to produce smaller-area cells. Due to the very low temperature coefficient of the III-V multi-junction concentrator solar cell, the performance of CPV systems is much less affected by temperature than any other PV technology, i.e., the loss of efficiency is approximately one third that of c-silicon modules.

This characteristic is extremely important for the best solar sites in the world, which are generally located in the equatorial regions where 'high temperatures' and 'direct solar radiation' with virtually perfect incidence angles with values up to 2800 kWh/(m²a) are available. Because of the low temperature coefficient, the efficiency and the electricity production of CPV systems are only slightly affected by high ambient temperatures in comparison to other PV technologies. The drop in efficiency with rise in temperature is by far the smallest for CPV systems as can be seen in Figure 5.8; the efficiency drop for a temperature difference of 40°K (e.g., from 25°C at standard testing conditions to typical operating cell temperatures of 65°C) is by far the smallest for CPV systems. The CPV technology, therefore, guarantees energy production during the entire day as well as maximum energy yield.

The other benefits that make CPV technology most cost-effective are: lower environmental effects, high potential for recycling, short energy payback time, requirement of less water for cooling compared to nuclear power plants and concentrating solar thermal systems, and most impressive potential for cost reduction.

There are, on the other hand, several drawbacks in using concentrators. The concentrating optics, for example, are significantly more expensive than the simple covers needed for flat-plate modules, and most concentrators must track the sun throughout the day and year to be effective. Thus, higher concentration ratios mean using not only expensive tracking mechanisms but also more precise controls than flat-plate systems with stationary structures. High concentration ratios are a particular problem, because the operating temperature of cells increases when excess radiation is

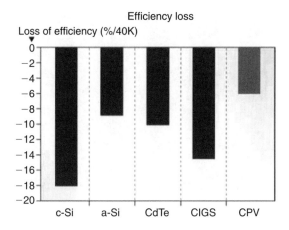

Figure 5.8 Efficiency loss for different PV technologies due to a rise of temperature of 40K
(*Source*: Gombert and Desrumaux 2010, reproduced with permission from Montgomery, J, News Editor, PVWorld)

concentrated. This, in turn, generates heat that disturbs the long-term stability of PV cells. Hence, the cells have to be cooled.

5.3.4 · Status of CPV

An excellent review was given by R.M. Swanson (Swanson 2000) analyzing the situation including commercial activity. There has been an intense development work on CPV system technology. In recent years, expensive multijunction III-V concentrator solar cells with efficiencies >40% under concentration are available in the market, and several of the first CPV products have been installed in power plants in 2008 (Gombert & Desrumaux 2009). These plants have helped to verify the maturity of this technology with very satisfying field data over the period of operation (Martinez *et al.* 2009; Gombert *et al.* 2009). The first example of a large scale project is the 1 MW project in Questa, New Mexico – installed by Chevron in 2010 – which demonstrates the self-assurance of major companies in CPV technology.

The major manufacturers of CPV for more than a decade are Amonix (USA) and Solar Systems (Australia) though the later is presently not in operation. The other prominent companies that emerged meanwhile are Isofoton, and Sol3G (Spain), Concentrator technologies, and Pyron Solar (USA), Daido Steel (Japan), Green and Golden Energy (Australia), Opel Solar (USA), Arima EcoEnergy (Taiwan), Zenith Solar (Israel) so on.

5.3.5 Overview of HCPV modules

Several designs of HCPV modules are developed using both planar Si cells as well as multijunction solar cells. A few initial models demonstrated on the field are explained.

In the design of HCPV module, the concentration is achieved via reflective or refractive optics. A reflective optics system is likely to use a central receiver, where an

Figure 5.9   Solar System's Concentrator receiver designed for applications at 400×–500×
(*Source*: Solar Systems, reproduced with the permission of Sam Carter, Solar Systems)

array of cells densely packed close to each other receives the concentrated sunlight. The Dish concentrator developed by Solar Systems of Australia is shown in Figure 5.9. These concentrators utilize reflective optics to concentrate the light at 500 times the suns energy. The CS500 dish concentrator PV system has 112 curved reflecting mirrors mounted on a steel frame, which tracks the sun throughout the day. The combination of mirror profile, mounting framework and solar receiver are carefully designed to deliver concentrated sunlight energy to each PV module. The tracking mechanism maximises the amount of electricity produced. The critical part of the system is an array of close-packed high efficiency Triple Junction Solar Cells packed into a 36 cm^2 actively cooled package that are located in the solar receiver, suspended above the focus of the mirrors. The cells are mounted in a way that allows efficient dissipation of thermal energy as well as extraction of electricity. Since PV performance falls by around 1.7% for every 10°C rise in cell temperature and the sunlight is concentrated 500x, effective cooling is critical to achieve efficient performance. The module also incorporates electrical connections to deliver DC output as well as current and temperature sensors for real-time monitoring.

The control system keeps each dish pointing to the sun, monitors performance and adjusts the DC voltage to maximise electricity output. It also incorporates several failsafe systems to protect the CS500 from damage. This configuration is modular allowing Solar Systems to design different receiver configurations and sizes for both Dish and Heliostat solar applications. With active cooling, this design increases the reliability of the cells and produces the highest output power compared to other technologies, increasing cell life and providing more reliable operation. Further, this technology offers the advantage of lower capital investment cost to increase production capacity compared to conventional thin film or crystalline silicon production. This manufacturer, however, is presently not in production.

In the module developed by Amonix, a parquet of Fresnel lenses concentrates the light on individual cells mounted on a heat sink, as shown in Figure 5.10. This design allows passive cooling of the cells without raising the temperature much above the ambient, as the cells are spaced out from each other. The operating temperature of the

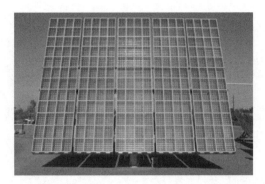

Figure 5.10 Amonix's Concentrator receiver designed for applications at 250×–400× (*Source:* Amonix, Inc., reproduced with the permission of Bob McConnell)

cell is typically 20–25°C above the ambient air. As with any passive cooling system, the temperature rise depends on the wind speed at the location. The Amonix module also uses the co-planar Si cell technology. There are currently over 700 kW of installed capacity operating in the states of Arizona and Nevada.

Other one is the micro dish module demonstrated by Concentrating Technologies (CT) which has utilised the multijunction cell technology in a grid-connected, high concentration module for the first time (Sherif 2005, Bett *et al.* 2006). The approach to the design of this module is a combination of reflective optics (Cassegrain like mirror optics) and the distributed location of small cells typical of refractive concentrators, thus avoiding the use of active cooling in a central receiver. This is accomplished by having a micro array of mirrors focusing the light on individual Power Conversion Units (PCUs). Each PCU has a secondary optical element to homogenize the light before it falls on the cells, and a heat sink attached to the back of the PCU to radiate heat to the ambient air.

Daido Steel of Japan has been active in the development of CPV systems since several years (Araki *et al.* 2003, 2004), and has developed a concentrating optic which uses a PMMA Fresnel dome lens and a glass kaleidoscope. Triple-junction cells manufactured at Sharp are used in the modules. Two types of modules with a concentration ratio of 400 and 550 are designed and are subjected to intensive testing. Module efficiencies as high as 30% have been reported (Araki *et al.* 2005). These results are very promising for any III-V solar cell based CPV system. Further, this manufacturer considers reliability issues seriously which are not often addressed.

Several developers such as Pyron Solar, Green & Gold Energy, Isofoton and Sol3G have come up with CPV system structures which are briefly explained by Bett *et al.* (2006).

Recently, Zenith Solar (Israel) has developed a system which generates electric power as well as heat (Figure 5.11). The system consists of a mirrored dish that concentrates the equivalent of 1000 suns onto a III-V multijunction solar cell. It will produce over 2 kW of electricity and the equivalent of 5 kW of solar hot water. The concentrator dish covers around 11 sq. m. of area, and the cell measures around 10 cm in each side. The high temperature created by the solar dish necessitate cooling the

Figure 5.11   Zenith Solar's HCPV system (Photo courtesy: Prof. Ezri Tarazi, reproduced with the permission of Roy Segev)

solar cell with water instead of with passive metal heat sinks. The water is then run through heat exchangers to provide hot water to an industry or a commercial site. It is a modular and easily scalable HCPV.

The multijunction solar cell that can convert more than 40 percent of the sun's energy into electrical energy is the heart of CPV technology. As seen already, the lattice-matched three-junction GaInP/GaInAs/Ge cell used in CPV is the result of innovation and refinement over the past decade. This cell structure is created by growing as many as 20 thin, single-crystal layers of III-V materials onto a single-crystal Ge wafer that is 100 mm in diameter using Metallorganic vapor phase epitaxy. The requirements for crystal growth quality, thickness and doping concentration uniformity are extraordinarily high. With continued global research on advanced multijunction cell architectures by several ways such as incorporation of metamorphic semiconductor materials, increasing numbers of junctions and so on, it can be expected to increase practical concentrator cell efficiencies to 45 or even 50 percent. It will then become economical to deploy solar cell technology on a larger scale.

5.3.6   Research and development

Several research groups have been studying the development of concentrator PV modules based on Si and III-V solar cells since 1990s.

Under DOE's High performance Photovoltaic project (HiPerfPV), several institutions, JX Crystals, CalTech, Fraunhofer ISE, Ohio State University, Amonix, University of Delware, Arizona State University, Spectrolab, SunPower, so on have participated in the development of CPV.

JX Crystals, under the project subcontract, has focused on demonstrating a 40%-efficient hybrid InGaP/GaAs – GaSb multijunction cell in a 33%-efficient Cassegrainian PV concentrator panel, shown in Figure 5.12.

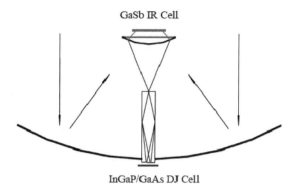

GaSb IR Cell

InGaP/GaAs DJ Cell

Figure 5.12 Cassegrain PV concentrator module with dichroic secondary optics separating the solar infrared from the visible spectrum
(*Source*: McConnell and Symko-Davies 2005: Used with permission by The Alliance for Sustainable Energy, LLC (Alliance). Alliance is the Manager and Operator of the NREL)

Recently, they have demonstrated a 30%-efficient InGaP/GaAs double-junction cell and an 8%-efficient GaSb infrared cell for incorporation into their 25 cm × 25 cm dichroic Cassegrain module (Fraas *et al.* 2005a).

California Institute of Technology, under a HiPerf PV subcontract, has undertaken the development of a 4-junction solar cell with 40% target efficiency fabricated by wafer bonding and layer transfer. The studies at Ohio State include the developing and demonstrating the use of novel 3-D substrate engineering in Si and Ge to achieve spectrum-optimized bandgap profiles leading to >40% III-V multijunction concentrators on Si or Ge. The University of Delaware focuses on demonstrating high-performance tandem solar cells based on the InGaN material system (Jani *et al.* 2005). The approach is with high-bandgap GaN and Indium-lean InGaN p-i-n and quantum well solar cells, and with low-bandgap Indium-rich InGaN grown on Ge. Sunpower Corporation focuses on low-concentration flat-plate module designs using high-efficiency silicon cells (Fraas *et al.* 2005b). The modules with this design are estimated to be available at $1.50/watt. The advantage is that there is no need for a solar tracker, and has the ability to compete in the PV rooftop market (McConnell & Symko-Davies 2005).

Amonix Corporation has focused on large-scale modules that will use >40% efficient multijunction solar cells under the same HiPerf PV Project. It is expected that such a module would achieve 33% efficiency. Amonix's approach includes: designing and packing for the Ge substrate-based multijunction solar cells, testing packaged cells under a Fresnel lens, characterizing different cells/package/lens configuration, and designing and constructing 33%-efficient modules, to be tested on-sun and not in the laboratory (McConnell & Symko-Davies 2005).

The design of a Fresnel lens concentrator system is an important aspect because it must account for several optical issues such as nonuniformity, chromatic aberration, lens absorption, and matching to multijunction solar cell response. The top cell is identified as the significant efficiency-limiting feature of the current design, and Amonix redesigned the optics with a yield of >26% lens/cell efficiency (without external cooling). Amonix sees a major inpact on the industry with the successful commercialization of modules exceeding 33%. They expect a standard shift in installed system cost and a new PV market ideally suited for PV concentrators (Garboushian 2005).

Fraunhofer ISE has developed cover and bottom plate made out of glass, and a relatively small aperture of each of the primary lenses, which are assembled in an array. The cells are mounted on heat spreaders which also serve as contact pads for the internal electrical connection of the module. The major reasons for using two glass panes are high durability, low cost, and the low coefficient of thermal expansion that ensures that the foci remain on the cell at varying operating temperatures. As the thermal expansion coefficient of glass is 3 times lower than that of aluminum for example, it is possible to keep the focal position on the cell within $100\,\mu m$ at all operating temperatures. The bottom plate need not be thermally conductive as the heat spreading is already efficiently done by the heat spreader, made of highly thermally conductive materials. The glass also serves as a scratch-resistant cover plate. Fresnel lens array is replicated in one piece into a silicone rubber on glass (SOG), allowing the usage of extremely UV stable materials in a cost effective mass production. The reasons for using a relatively small lens aperture are thermal management and low module depth. In the case of this primary lens, a simple heat spreader made out of a metal with an appropriate thermal conductance is sufficient for the thermal management. It was shown that the cell temperature in a CPV module does not exceed 40K above ambient temperature on average (Seifer & Bett 2006). Furthermore, a small lens allows for small cells, which help to achieve the highest efficiency because of the low resistance losses in the cell.

Soitec, Concentrix Solar, Fraunhofer ISE and CEA-Leti have collaborated on the development of the next generation of very high efficiency CPV solar cells based on Soitec's proprietary technologies. The newly developed cell is code named 'Smart Cell,' which will be integrated into the FLATCON CPV system. The modularity of the CPV approach, i.e., the independence of the module optimization from the cell optimization, allows for a very rapid and focused development. A complete system consists of the modules, the tracker, the inverter, the auxiliaries, and the control and monitoring hardware and software.

The concept of FLATCON was developed in the late 90s as a collaborative effort between Fraunhofer ISE and the Russian Ioffe Institute (Rumyantsev et al. 2000, Bett et al. 2003). The main ideas behind this concept are: (i) small sized III-V based cells, (ii) passive cooling, (iii) a small focus length and a Fresnel lens structure, and (iv) cheap and long term proved materials like silicone or glass. The development of this technology over the decade starting with the use of single-junction cells initially to the present triple junction cells, and from manual manufacturing to automated processing are dealt in several papers published from Fraunhofer ISE (for example, Bett et al. 2006).

Basically a 'unit' element of a FLATCON module consists of a primary lens and a solar cell plus bypass diode mounted on a small planar heat spreader. This planar design can be manufactured by using standard semiconductor assembly and printed circuit board machines. The lens and bottom plate are mounted by using standard technologies used in architectural glazing industry. The first fully operational and grid connected FLATCON®CPV system (Figure 5.13) was installed at Fraunhofer ISE. Currently, the FLATCON modules CX-75 have an average efficiency of 27% (Gombert & Desrumaux 2009).

Figure 5.13 Photograph of the grid-connected CPV system using FLATCON®-type modules on top of the roof of Fraunhofer ISE. In Europe, this is the first CPV demonstration system using III-V cells
(*Source*: Bett *et al.* 2003, reproduced with the permission of Dr. Andreas Bett)

At Ioffe Physical Technical Institute (Russia), development of HCPV modules with small aperture area glass-silicone Fresnel lenses and multi junction III-V solar cells is undertaken in the recent years (Alferov *et al.* 2007; Rumyantsev 2007, 2010a). The research group has demonstrated that using $40 \times 40 \, mm^2$ or $60 \times 60 \, mm^2$ Fresnel lenses and solar cells of 1.7 or 2.3 mm in diameter mounted on passive copper heat spreaders, a high concentration and proper thermal regimes can be realized in a very simple design with low consumption of the materials in fabrication of the modules. In this design, proper choice of the optical parameters of the Fresnel lenses and structural features of the heat spreaders are very vital.

The use of the lense-type secondary elements decreases the accuracy demands on assembly, alignment, and tracking technology for the HC PV modules (Rumyantsev *et al.* 2002, 2010b). Further, the secondary lenses help to increase the sun concentration ratio significantly, leading to the more effective use of the semiconductor materials. The disadvantage of the secondary elements is the inherent optical losses due to reflection from the air-glass interface, while the positive feature is the possibility to arrange without a mechanical contact with the cell surface, so that the problems related to difference in thermal expansions as well as that of the long-term stability of the optical lens-cell contact do not arise. Figure 5.14 shows design of the small-aperture area HCPV sub-module under investigation.

The concentrator module is made in a form of a front lens panel connected with a rear energy generating panel by aluminum walls. Each lens focuses solar radiation onto an underlying solar cell (InGaP/GaAs/Ge cells) with a photosensitive surface area diameter of 1.7 or 2.3 mm. The heat-distributing plates are glued to the outer surface of the rear module plate, which is made of silicate glass. The heat is removed from the cells by spreading along the copper plates and glass base. The heat sink plates are protected from environment by a laminating film on the rear side of the module. An ordinary

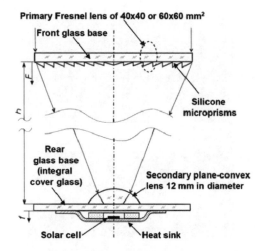

Primary Fresnel lens of 40x40 or 60x60 mm²

Front glass base

Silicone microprisms

Rear glass base (integral cover glass)

Secondary plane-convex lens 12 mm in diameter

Solar cell

Heat sink

Figure 5.14 Optical scheme of an individual solar concentrator submodule. Shown are the planes fromwhich the focal distances F and f are measured. A full-size module contains a panel of the primary lenses (144 lenses in arrangement of 12×12 for 40×40 mm lenses) and a panel of the cells with secondary optical elements (Credit: Rumyantsev *et al.* 2010d, reproduced with permission of Dr. Rumyantsev)

convex lens as a secondary element can distinctly improve the misorientation curve of a module. Also, the rear glass plate may serve as a common base for these lenses as well as protective cover for all the cells in a module. In the case of illumination with increased local sun concentration ratio, the cells would certainly improve parameters regarding the low internal ohmic resistance and the high enough peak current in the built-in tunnel junctions. This performance is fortunately realized in commercially available triple-junction cells.

To reduce losses due to reflections, an ARC may be applied, so that residual losses for two lens sides could be held at about 2 to 3%. The cost contribution of an ARC deposition to the cost of the lenses may be of the same order as that of highly reflective coatings in the case of the reflective pyramids. The capacity of the cells operating at very high local concentration was confirmed for practical structures of the high efficiency triple-junction cells. In fact, the fill factor was around 85 to 86% at local concentration ratios of about $5000\times$.

An elaborative process is required for assembling the concentrator modules consisting of a large number of lenses and cells, for checking and precision positioning all the elements in the module. The success of multistage module assembling procedure depends on proper quality of the solar cells. For instance, a cell with current leakage may result in shortening the cell string as a whole. On the other hand, dispersion of the photo-generated voltages in the cells within a string leads to reduction in PV conversion efficiency. It means that preliminary testing of the solar cells before their mounting is a very important step.

A contactless cell characterization has been developed at Ioffe Institute. It is based on the fact that the triple-junction InGaP/GaAs/Ge cells are characterized by intensive photo- and electro-luminescence (PL and EL) arising in both top and middle sub-cells

 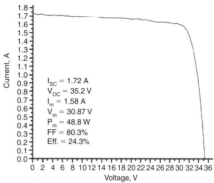

$I_{SC} = 1.72$ A
$V_{OC} = 35.2$ V
$I_m = 1.58$ A
$V_m = 30.87$ V
$P_m = 48.8$ W
FF $= 80.3\%$
Eff. $= 24.3\%$

Figure 5.15 (left) PV installation with concentrator modules for 1 kWp of output power on the roof of the Ioffe Institute; (right) illuminated I-V curve for one of the full-size modules (50×50 cm^2) measured outdoors (Efficiency $= 24.3\%$)
(Credit: Rumyantsev 2010d, reproduced with the permission of Dr. Rumyantsev)

under a local photo-excitation (Rumyantsev *et al.* 2010c). In general, it is possible to estimate the main cell parameters (collection efficiency, open circuit voltage, sheet resistance etc) of a direct bandgap p-n junction cell by the contactless methods, by analyzing only PL- and EL-signals from a cell wafer under photoexitation (Andreev *et al.* 1997). Scanning the wafer with separated cells by a green laser beam and analyzing the intensity of the EL signal from each cell is an effective tool for sorting the cells by photo-generated voltage.

5.3.7 Installations with HCPV modules

A complete system consists of the modules, the tracker, the inverter, the auxiliaries, and the control and monitoring hardware and software. High accuracy of module alignment to the sun is a specific feature of the concentrator installations.

In Figure 5.15, a stair-step principle of HCPV module arrangement on a solar tracker developed at Ioffe Institute is shown.

Advantage of such an arrangement is reduction of wind pressure on a frame with modules in different tracker positions during a day. Also, symmetry of two outermost positions of the frame (in directions to sunrise/sunset and to zenith) enables to make this frame more rigid. The tracker is equipped with a digital circuit for programmable rotation in both daytime during cloudy periods and at night from sunset to sunrise position. Analog sun sensor is used for positioning the frame with modules in direction to the Sun with accuracy better than 0.1° of arc.

The illuminated I-V curve for one of the developed HCPV modules, measured outdoors, is shown in Figure 5.15. Overall conversion efficiency of 24.3% is not temperature corrected value. If measured at standard cell temperature of 25°C, it would be 26.5%. The cell efficiency in these modules was at 33%; this means, one may expect a definite increase in module efficiency if the cell efficiency approaches around 40%, the value characterizing the best cells today (Renewable energy world Report 2009).

5.3.8 Cost benefits

CPV has the highest efficiency and energy output, i.e., the lowest levelized cost of electricity (LCOE). Until now, most of the industry has been focusing on PV module cost. Being a new and maturing industry, a great deal of the effects of scaling, productivity, and the cost reduction learning curve had to be achieved. Now, giga watt-level production has been achieved by top PV module suppliers, meaning that most of this learning has been integrated. So, cost reduction and efficiency improvement move now into a continuous improvement mode, with a few percent maximum gains per year. Module cost reduction will be limited by the cost of raw materials (glass, metals, semiconductor, etc.). CPV has two fundamental cost advantages: firstly, it minimizes the semiconductor content in the module due to high concentration; and any semiconductor material is far more expensive than any other material used in a PV solar plant. Secondly, the higher the efficiency (CPV has 2X higher efficiency than multi-crystalline silicon technology), the better the usage of all the other materials (glass, frame, etc.). That's why, for similar volumes, a CPV module (system) will be cheaper than any other PV technology.

From a mid- to long-term perspective, PV electricity costs in general will also be dominated by fixed costs associated with a PV plant, such as project development, installation, and O&M costs. In order to drive LCOE down to a very low level, energy output (kWh) has to be as high as possible to compensate for this fixed-cost structure. There are two ways to increase energy output, either by enhancing module efficiency or by sun tracking. Knowing that sun tracking only makes economic sense if module efficiency is high enough (because the tracker cost, which is measured in $/m^2$, has to be compensated for by the energy output 'boost') everything comes back again to the importance of efficiency. So far, there is no alternative to CPV with respect to the highest efficiencies, making it the technology of the future in sunny regions (Gombert and Desrumaux 2009).

Five MW CPV Systems: 5 MW CPV System shown in Figure 5.16 has been installed by Amonix, the American supplier in the village of Hatch in New Mexico. It has been generating 4.8 MW power since July 2011 and is the biggest CPV system so far installed. The system uses 84 Amonix's trackers installed over an area of 39 acres and is built and operated by NextEra Energy. Each of Amonix modules produces about 60 kilowatts. The electricity produced will be sold to El Paso Electric under a Purchase Power Agreement (Greentech Solar News, August 2011).

5.3.9 CPV and cogeneration

There have been trials where CPV activity involves extraction of heat energy also. For example, Zenith Solar based in Israel has developed modular and easily scalable high-concentration photovoltaic systems (HCPV) in collaboration with Fraunhofer Institute (Germany) and Ben Gurion University. The technology is based on unique optical design that extracts maximum energy. The system provides both high electrical output and heat at wide range of temperatures suitable for many cogeneration applications. Presently, the manufacturer utilizes the heat generated at the solar cell receiver to provide usable hot water, thereby improving the overall power conversion efficiency to 75%. Hotels, hospitals, commercial buildings, resorts so on can be benefited by

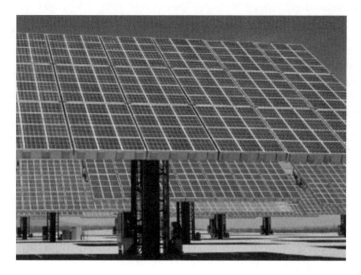

Figure 5.16   5 MW CPV System installed by Amonix trackers
(*Source:* Amonix, Inc, reproduced with permission)

these systems due to their high electrical output and high temperature output (Sun &
Wind Energy 3/2010, p. 105).

REFERENCES

Alferov, Zh. I., Andreev, V.M. & Rumyantsev, V.D. (2007): III-V hetero-structures in photovoltaics, In: Luque, A. & Andreev, V. (eds.), *Concentrator Photovoltaics,* Springer Series in
Optical Sciences, **130**, pp. 25–50

Algora, C. (2005): *Proc. 20th EU Photovoltaic Solar Energy Conf.* June 2005, Barcelona, Spain,
pp. 82–85

Andreev, V.M, Grilikhes, V.A. & Rumyantsev, V.D. (1997): *Photovoltaic Conversion of
Concentrated Sunlight,* John Wiley & Sons, Chichester, Chapter 4

Araki, K., Kondo, M., Uozumi, H. & Yamaguchi, M. (2003): Development of a Robust and
High efficiency Concentrator PV Systems in Japan, *Proc. 3rd WC PV Energy Conversion,*
May 2003, Osaka, Japan, pp. 630–633

Araki, K., Kondo, M., Uozumi, H., Kemmoku, Y., Egami, T., Hiramatsu, M., Miyazaki, Y.,
Ekins-Daukes, N.J., Yamaguchi, M., Siefer, G. & Bett, A.W. (2004): A 28% efficient 400x
and 200 Wp concentrator module, *Proc. 19th EU PV Solar Energy Conference*, June 2004,
Paris, France, pp. 2495–2498

Araki, K., Emery, K., Siefer, G., Bett, A.W., Sakakibara, T., Kemmoku, Y., Ekins-Daukes, N.J.,
Lee, H.S. & Yamaguchi, M. (2005): Comparison of efficiency measurement for a
HCPV module with 3J cells, *31st IEEE PV Specialists Conf,* June 2005, Orlando, FL,
pp. 846–849

Bailey, S.G., Castro, S.L., Raffaelle, R.P., Fahey, S.D., Gennett, T. & Tin, P. (2003): Nanostructured materials developed for solar cells, *NASA Glenn Research Center* Report.

Barnham, K.W.J. (2005): Future applications of low dimensional structures in Photovoltaics,
Proc. Photovoltaics for the 21st Century, Vol. **10**, 2005

Bett, A.W., Dimroth, F., Stollwerck, G. & Sulima, O.V. (1999): III-V Compounds for Solar Cell Applications, *J. Appl. Phys.*, **69** (2), pp. 119–129

Bett, A.W., Baur, C., Dimroth, F., Lange, G., Meusel, M., van Riesen, S., Siefer, G., Andreev, V.M., Rumyantsev, V.D. & Sadchikov, N.A. (2003): FLATCON Modules: technology and Characterisation, *3rd WC PV Energy Conversion*, May 2003, Osaka, Japan, pp. 634–637

Bett, A.W., Baur, C., Dimroth, F. & Schöne, J (2005): Metamorphic GaInP-GaInAs Layers for Photovoltaic Applications, *Materials Research Society Symp. Proceedings*, **836**, p. 223

Bett, A.W., Burger, B., Dimroth, F., Siefer, G. & Lerchenmüller, H. (2006): High Concentration PV using III-V solar cells, *4th WC PV Energy Conversion*, May 2006, Waikoloa, HI, pp. 615–619

Bosi, M. & C. Pelosi (2007): The Potential of III-V semiconductors Semiconducto as Terrestrial Photovoltaic Devices, *Prog. Photovolt: Res. Appl.* **15**, 51–68, doi:10.1002/pip.715

Burnett, B. (2002): *The Basic Physics and design of III-V multijunction solar cells*, at htttp://www.nrel.gov/ncpv/pdfs/11_20_dga_basics_9-13.pdf

Cotal, H.L., Lillington, D.R., Ermer, J.H., King, R.R. & Karam, N.H. (2000): Triple junction solar cell efficiencies above 32%: The promise and system challenges of their application in high-concentration-ratio PV systems, *Proc. 28th IEEE PV Specialists Conf.*, September 2000, Anchorage, AK, pp. 955–960

Das, B. (2001): Multijunction solar cells based on nanostructure arrays, *Electrochemical Society Meeting*, September 2001

Delta commercializing Spectrolab's high efficiency cells, March 9, 2007, http://www.insidegreen tech.com/node/863

Dimroth, F., Schubert, U. & Bett, A.W. (2000): 25.5% efficient $Ga_{0.35}$ $In_{0.65}$ $P/Ga_{0.83}In_{0.17}As$ tandem Solar Cells Grown on GaAs Substrates, *IEEE Electron Device Letters*, **21**, 209–211

Dimroth, F. & Bett, A.W. (2001): Next generation GaInP/GaInAs/Ge multi-junction space solar cells, *Proc. 17th EU PV Solar Energy Conf.*, (WIP-Munich and ETZ-Florence), October 2001, pp. 2150.

Dimroth, F. (2005): 3–6 junction photovoltaic cells for space and terrestrial applications, *Proc. 31st IEEE Photovoltaic Specialists Conf*, January 2005, Lake Buena Vista, FL. pp. 525

Dimroth, F., Baur, C., Bett, A., Köstler, W., Meusel, M. & Strobl, G. (2006): Thin 5-Junction Solar Cell Structures with Improved Radiation Hardness, *Proc. 4th WC PV Energy Conversion*, pp. 1777–1780

Dimroth, F. & Kurtz, S. (2007): High-efficiency multijunction solar cells, *MRS Bulletin*, **32**, 230–235

Ekins-Daukes, N.J. (2002): Strain-balanced quantum well solar cells, *Physica E*, **14**, 132

Fetzer, C.M., Yoon, H., King, R.R., Law, D.C., Isshiki, T.D. & Karam, N.H. (2005): 1.6/1.1 eV metamorphic GaInP/GaInAs solar cells grown by MOVPE on Ge, *J. Crystal Growth*, **276**, 48–56

Fraas, L.M., Avery, J.E., Sundaram, V.S., Kinh, V.T., Davenort, T.M., Yerkes, J.W., Gee, J.M. & Emery, K.A. (2005a): Toward 40% and Higher Solar Cells in a New Cassegrainian PV Module, *Proc. 31st IEEE PV Specialists Conference*, January 2005, Lake Buena Vista, FL., pp. 751–753

Fraas, L.M. (2005b): Efficient Solar Photovoltaic Mirror Modules for Half the Cost of Today's Planar Modules, *3rd Intl. Conf. on Solar Concentrators for the generation of Electricity or Hydrogen*, May 2005, Scottsdale, Az.

Garboushian, V. (2005): Marketing Concentrating Solar Systems–Challenges and Opportunities, *3rd Intl. Conf. on Solar concentrators for the generation of Electricity or Hydrogen*, May 2005, Scottsdale, Az.

Geisz, J.F., Kurz, S., Wanlass, M.W., Ward, J.S., Duda, A., Friedman, D.J. & Olson, J.M. (2007): High-efficiency GaInP/GaAs/InGaAs triple-junction solar cells grown inverted with a metamorphic bottom junction, *Appl. Phys. Lett.* **91**, Article ID 023502, 3 pages

Gombert, A., Hakenjos, A., Heile, I., Wüllner, J., Gerstmaier, T. & van Riesen, S. (2009): FLATCON CPV Systems: Field data and new developments, *Proc. 24th EU PV Solar Energy Conf. and Exhibition*, September, 2009, Hamburg, Germany, pp. 156–158

Gombert, A. & Desrumaux, C. (2010): Concentrator photovoltaics: a mature technology for solar power plants, *PVWorld, Vol. 2010, Issue 3*, May–June 2010; http://www.electroiq.com/ articles/pvw/print/volume-20100/issue-3/features/ concentrator.

Green, M.A. (2006): *Third generation Photovoltaics: Advanced solar energy conversion*, Berlin: Springer

Green, M.A., Emery, K., Hishikawa, Y. & Warta, W. (2007): Solar cell efficiency tables (Version 29), *Prog. in Photovoltaics*, **15**, 425

Gur, I. (2005): Air-stable all-inorganic nanocrystal solar cells processed from solution, *Science Reports,* Vol. **310**

Holland, D. (2003): Overview of Solar Systems, Oral *Presentation at the International Solar Concentrator Conference*, Alice Springs, AU, November 2003.

Honsberg, C.B., Corkish, R. & Bremner, S.P. (2001): A new generalized detailed balance formulation to calculate solar cell efficiency limits, *Proc. 17th EU PV Solar Energy Conf.*, October 2001, Munich, Germany

Honsberg, C.B. (2005): Paths to ultra-high efficiency (>50% efficient) photovoltaic devices, *20th EU Photovoltaic Solar Energy Conference*, June 2005, Barcelona, Spain

Honsberg, C.B. (2006): Nanostructured solar cells for high efficiency photovoltaics, *Proc. 4th WC on Photovoltaic Energy Conversion*, May 2006, Waikoloa, HI

Hung, H.A. (1988): IEEE Trans. Microwave Theory and Techniques **36**, p1966, doi:10.1063/ 1.117995.

Industrial strength solar energy gets some good energy cash, March 03, 2006, www. thealarmclock.com.

Jayarama Reddy, P. (2010): *Science and Technology of Photovoltaics*, 2nd Edn, CRC Press, Leiden, The Netherlands, ISBN 13: 978-0-415-57363-4

Kasap, S.O. (2001): *Optoelectronics and photonics: Principles and Practices*, New York: Prentice Hall

Karam, N.H., King, R.R., Cavicchi, B.T., Krut, D.D., Ermer, J.H., Haddad, M., Cai, L., Joslin, D.E., Takahashi, M., Eldredge, J.W., Nishikawa, W.T., Lillington, D.R., Keyes, B.M. & Ahrenkiel, R.K. (1999): *IEEE Transactions on Electron Devices*, **46**, 2116

King, R.R., Haddad, M., Isshiki, T., *et al.* (2000): Metamorphic GaInP/GaInAs/Ge Solar Cells, *Proc. 28th IEEE Photovoltaic Specialists Conf*, September 2000, Anchorage, AK, pp. 982–985

King, R.R., Fetzer, C.M., Colter, P.C., Edmondson, K.M., Ermer, J.H., Cotal, H.L., Yoon, H., Stavrides, A.P., Kinsey, G., Krut, D.D. & Karam, N.H. (2006): High-Efficiency Space and Terrestrial Multijunction Solar Cells Through Bandgap Control in Cell Structures, *Proc. 29th IEEE PV Specialists Conference,* May 2006, New Orleans, LA,(IEEE, New York), pp. 776

King, R.R., Fetzer, C.M., Edmondson, K.M., Law, D.C., Colter, P.C., Cotal, H.L., Sherif, R.A., Yoon, H., Isshiki, T., Krut, D.D., Kinsey, G.S., Ermer, J.H., Kurtz, S., Moriarty, T., Kiehl, J., Emery, K., Metzger, W.K., Ahrenkiel, R.K. & Karam, N.H. (2004): Metamorphic III-V Materials, Sublattice Disorder, and Multijunction Solar Cell A pproaches with over 37% Efficiency, *Proc. 19th EU Photovoltaic Solar Energy Conf,* June 2004, Paris, France, pp. 3587

King, R.R., Law, D.C., Fetzer, C.M., Sherif, R.A., Edmondson, K.M., Kurtz, S., Kinsey, G.S., Cotal, H.L., Krut, D.D., Ermer, J.H. & Karam, N.H. (2005a): "Pathways to 40%-Efficient

Concentrator Photovoltaics, *Proc. 20th EU PV Solar Energy Conf. and Exhib*, June 2005, Barcelona, Spain, pp. 118

King, R.R. (2005): Bandgap engineering in high-efficiency multijunction concentrator cells, *Proc. Intl. Conf. on Solar Concentrators for the Generation of Electricity or Hydrogen*, May 2005, Scottsdale, Az.

King, R.R., Law, D.C., Edmondson, K.M., Fetzer, C.M., Sherif, R.A., Kinsey, G.S., Krut, D.D., Cotal, H.L. & Karam, N.H. (2006a): Metamorphic and Lattice-Matched Solar Cells under Concentration, *Proc. 4th WC PV Energy Conversion*, May 2006, Waikoloa, HI 1, pp. 760–763

King, R.R., Sherif, R.A., Law, D.C., Yen, J.T., Haddad, M., Fetzer, C.M., Edmondson, K.M., Kinsey, G.S., Yoon, H., Joshi, M., Mesropian, S., Cotal, H.L., Krut, D.D., Ermer, J.H. & Karam, N.H. (2006b): New Horizons in III-V Multijunction Terrestrial Concentrator Cell Research, *Proc. 21st EU PV Solar Energy Conf and Exhib*, September 2006, Dresden, Germany, pp. 124–128

King, R.R., Law, D.C., Edmondson, K.M., Fetzer, C.M., Kinsey, G.S., Krut, D.D., Ermer, J.H., Sherif, R.A. & Karam, N.H. (2007a): 40% efficient metamorphic GaInP/GaInAs/Ge multi junction solar cells, *Appl. Phys. Lett*, **90**, 183516/183511-183513

King, R.R., Law, D.C., Edmondson, K.M., *et al.* (2007b): Metamorphic Concentrator solar cells with over 40% conversion efficiency, *Proc. 4th International Conf. on Solar Concentrators (ICSC-4)*, El Escorial, Spain, 12–16 March 2007

King, R.R., Law, D.C., Edmondson, K.M., Fetzer, C.M., Kinsey, G.S., Yoon, H., Krut, D.D., Ermer, J.H., Sherif, R.A. & Karam, N.H. (2007c): Advances in high efficiency of III-V Multi-junction solar cells, *Advances in Opto-Electronics*, Volume 2007, Article ID 29523, 8 pages, Hindawi Publishing Corporation

King, R.R. (2007d): *DOE Solar Program Review meeting*, Denver, CO, April, 2007

Kurtz, S. (2009): Opportunities and Challenges for Development of a Mature Concentrating Photovoltaic Power Industry, *Technical Report* NREL/TP-520-43208 Revised Nov. 2009.

Luque, A. & Hegedus, S. (Eds.) (2003): *Handbook of Photovoltaic Science and Engineering*, New York, Wileys

Martí, A. & Luque, A. (2004): Next generation PV: high efficiency through full spectrum utilization

Martínez, M., Sánchez, D., Perea, J., Rubio, F. & Band, P. (2009): ISFOC Demonstration Plants: Rating and production data analysis, *Proc. 24th EU PV Solar Energy Conf. and Exhibition*, June 2009, Hamburg, Germany, pp. 159–164

Mawst, L.J., A. Bhattacharya, J. Lopez, D. Botez, D.Z. Garbuzov, L. DeMarco, J.C. Connolly, M. Jansen, F. Fang & R.F. Nabiev (1996): 8 W continuous wave front-facet power from broad-waveguide Al-free 980 nm diode lasers, *Appl. Phys. Lett.* **69**, 1532–1534

McConnell, R. & Symko-Davies, M. (2005): DOE High performance Conc. PV Project, Paper: NREL/CD- 520-38172, Presented at *3rd Intl. Conf. on Solar Concentrators for the generation of Electricity or Hydrogen*, May 1–5, 2005, Scottsdale, Az.

McConnell, R. (2005): Representative Samples for Concentrator PV Module Qualification Test-ing, *3rd International Conference on Solar Concentrators for the Generation of Electricity or Hydrogen*, May 1–5, 2005, Scottsdale, AZ.

Meusel, M., Dimroth, F., Baur, C., Siefer, G., Bett, A.W., Volz-Koch, K., Stolz, W., Strobl, G., Signorini, C. & Hey, G. (2004): European roadmap for the development of III-V multi-junction solar cells, *Proc. 19th EU Photovoltaic Solar Energy Conference*, June, 2004, Paris, France, pp. 3581–3586

Meusel, M., Baur, C., Guter, W., Hermle, M., Dimroth, F, Bett, A.W., Bergunde, T., Dietrich, R., Kern, R., Köstler, W., Nell, M., Zimmermann, W., LaRoche, G., Strobl, G., Taylor, S., Signorini, C. & Hey, G. (2005): Development Status of European Multi-Junction Space Solar

Cells with High Radiation Hardness, *Proc. 20th EU Photovoltaic Solar Energy Conference*, June 2005, Barcelona, Spain.

Meusel, M., Bensch, W., Bergunde, T., *et al.* (2007): Development and production of European III-V Multi junction solar cells

Morf, R.H. (2002): Unexplored opportunities for nanostructures in photovoltaics, *Physica* E, **14**, 78.

Myong S.Y. (2007): Recent progress in inorganic solar cells using quantum structures, Recent patents on nanotechnology, February 2007, pp. 67–73

Nosova, L. *et al.* (2005): Design of semiconductor nanostructures for solar cell application, Springer, 2005.

Nozik, A.J. (2002): Quantum dot solar cells, *Physica E*, **14**, 115.

Olson, J.M., Kurtz, S.R., Kibbler, A.E. & Faine, P. (1990): A 27.3% Efficient Ga0.5In0.5P/GaAs Tandem Solar Cell, *Appl. Phys. Lett.* **56**, 623

Poortmans, J. & Arkhipov, V. (2006): *Thin film solar cells: fabrication, characterization and applications*, Hoboken, NJ: Wiley

Raffaelle, R.P. (2006): Multi-junction solar cell spectral tuning with quantum dots

Renewable energy World report (2009): Spectrolab Hits 41.6% PV Cell Efficiency record (Renewable energy world, 2009) at http://www.renewableenergyworld.com/rea/news/article/2009/08/spectrolabsets- solar-cell-efficiency-record-at-41-6

Román, J.M. (2004): State-of-the-art of III-V solar cell fabrication technologies, device designs and applications, *Advanced Photovoltaic Cell Design,* 2004, available at jmroman_solarcells3-5.pdf

Rumyantsev, V.D., Chosta, O.I., Grilikhes, V.O., Sadchikov, N.A., Soluyanov, A.A., Shvarts, M.Z. & Andreev, V.M. (2002): Terrestrial and space concentrator PV modules with composite (glass-silicone) Fresnel lenses, *Proc. 29th IEEE PV Specialists Conf,* May 2002, New Orleans, LA, pp. 1596–1599

Rumyantsev, V.D. (2007): Terrestrial concentrator PV systems, In: A. Luque, A. & Andreev, V. (eds.) *Concentrator Photovoltaics*, Springer Series in Optical Sciences, **130**, pp. 151–174.

Rumyantsev, V.D. (2010): Solar concentrator modules with silicone-on-glass Fresnel lens panels and multijunction cells, *Optics Express*, **18**, A17–A24

Rumyantsev, V.D., Davidyuk, N.Y., Ionova, E.A., Malevskiy, D. Pokrovskiy, P.A., & Sadchikov, N.A. (2010b): HCPV Modules With Primary And Secondary Minilens Panels, *Proc. 6th International Conf. on CPV Systems*, April 2010, Freiburg, Germany

Rumyantsev, V.D., Davidyuk, N.Y., Ionova, E.A., Pokrovskiy, P.V., Sadchikov, N.A. & Andreev, V.M. (2010c): Thermal Regimes of Fresnel Lenses and Cells in "All-Glass" HCPV Modules, *Proc. 6th International Conf. on CPV Systems,* April 2010, Freiburg, Germany.

Rumyantsev, V.D., Ashcheulov, Y.V., Davidyuk, N.Y., Ionova, E.A., Pokrovskiy, P.V., Sadchikov, N.A. & Andreev, V.M. (2010d): CPV modules based on Lens Panels, *Advances in Science and Technology*, **74**, 211–218 (Trans Tech Publications, Switzerland, www.ttp.net)

Sala, G.: Classification of PV concentrators (IES report, Spain) http://www.ies-def.upm.es/ies/CRATING/Chapter1.pdf

Sharps, P.R., Stan, M.A., Aiken, D.J., Newman, F.D., Hills, J.S. & Fatemi, N.S. (2004): High Efficiency Multi-Junction Solar Cells – Past, Present, and Future, *Proc. 19th EU PV Solar Energy Conference*, June 2004, Paris, France, pp. 3569–3574

Sherif, R.A., King, R.R., Karam, N.H. & Lillington, D.R. (2005): First Demonstration of Multi- Junction Receivers in a Grid-Connected Concentrator Module, *Proc. 31st IEEE PV Specialists Conf.*, January 2005, Lake Buena Vista, FL, pp. 17–22

Shockley, W. & H.J. Queisser (1961): Detailed balance limit of efficiency of p-n junction solar cells, *J. Appl. Phys,* **32** (3), p. 510

Siefer, G. & Bett, A.W. (2006): Calibration of III-V Concentrator Cells and Modules, *Proc. 4th WC on PV Energy Conversion*, May 2006, Waikoloa, HI, pp. 745–748

Sinharoy, S., King, C.W., Bailey, S.G. & Raffaelle, R.P. (2005): InAs QD development for enhanced InGaAs space solar cells, *Conf. record of 31st IEEE PV Specialists Conf,* January 2005, Lake Buena Vista, Fl. pp. 94–97

Smestad, G.M. (2002): *Optoelectronics of solar cells,* Bellingham, WA: SPIE Press.

Sobolev, N.A. (2001): Enhanced radiation hardness of InAs/GaAs quantum dot structures, *Phys. Stat. Sol. B,* **224,** 93

Strobl, G.F.X. (2006): EU roadmap of MJ solar cells and qualification status

Suwaree, S., Thainoi, S., Kanjanachuchai, S. & Panyakeow, S. (2006): Quantum Dot Integration in Heterostructure Solar Cell, *Sol. Energy Mater. & Sol. Cells,* **90,** 2968–2974

Swanson, R.M. (2000): The Promise of Concentrators, *Prog. Photovolt: Res. Appl.* 8, 93–111

Spectrolab solar cell breaks 40% efficiency barrier, December 7, 2006, at http://www.insidegreentech.com/node/454

Takamoto, T., Agui, T., Kamimura, K., Kaneiwa, M., Imaizumi, M., Matsuda, S. & Yamaguchi, M. (2003): Multijunction solar cell technologies – high efficiency, radiation resistance, and concentrator applications, *Proc. 3rd WC on PV Energy Conversion,* May 2003, Osaka, Japan, **1,** pp. 581–586

University of Strathclyde: *Energy and the Environment* course, at www.esru.strath.ac.uk.

Walters, R.J., Messenger, S.R., Summers, G.P., Freundlich, A., Monier, C. & Newman, F. (2000): Radiation hard multi-quantum well InP/InAsP solar cells for space applications, *Prog. in Photovolt,* 8, 349–354

Wanlass, M.W., Ahrenkiel, S.P., Ahrenkiel, R.K., Albin, D.S., Carapella, J.J., Duda, A., Geisz, J.F., Kurtz, S. & Moriarty, T. (2005): "Lattice-Mismatched Approaches for High-Performance, III-V, Photovoltaic Energy Converters," *Proc. 31st IEEE PV Specialists Conf,* January 2005, Lake Buena Vista, FL, pp. 530–535

Weiss, P. (2006): Quantum-dot leap: Tapping tiny crystals' inexplicable light-harvesting talent, *Science News,* June 03, 2006

Wolf, M. (1960): Limitations and possibilities for improvement of photovoltaic solar energy converters, *Proc. Inst. Radio Engineers,* vol. **48**

Würfel, P. (2005): *Physics of solar cells: From principles to New concepts,* Wiley-VCH, Weinheim

Yamaguchi, M. (2001): Present status of R&D super-high-efficiency III-V compound solar cells in Japan, *Proc. 17th EU PV Solar Energy Conference* (WIP-Munich and ETZ-Florence), October 2001, pp. 2144

Yamaguchi, M. (2005): Super-high-efficiency multi-junction solar cells, *Prog. Photovolt: Res. and Appls,*13, 125

Yang, J., Banerjee, A., Glatfelter, T., Hoffman, K., Xu, X. & Guha, S. (1994): *Conference Record, 1st WC on PV Energy Conversion,* Hawaii, December 1994, pp. 380–385

Yastrebova, N.V., (2007): High-efficiency Multi junction solar cells: Current Status and Future Potential, Centre for Research on Photonics, University of Ottawa, 2007 available at HiEfficMjSc-CurrStatus&FuturePotential.pdf

Yoon, H., Granata, J.E., Hebert, P., King, R.R., Fetzer, C., Colter, P.C., Edmondson, K.M., Law, D., Kinsey, G.S., Krut, D.D., Ermer, Gillanders, J.H & Karam, N.H. (2005): Recent advances in efficiency III-V Multi junction solar cells for Space applications: Ultra triple junction Qualification, *Paper: NASA/CA- 2005-213431,* available at 10YoonMJ3-5cells.pdf

Yoshimi, M., Sasaki, T., Sawada, T., Suezaki, T., Meguro, T., Ichikawa, M., Nakajima, A., Yamamoto, K., Matsuda, K., Wadano, K. & Santo, T. (2003): High efficiency Thin film Si hybrid solar cell module on 1m2-Glass large area substrate, *Conf. Rec, 3rd WC on PV Energy Conversion,* May 2003, Okasa, Japan, p. 1566

Zweibel, K. (1984): *Basic Photovoltaic Principles and Methods,* New York: Van Nostrand Reinhold, 1984

Chapter 6

New concepts based solar cells

There has been increasing interest in the last couple of decades, in materials with nanometer-scale dimensions. 'Quantum dots, nanowires and nanotubes' have received special attention for their distinctive properties and intricate structures. Many nanowire-based materials are promising candidates for energy conversion devices. The quantum dots, nanowires and nanotubes have been incorporated in solar cells based on silicon and compound semiconductor thin films to advance the performance. There have been tremendous efforts in this field showing promising results indicating high potential for the fabrication of low cost high efficiency PV devices. These approaches are discussed in this chapter.

Progress on CSG technology, another promising and inexpensive technology, and other novel approaches (hot carrier solar cells, plasmonic photovoltaics, solar cells based on nano materials) where there has been enormous research activity is broadly covered.

Earlier, Polymer solar cells (Organic and Dye-sensitized solar cells), also third-generation technologies which have matured to a stage of module development/production are discussed.

6.1 QUANTUM DOT SOLAR CELLS

In recent years, the quantum dot (QD) solar cell which may exceed the theoretical limitation of the conversion efficiency has gained attention.

The quantum dot is a nano crystalline structure ranging from a few nm to about 10 nm in size, and is achieved by epitaxial growth on a substrate crystal. The quantum dots are surrounded by high potential barriers in a three dimensional shape, and the electrons and electron holes in the quantum dot become a discrete energy as it is confined in a small space. Consequentially, the energy state of the ground-state energy of the electrons and holes in the quantum dot would be subject to the size of the quantum dot.

The concept of QD solar cell structure is shown in Figure 6.1.

The physical characteristics of the quantum dot that are important from the point of application to the solar cell (Okada *et al.* 2006, 2007; Kanama & Kawamoto 2008) are 'Quantum size effect' and 'Energy relaxation time.' By adjusting the size of the quantum dot, the optical absorption wavelength that can be more consistent with the solar spectrum can be selected (Okada *et al.* 2006; Barnham & Duggan 1990).

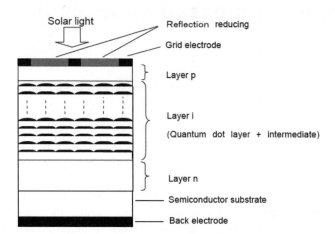

Figure 6.1 Concept of QD solar cell structure (Credit: Kanama & Kawamoto 2008, reproduced with permission of the authors)

The energy relaxation time of the electron slows down in the quantum dot; and this may provide a possibility to remove the electrons in a state of high energy by phonon emissions, before energy relaxation occurs. By realizing these, a theoretical efficiency exceeding 60% (Luque & Marti 1997; Green 2003) is expected to achieve.

These two characteristics are considered vital to overcome the two main losses: transmission loss of light and loss of thermal energy by phonon emission. The formation of 'miniband' is another physical characteristic that contributes to high theoretical efficiency. When there is a bonding of quantum dots, a miniband is formed at the superconductor and valence band (Ezaki & Sasaki 1988). If this middle layer between the quantum dots is very thick, the energy band diagram of the QD cell structure becomes as shown in Figure 6.2(a). In this case, the electrons excited by sunlight slip out of the well of the quantum dots by further optical excitation or thermal excitation giving rise to an electric current. Meanwhile, if the middle layer is several nm thin, a miniband is formed between quantum dots, and the electrons and the holes can move with little energy loss (Figure 6.2(b)) (Kanama & Kawamoto 2008).

However, there are many vital issues to be understood, for example, arranging the quantum dots in a regular and stable manner using general materials and so on.

6.1.1 Silicon-QD solar cell

A typical example of a Si nanostructure is a Si quantum dot superlattice (Si-QDSL). In the Si-QDSL, Si quantum dots are periodically arranged with an interval of 2–3 nm or less in a material with a band gap wider than that of Si. When several Si-QDSL thin films with different sizes of Si quantum dots are stacked, a wider wavelength range of the solar spectrum can be absorbed, and a marked increase in conversion efficiency of >40% is likely to be achieved by using a multi junction structure. SiO_2, Si_3N_4, and SiC are the wide bandgap materials considered for the barrier layers. Thus, Si-QDSL thin films can be made of Si-based materials only, a favourable factor from the point of abundant availability of raw materials and their non-toxic nature.

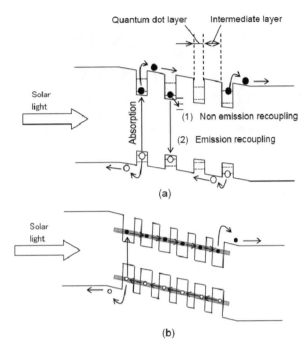

Figure 6.2   QD solar cell (a) energy band diagram, and (b) mini band
(*Source*: Kanama & Kawamoto 2008, reproduced with the permission of the authors)

It used to be difficult to fabricate Si-QDSL thin films containing Si quantum dots with a diameter of 10 nm or less, the optimum size for demonstrating their quantum effect, using conventional technology. In 2002, Zhacharias and colleagues at the Max-Planck Institute succeeded in fabricating a Si-QDSL by a simple method, as shown in Figure 6.3 (Zacharias *et al.* 2002).

In their method, a SiO_x/SiO_2 multi-layered film is deposited by reactive evaporation and thermally annealed to form a size-controlled Si-QDSL. The TEM of such a cell is shown in Figure 6.4. By sandwiching a thermally unstable SiO_x layer between thermally stable SiO_2 layers, the diameter of Si quantum dots, which separate out during annealing, is restricted to the separation of the two SiO_2 layers, thus enabling to limit the diameter to 10 nm or less. The photoluminescence measurement had shown that the band gap was successfully controlled to ~1.3–1.5 eV for quantum dots with a diameter of 3.8–2.0 nm.

Applying the plasma CVD technique, Miyajima made a silicon quantum dot super grid in a self-organizing manner, by making an amorphous super grid consisting of a-SiC with the stoichiometry and silicon excess composition (Kanama & Kawamoto 2008). This is a method for forming a quantum dot of its own accord, during the process of crystallization by making use of the phenomenon in which grid distortion is caused due to difference in the grid parameters between the materials used as a quantum dot and as an energy barrier layer.

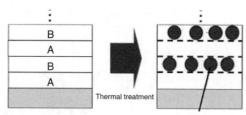

Multi–layered structure   Size–controlled Si quantum dot

Figure 6.3 A layer: Stoichiometric a-Si alloy (a-SiC, a-Si$_3$N$_4$, a-SiO$_2$); B layer: Si-rich a-Si alloy (a-SiC$_x$, a-Si$_y$N$_x$, a-SiO$_x$): Method of fabricating size-controlled Si-QDSL by thermal treatment (*Source:* Zacharias *et al.* 2002, reproduced with permission, Copyright © 2002, American Institute of Physics)

Figure 6.4 TEM image of size-controlled Si-QDSL
(*Source:* Zacharias *et al.* 2002, reproduced with permission, Copyright © 2002, American Institute of Physics)

Another technique developed by Sasaki that may be applicable to the quantum dot structure is the synthesis of nanoporous silica (SiO$_2$), which contains a laminar silica salt precursor (Kanama & Kawamoto 2008). At present, this is not for producing quantum dots supergrid. The high-resolution transmission electron's mirroring image (Figure 6.5) shows that there is a structure of silica pore controlled by a nano scale.

If a technology to enclose the semiconductor nano particle in each nano hole is developed, it is possible to use the silica itself as the interlayer. It may even enable the creation of a three dimensional quantum-dot structure containing thin films with regular layouts of the semi-conductor nano particles.

Green and his group at the UNSW was the first to apply a Si-QDSL to solar cells. They succeeded in fabricating a SiO$_2$-based Si-QDSL and a Si$_3$N$_4$-based Si-QDSL in 2005 (Green *et al.* 2005; Cho *et al.* 2005). Recently, they fabricated a crystalline Si hetero junction solar cell using a p-doped SiO$_2$-based Si-QDSL as n-layer, demonstrating that the quantum dot layer can function as a doping layer (Cho, EC *et al.* 2008). Konagai's group could successfully fabricate a Si-QDSL using SiC, considered to be useful in terms of the electrical conductivity of carriers (Kurokawa *et al.* 2006). A solar cell structure constructed using a SiC-based Si-QDSL as a power generation layer demonstrated its photovoltaic effect for the first time (Kurokawa *et al.* 2008).

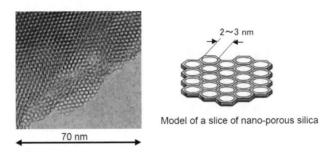

Model of a slice of nano-porous silica

Figure 6.5 High resolution TE mirroring image of nanoporous silica
(Credit: Kanama & Kawamoto 2008, reproduced with the permission of the authors)

Later, Green's group fabricated a p–i–n homo junction solar cell using a SiO_2-based Si-QDSL and achieved an open circuit voltage of 492 mV (Conibeer *et al.* 2009).

These quantum-dot structure fabrication methods explained are all in the process of development, and have only been able to produce small-scale samples. There are still several problems to be solved for arranging the quantum dots to a uniform nominal size in a three dimensional order, while achieving highly dense and thin inter layer. The main issue, in particular, is how to control the quantum dot size fluctuation. For large scale applications in the future, it may be essential to develop a technique not requiring an ultra-air-tight-vacuum.

In early research of quantum dot solar cells, the compound semiconductor was the main focus. In the case of production methods, producing the quantum dots by self-organization using the difference of the grid parameter and the high index plane substrate are considered to be the most feasible process. For example, if a material with a higher grid parameter than that of the substrate needs to be grown epitaxially, the quantum dot grows as islands on a regular basis to reduce the strain energy which accompanies the growth process. Typical examples include InAs/GaAs, where a grid parameter of InAs (7.2%) is larger than that of GaAs. A conversion efficiency of 8.54% was achieved by Okada *et al.* with this system (Oshima *et al.* 2007). These trends are possible with compound semiconductors only.

6.1.2 III-V multi junction QD solar cells

Bandgap engineering of multijunction solar cells (MJSCs) using quantum dots has been the approach of Sharps *et al.* (2008) and Hubbard *et al.* (2009a) to improve the efficiency of III-V multi junction solar cells. Luque and Marti (1997) also proposed a novel extension of the bandgap engineering approach. It uses multiple QD superlattices to form an optically isolated intermediate band (IB) within the bandgap of a standard single-junction solar cell. Photons with energy below the host bandgap are absorbed from the host valence to the intermediate band and from the intermediate to the host conduction band. As these lower-energy photons are normally lost in transmission in single-junction solar cells, the IB approach may result in a high limiting efficiency. Recently, several experiments have established the key operating principles of the IB solar cell using both Indium arsenide (InAs) (Marti *et al.* 2006; Hubbard *et al.* 2008)

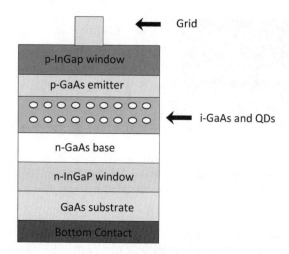

Figure 6.6 Schematic representation of QD-enhanced solar-cell design (Redrawn from Hubbard
et al. 2009)

and Gallium antimonide (Laghumavarapu *et al.* 2007) QDs in a gallium arsenide
(GaAs) host. For example, effective bandgaps of InAs quantum dot with sizes 5 nm,
10 nm and 12 nm are respectively 1.071 eV, 0.553 eV and 0.045 eV (Sobolev *et al.*
2001). Theoretically, a single intermediate electronic band created by QDs would
offer a 63.2% efficiency of an ordinary solar cell, which greatly exceeds the maximum
conversion efficiency of 31% for even a single-junction device (Bailey *et al.* 2003).
A system with an infinite number of sizes of QDs has the same theoretical efficiency
as an infinite number of band gaps or 86.8% (Luque & Hegedus 2003).

The NanoPower Research Laboratories of Rochester Institute of Technology have
engineered III-V-type solar cells to get the benefit of the full absorption spectrum
of lower-bandgap heterostructures (such as QDs) inserted into the current-limiting
junction of a multi junction solar cell (Hubbard *et al.* 2009a). The larger absorption
spectrum of the nanostructures enhances the overall short-circuit current and global
efficiency of the cell. Models of InGaP/InGaAs/Ge triple-junction solar cell, in which
QDs extend the middle junction's absorption spectrum, indicate that the theoretical
limiting efficiency could be improved to 47% under one sun illumination. These devices
have additional merits, such as enhanced radiation tolerance and temperature coefficients, which are very essential for space applications (Sinharoy 2005; Walters *et al.*
2000; Sobolev *et al.* 2001; Cress *et al.* 2007).

The feasibility of the QD tuning approach using strain-balanced InAs dots inserted
in the intermediate i – region of a GaAs p-i-n solar cell has been demonstrated (Hubbard
et al. 2008; Bailey *et al.* 2009). The schematic of solar-cell structure (grown by metal
organic vapor-phase epitaxy) is shown in Figure 6.6. These cells are fabricated using
standard III-V processing technology and a concentrator grid designed with 20% metallization coverage. The control results are equal or close to the standard values for
GaAs solar cells. Illuminated J-V curves for a control GaAs cell without QDs and
for cells with 5×, 10× and 20× stacked layers of InAs QDs under one sun global

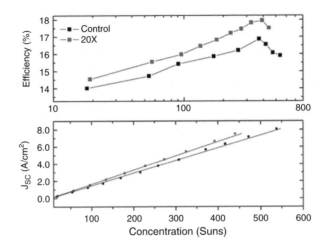

Figure 6.7 Direct air mass 1.5 (AM1.5d) efficiency and short-circuit current (J_{sc}) versus concentration for the baseline and QD cells. The enhanced cells show improved efficiency at higher concentration
(*Source:* Hubbard & Raffaelle 2010, reproduced with the permission of Professor Raffaelle)

air mass 1.5 (AM1.5 g) are measured. These measurements have clearly revealed the improvement in short-circuit current using QDs. This is a direct result of photo-generated current contributed by the nanoparticles. This trend is expected to continue with addition of QDs, because a greater portion of the sub-GaAs bandgap solar spectrum should be absorbed with increasing number. In addition, the strain-balancing technique used for QD growth led to higher material quality in layers deposited on top of the InAs nanoparticles. This has enabled short-circuit current enhancement and minimal degradation of open-circuit voltage (Hubbard *et al.* 2008).

EQE measurements on 5× to 20× QD samples show an increased response at wavelengths greater than the GaAs band edge, indicating that a portion of the short-circuit current of these cells is generated by QD-related absorption processes. Also, increasing the number of layers raises the EQE at all QD-related transitions. At 909 nm (1.36 eV), this amounted to increases from 2.5 to 9.2% by increasing the stacking from five to 20×. This was clear evidence that raising the volume of the quantum-dot absorbers successively increases short-circuit current (Hubbard & Raffaelle 2010).

To assess the spectral enhancement of the QDs at higher concentration, the control and 20-layer cells are studied under high-intensity illumination (Hubbard *et al.* 2009b). The trends observed at one sun continued; with the enhanced cells showing improved short-circuit currents (Figure 6.7). All cells, including the baseline, show a slight super-linear trend in current, as has been observed in GaAs because of high-level injection effects. The 20-layer QD cell generated the largest short-circuit current (7.53 A/cm²) at 440 suns – an 11% increase compared to the control-cell current at the same concentration. The open-circuit voltage for all cells rose logarithmically with concentration.

The efficiency for each type of cell under direct AM1.5 g conditions is shown in Figure 6.7. The efficiency peaked near 400 suns, consistent with their concentrator-grid design. The QD-enhanced cell showed almost 18% efficiency at 400 suns, representing absolute efficiency improvement by ~1% compared to the control-cell's 6% relative improvement. Under high-illumination intensity, the reduced (longer-wavelength) effective bandgap of the cell, as well as increased optical and thermal extraction of QD-generated carriers, leads to direct improvement in cell efficiency. This increase was due to higher short-circuit current combined with minimal loss of open-circuit voltage.

Suwaree (2006) had shown, by the study of InAs self-assembled QDs incorporated into GaAlAs/GaAs hetero structure, that an integration of QD into photovoltaic cell can provide an additional spectral response and stronger light absorption at long wavelength.

The research team from US Army Research Lab., University at Buffalo, and US Air Force Office of Scientific Research Physics and Electronics directorate has recently studied (Sablon et al. 2011) the possibility of using doping to improve QD solar cell performance by adding charge to the dots. Such doping has been found to avoid decrease of V_{oc} in normal hetero junction solar cells (i.e., without dots). Since the energy-level spacing for electrons in QDs is relatively large compared to the spacing for holes and the thermal energy, the electron intradot processes are precisely the ones that limit the electron–hole escape from QDs. The researchers have, therefore, preferred to enhance the photo-excitation of electrons rather than holes. It is further expected that doping-induced light absorption processes would aid energy harvesting and the extension of infrared response.

Using molecular beam epitaxy (MBE), the quantum dots with built-in charge (Q-BIC) solar cell was prepared on heavily doped n-type gallium arsenide substrates (Figure 6.8). The n-type buffer layer was grown at 595°C, while the quantum dot layers were grown at 530°C. Quantum dots were stacked in 20 layers consisting of 2.1 monolayers of indium arsenide (InAs) separated by 50 nm gallium arsenide spacers. The effect of putting a very thin layer of InAs in GaAs is for phase separation to occur, resulting in self-assembled InAs quantum dots. The distance between the layers was made relatively large in order to dissipate strain and hence suppress dislocations. The large spacing, in addition, reduces tunneling and prevents formation of an intermediate band.

Delta doping between the layers was used to provide electrons to occupy the quantum dots with up to six electrons on average. As expected, Delta doping with p-type material was found to reduce conversion efficiency. The structure was completed with three p-doped layers.

The epitaxial material was used to create circular solar cells of 250 μm diameter. The n-type contact consisted of annealed tin-gold-tin evaporated on to the back side of the substrate. A ring p-type contact made of chromium-gold with 100 μm opening was used. The devices were mounted in 68-pin leaded chip carrier (LDCC) packages using indium. The results are shown in Table 6.1.

The most heavily doped sample records an improved photovoltaic efficiency by as much as 50% compared with the undoped one. Since no evidence of the effect saturating is observed, an even stronger increase of the efficiency for further increase of the doping level can be expected. But, the occurance of Auger recombination may limit the doping levels.

Spectral studies showed that the response to longer-wavelength photons (880–1150 nm) was increased, at the expense of short wavelengths (less than 880 nm).

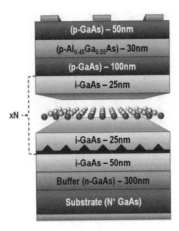

Figure 6.8 Schematic representation of Q-BIC Solar cell structure
(*Source*: reproduced with permission from Sablon *et al.* 2011, Copyright © 2011 American Chemical Society)

Table 6.1 QD solar cell parameters under 1 sun (AM1.5 G) at 100 mW/cm² illumination for as a function of doping designed to provide 0, 2, 3, and 6 electrons per dot on average

Ave. Dot electron population	J_{sc} (mA/cm^{-2})	V_{oc} (volt)	FF (%)	Efficiency (%)
0	15.1	0.77	77	9.31
2	17.3	0.74	76	9.73
3	18.5	0.79	75	12.1
6	24.3	0.78	72	14.0

This response above 920 nm was associated with QD excited and ground states. Beyond this, a rise in the region 4.8 μm (4800 nm) was found which was attributed to a transition from the dot ground state to the low-energy resonance conducting state. A broad weak peak was also seen above this up to ~8.0 μm, where the experimental measurements were close to the edge of their capability. The strength of the peak depended on the amount of delta doping and was totally absent in undoped GaAs reference samples (i.e., without QDs).

The improvement of IR absorption is expected to be even stronger at higher radiation intensities due to optical pumping effects. This feature makes the Q-BIC solar cells promising candidates for use with concentrators of solar radiation.

The key challenges in inclusion of nanostructures in photovoltaic devices is that the design considerations for achieving high efficiency nanostructured solar cells are substantially different compared to those of conventional devices, and many design parameters do not have sufficient theoretical or experimental support (See Chapter 2). However, the recent progress in the study of nanomaterials and nanostructures (for example, the current and efficiency enhancement in QD-based solar cells) is significantly impacting on the developments in photovoltaics.

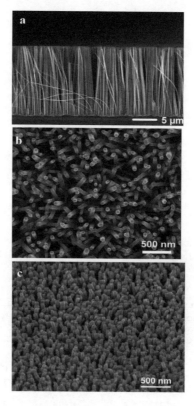

Figure 6.9 SEM images of (a) Silicon, (b) ZnO and (c) InGaN nanowires
(*Source*: Hochbaum & Yang 2010, reproduced with permission of Prof. Peidong Yang 2010,
Copyright © 2010 American Chemical Society)

6.2 NANOWIRE (NW) SOLAR CELLS

Semiconductor nanowires (NWs) have gained much attention over the course of the past decade. The term 'nanowire' is generally used to describe a large aspect ratio rod, 1–100 nm in diameter (Hochbaum & Yang 2010, Figure 6.9). The radial dimension of a nanowire is comparable to typical scale of fundamental solid-state phenomena (Hochbaum & Yang 2010) and the unique geometry leads to a large surface area-to-volume ratio which allows for distinct structural and chemical behavior. The unrestrained dimension of nanowire allows the conduction of particles such as electrons, phonons and photons. This fundamental advantage and control over transport of various forms of energy makes it an ideal form of material to fabricate different solid-state devices. These properties have been exploited in devices such as biological and chemical sensors (Cui *et al.* 2001; Zheng *et al.* 2005; Zhang *et al.* 2004; Hsueh *et al.* 2007).

Nanowire (NW) based solar cells are promising photovoltaic devices due to their several performance and processing benefits, including a direct path for charge transport provided by the geometry of such nanostructures.

The application of nanowires (and nanorods, defined as having an aspect ratio $\leq$ 5:1) to solar cells has been experimented in several device configurations and materials systems (Tsakalakos *et al.* 2007, Tian *et al.* 2007, Garnett & Yang 2008, Peng *et al.* 2002). Nanowire/rod-enabled solar cells demonstrated to date have been primarily based on hybrid organic-inorganic materials or that have utilized compound semiconductors such as CdSe. Huynh *et al.* (2002) have used CdSe nanorods as the electron-conducting layer of a hole conducting polymer-matrix solar cell and achieved an efficiency of 1.7% for AM1.5 irradiation. Similar structures have been demonstrated for dye-sensitized solar cells using titania or ZnO nanowires, with efficiencies ranging from 0.5% to 1.5% (Law *et al.* 2005; Baxter & Aydil 2005). These as well as other recent studies (Gur *et al.* 2005; Tian *et al.* 2007; Andra *et al.* 2007) demonstrate the benefits of using nanowires in improving charge transport in nanostructured solar cells.

NW solar cells can be in the form of a single NW or an array of NWs, which are usually aligned vertically. The p-n junctions in NW solar cells can be formed in radial or axial directions. For axial NW PVs, the electrode contacts are placed on the p- and n-regions, respectively, while the contacts of the core/shell NWs are placed at the bottom and top of NWs. NW arrays for solar-cell applications are normally unconnected NWs with abrupt junctions for charge separation. Vertical NW arrays enhance light absorption and improve both charge separation and collection efficiencies. The formation of axial and radial multijunction nanowire heterostructures enable enhanced light absorption by solar spectrum and through the formation of stacked heterojunction solar cell devices. The axial p-n junction, radial (core/shell) p-n junction, and heterojunction structures are the other configurations studied (Hochbaum & Yang 2010; Ke Sun *et al.* 2011).

These devices offer several advantages: (a) large surface-to-volume ratio for effective chemical and catalytic reactions and a large number of surface states to minimize dark currents (Soci *et al.* 2007), (b) allow the use of low-cost materials and substrates, because NWs can be grown epitaxially on lattice mis-matched or amorphous substrates, and (c) can be produced via sufficiently controllable top-down and bottom-up fabrication techniques that can potentially lead to low cost manufacturing (Zhu & Cui 2010).

6.2.1   Silicon NW solar cells

Silicon is the dominant semiconducting material in today's technology and has been used extensively for PV applications (Boettcher *et al.*, 2010; Peng *et al.* 2010; Shu *et al.* 2009) because it is an abundantly available low cost material, and its doping and morphology are easily tunable (Tian *et al.* 2007; Kempa 2008). Si NWs have been applied to solar cell applications with different structures, single NW or an array of NWs.

Ever since the fabrication of Silicon NWs in 1964 at Bell Laboratories of USA using a vapor-liquid-solid mechanism (Wagner & Ellis 1964), there has been tremendous progress in the fabrication and study of Si nanowire cells. The Si NWs are simple to grow: Au dots are formed on a Si substrate, which is heated to generate a liquid Au/Si alloy, causing Si atoms to be absorbed by the liquid formed from tetrachlorosilane in vapor phase. When the liquid is Si-supersaturated, Si atoms are deposited on the substrate, the melting point of the liquid increases, and solid crystals are precipitated.

In the early period of development, it was necessary to heat the Si substrate to 950°C, to obtain Si nanowires with a minimum diameter as large as 50 nm. The growth temperature has since been reduced to near the eutectic temperature of Au/Si (370°C); and using SiH_4 as a precursor, Si nanowires of diameter, 10 nm or less, have been realized (Cui 2001).

Si nanowire-based solar cells with vertical array geometry scatter light efficiently, especially at blue wavelength region. Further, they can absorb more light compared to thick crystalline films depending on the nanowire dimensions (Hu & Chen 2007). Since the NWs do not span the junction in these devices, the cells behave like conventional mono-crystalline p-n junction solar cells, but with greater absorption of incident light. Since the charge extraction occurs through the nanowires resulting in decrease of the current density (because of large series resistance of the NWs and their contacts), the efficiency of these cells are less than that of conventional cells. This effect is more distinct in VLS-grown nanowires (Stelzner et al. 2008; Tsakalakos et al. 2007).

Garnett and co-workers (2008) fabricated radial p-n junctions on nanowire arrays synthesized by a scalable, aqueous solution etching synthesis (Peng et al. 2002). The radial heterostructures are the optimal design for efficient charge collection. These cells have exhibited cell efficiency of around 0.5%. This poor efficiency is considered due partly to interfacial recombination losses as seen by a significant dark current, and partly to a large series resistance in the polycrystalline shell.

Single coaxial p-i-n silicon NW solar cells (Figure 6.10), consisting of a p-type core with intrinsic n-type shells have been fabricated by Lieber and his group (Tian et al. 2007). Silicon nanowire p-cores were synthesised by the nanocluster-catalysed vapor–liquid–solid (VLS) method and subsequent i- and n-type nano-crystalline Si shells by using chemical vapour deposition. After nanowire growth, SiO2 was deposited conformally by plasma-enhanced chemical vapour deposition (Tian et al. 2007). Under one-sun illumination, these p-i-n silicon nanowire elements yield a maximum power output of up to 200 pW per nanowire device. They have demonstrated an open circuit voltage of 260 mV, a short circuit current of up to 15 mA/cm², a fill factor of 0.55, and conversion efficiency of 3.4%. This performance is mainly due to a high J_{sc} caused by highly improved absorption in the thin shell due to its nanocrystalline feature and intrinsic layer. This group has further demonstrated a single nanowire PV device by selective chemical etching and contacting the p- and n-type regions separately which could power a logic circuit.

Kelzenberg et al. (2008) have fabricated single-nanowire silicon solar cells using VLS method by creating rectifying contacts which have shown an efficiency of 0.46%. When compared to single radial p-i-n NW solar cells, the conversion efficiency is low, while the carrier diffusion lengths are long (>2 μm).

Several studies have been reported in recent years on silicon NW array-based solar cells (Kayes et al. 2005; Stelzner et al. 2008; Tsakalakos et al. 2007; Kelzenberg et al. 2009; Garnett & Yang 2010; Gunawan & Guha 2009; Peng et al. 2005; Shu et al. 2009; Fang et al. 2008). Despite high anti-reflecting characteristics of silicon NW arrays, conversion efficiency of 9.31% has been achieved for the solar cells consisting of n-type Si NW arrays on p-type Si substrates. This is not, however, as high as efficiencies of conventional solar cells (Peng et al. 2005; Green 2003a). The lower conversion efficiency is believed to be due to the limited junction area (axial junction between n-NW and p-substrate) as well as surface states in etched Si NWs.

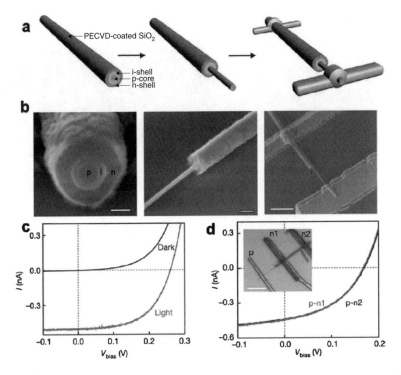

Figure 6.10 (a) Schematics of single nanowire PV device fabrication, (Left) The p-core, i-shell, n-shell, and PECVD-coated SiO₂ are shown; (Middle) Selective etching to expose the p-core; (Right) Metal contacts deposited on the p-core and n-shell, (b) SEM images corresponding to schematics in a. Scale bars are 100 nm (left), 200 nm (middle), and 1.5 μm (right), (c) Dark and light *I-V* curves, (d) Light *I-V* curves for two different n-shell contact locations. (Inset) Optical microscopy image of the device, Scale bar, 5 μm
(*Source*: Tian *et al.* 2007, reproduced with permission of Professor Charles Lieber, Copyright © 2007 Nature)

These effects, compounded by the ultrahigh surface area of the NW arrays, eventually increase the surface recombination velocity and limit the carrier collection efficiency.

Tsakalakos *et al.* (2007) have fabricated NW solar cells by depositing a thin n-type amorphous Si layer on p-type silicon NW array to form a p-n junction through the VLS method (Figure 6.11). The p-type Si NWs of diameter 109 ± 30 nm and length ~16 μm are grown with silane, hydrogen, hydrochloric acid, and trimethylboron (Lew *et al.* 2004) at 650°C for 30 min. The NW array is then processed to produce a dielectric layer by oxidation in a dry oxygen atmosphere at 800°C, followed by spin coating with photoresist and partial etch back of the resist by reactive ion etching. The nanowires are then dipped in a buffered oxide etch (BOE) (6% HF in de-ionized water and buffered with ammonium fluoride) to remove the grown oxide on the exposed nanowire surfaces. The photoresist is then removed using acetone. The array is subsequently coated with a PECVD conformal n-type a-Si:H layer (40 nm thick) to create the photoactive p-n junction. Though this step is performed immediately (within ~30 min) after the

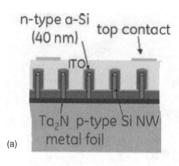

(a)

(b)

Figure 6.11   (a) Schematic cross-sectional view of the Si NW solar cell structure. The NW array is coated with a conformal a-Si:H thin film layer, (b) SEM of a typical Si NW solar cell on stainless steel foil, including a-Si and ITO layers; the insets show a cross-sectional view of the device and a higher magnification of an individual Si nanowire coated with a-Si and ITO (*Source:* Tsakalakos *et al.* 2007, Copyright © 2007 American Physical Society, reproduced with permission)

BO etch and photoresist removal steps, regrowth of a very thin native oxide could not be fully prevented.

After a-Si:H deposition, the array is sputter coated with a 200 nm-thick transparent conducting ITO layer to provide electrical contact to all the NW wires. Top finger contacts are created by shadow evaporation of Ti (50 nm)/Al (2000 nm). The stainless steel substrate is then spin coated with photoresist and diced into 1×1 or 1.2×1.5 cm pieces; and the photoresist is finally removed in acetone. The solar cells are then mounted using silver epoxy onto a Cu-coated printed circuit board with a notch, and a thin Au wire is manually attached to the top finger contact.

These cells prepared by Tsakalakos *et al.* were tested at NREL. The V_{oc} of the best devices was ~130 mV, less than that of the single NW devices demonstrated by Tian *et al.* (2007), and the fill factor was ~0.28. Both the high series and low shunt resistances of the cells seem to have limited their efficiency.

Despite low conversion efficiency (0.1%), these cells showed a spectrally broad EQE, indicating that the observed PV effect was due to absorption within the nanowire array. The processes employed are readily scalable, making this solar cell structure a promising candidate for future PV applications.

Sivakov *et al.* (2009) have demonstrated higher efficiencies with a solar cell based on Si NW arrays on glass substrates. This multijunction device is formed by applying a wet chemical etching process to form p^+, n, and n^+ polycrystalline silicon layers on glass; and it has exhibited a conversion efficiency of 4.4%. To improve the minority carrier collection in Si NW-array based solar cells, NWs with radial junctions are favoured. Si NW arrays consisting of an n-type core and a p-type shell have been fabricated and characterized (Garnett & Yang 2008). This NW cell which contains vertically aligned 18 μm long NWs with high packing density, is synthesized by a low-cost solution method and gives an overall cell efficiency of 0.46%. Recently, they have

(Garnett & Yang 2010) demonstrated strong light-trapping properties of NW arrays, which improve the conversion efficiency of Si NW-array solar cells. Their 5-μm NW-array radial p-n junction solar cells fabricated from 8 μm and 20 μm thin Si absorbing layers have achieved conversion efficiencies of 4.83% and 5.30%, respectively, under AM 1.5G illumination. The 4.83% conversion efficiency for 8 μm absorber Si-NW-array cells is about 20% higher than those on 8-μm-thick Si ribbon solar cells (4% higher J_{sc}). However, the 5.30% power conversion efficiency for the 20 μm absorber Si-NW-array solar cells is about 35% lower than the corresponding microfilm solar cells that have given an efficiency of 7.2% (14% lower J_{sc}) (Yoon et al. 2008). These devices demonstrated a significant light-trapping effect, above the theoretical limit for a randomizing system, indicating that there might be photonic crystal improvement effects. Nevertheless, the overall efficiency of these NW cells has not outdone that of planar cells due to increased junction and surface recombination. Even in comparison, the efficiency is more than ten times that of previous reported data (Garnett & Yang 2008). Therefore, such a kind of vertical NW array structure provides a feasible path toward high-efficiency and low-cost solar cells by reducing both the quantity and quality of the needed semiconductor materials.

One of the ways to enhance the power conversion efficiency of Si NW array-based solar cells is to decorate them with nanoparticles (NPs) or quantum dots (QDs). Materials such as insulators, metals, and semiconductors are widely used for this purpose, each contributing a different improvement due to a different mechanism (Huang et al. 2010; Kelzenberg et al. 2010; Peng et al. 2009; Hu et al. 2006).

Plasmon-enhanced solar cells are recognized as the next generation of solar cells (Atwater & Polman 2010). Also, Si NWs decorated with Pt NPs show exceptional catalytic activity at interfaces with liquids (Peng et al. 2009). Semiconductor QDs normally have high refractive index, which is believed to offer better light trapping performances. On the other hand, semiconductor NPs (example, PbS) provide additional absorption over a broad spectrum due to their high absorption coefficient (Lu et al. 2009). In addition, resonant excitons with high mobility transfer to the adjacent Si NW channel upon light absorption.

Ma et al. have demonstrated the possibility to widen the band gap of Si NW materials by decreasing their diameter to 3 nm or less (Ma 2003), indicating the possibility of developing Si nanowire materials with a controllable band gap. A key issue is the development of highly efficient surface passivation technology because Si nanowires have a large surface area (Konagai 2011).

6.2.2 Compound semiconductor NW solar cells

The potential of compound semiconductor NW solar cells is very high, mainly due to the rational control over NW and heterostructure growth. There has been enormous research activity on III-V and II-VI nanowire solar cells in recent years. III–V compounds are very significant for PV applications (Wei et al. 2009; Guo et al. 2009; Kempa et al. 2008; Yoon et al. 2008; Li et al. 2006; Dubrovskii et al. 2009) due to their outstanding electrical properties: (a) tunable energy band gaps and alloys; (b) a larger band gap effectively offering lower excess reverse saturation current with increased V_{oc} (Torchynska 2002); (c) excellent material qualities; and (d) high absorption coefficient.

III-nitrides are very attractive due to large energy band tunability, from ultraviolet to infrared. GaN, in particular, offers several advantages including high carrier mobility, p- or n-type selectivity (Sun *et al.* 2008), high stability, and a broad band gap for wide spectrum coverage (Ponce & Bour 1997). In addition, the dislocation defect density due to lattice mismatch between NW and substrate is potentially lower when compared to thin films. The fabrication of p-type GaN nanorod arrays on n-type Si substrates have been reported by Wagner and Shealey (1984) with good rectification characteristics, small reverse currents and highest conversion efficiency of 2.73%. Dong *et al.* (2009) have demonstrated the synthesis of n-GaN/i-In$_x$Ga$_{1-x}$N/ p-GaN core/shell NW solar cells using MOCVD. The single NW device with 27% of indium, simulated with AM 1.5G illumination, has shown a maximum efficiency of 0.19%.

GaAs is another attractive III-V material for solar cell application due to its large light absorption coefficient and ideal band gap (Chapter 5). LaPierre's group has fabricated GaAs NW radial p-n junction solar cells using MBE (Czaban *et al.* 2008), where Te and Be have been used for n-type shell and p-type core doping. The fabricated NWs show different morphologies due to the inclusion of dopants during NW growth. Although the cell exhibits rectifying characteristics, the leakage current is high, which can be explained based on the low breakdown voltage and possible tunneling effects. This cell shows a low efficiency of 0.83%, partly due to the formation of nonuniform core/shell p-n junctions. Colombo *et al.* (2009) have reported the fabrication of GaAs core/shell NW p-i-n junction solar cells, consisting of a p-type core, intrinsic and n-type shell using MBE. The conversion efficiency achieved for this single NW solar cell structure is 4.5%, which is probably the highest reported efficiency for III-V NW-based solar cells.

Goto *et al.* (2009) have reported high-quality core/shell InP nanowire arrays for photovoltaic applications grown by a selective-area MOCVD, with an overall conversion efficiency of 3.37%. This is attributed to InP's higher optical absorption coefficient and the band-gap, which offer finest match to the solar spectrum than Si (Yamaguchi *et al.* 1986). Wang's group has demonstrated heteroepitaxial growth of vertical InAs NW arrays on Si using catalyst-free MOCVD (Wei *et al.* 2009). This type of solar cell with multiple band gap absorber materials, InAs and Si, can efficiently harvest solar energy due to enhanced broad-band light absorption.

II-VI compound semiconductor NWs, such as CdS (Kwak *et al.* 2009; Ouyang *et al.* 2006; Cao *et al.* 2006), CdSe (Ouyang *et al.* 2006; Yang *et al.* 2002), CdTe (Yang *et al.* 2002), ZnSe (Fanfair & Korgel 2007), ZnTe (Dong *et al.* 2007), ZnO and PbSe (Hull *et al.* 2005) are prepared by low cost solution- based methods. This advantage of low-cost production has created interest in developing II-VI compound semiconductor NW-based PVs. Moreover, these materials carry the intrinsic benefits of inorganic nanomaterials, i.e., high carrier mobility, robust material stability, and high interfacial area.

Exploring the use of II-VI compound NWs in combination with semiconductor dyes and organic polymers in hybrid solar cells to substantially increase their efficiency and viability is highly desirable and extensive research effort is needed to realize the objective. However, the utility of II-VI materials is limited by the presence of native defects. Controlled doping in II-VI semiconductors is still technically challenging (Desnica *et al.* 1998; Chadi 1994), particularly for II-VI NWs grown by low-temperature synthesis methods.

6.3   NW-POLYMER HYBRID SOLAR CELLS

In the field of photovoltaics, inorganic–organic hybrid solar cell devices based on conjugated polymers have generated a lot of attention because of their low cost and scalability to large-area devices. NW–polymer hybrid solar cells derive advantage from the high electron affinity of inorganic semiconductors, and the low ionization energy of organic polymers to generate rapid charge transfer (Hyunh *et al.* 2002) which ensures high energy conversion efficiency. Large absorption coefficients of polymers and large electron mobilities in II-VI semiconductors are other features utilized by these cells to obtain high performance.

Nanorods (Huynh *et al.* 2002), nanoparticles (Arici *et al.* 2004; Plass *et al.* 2002), and carbon nanotubes (Kymakis & Amarathunga 2002; Miller *et al.* 2006) are dispersed in conjugated polymer hybrid solar cells as well as in nanowire-based dye-sensitized solar cells (Law *et al.* 2005; Baxter *et al.* 2006) to improve carrier mobility and collection efficiency. Compared to silicon or III-V substrates, the use of glass or plastic in these systems allows for a much higher cost-saving. However, one of the issues that many of these devices exhibit is poor carrier transport (Greenham *et al.* 1996). In the case of nanorods or nanoparticles embedded in a polymer, carriers must hop from one site to the next as they are transported through the host material. This random pathway leads to an increase in the probability of recombination and thus results in decrease of the overall photoconversion efficiency. To overcome this problem, attempts are made to improve the alignment of the NWs by using intermediate steps such as template-assisted growth (Kang *et al.* 2005) and NW growth on an intermediate layer on top of the metal oxide (Ravirajan *et al.* 2006).

Hybrid systems based on metal oxide, such as ZnO (Ravirajan *et al.* 2006; Beek *et al.* 2004, 2006; Peiro *et al.* 2006) and TiO$_2$ (Arango *et al.* 2000; Kim *et al.* 2006) nanostructures which improve cell performance are widely investigated. TiO$_2$/ZnO core–shell (Greene *et al.* 2007) and CdS/ZnO core–shell (Hao *et al.* 2010) heterostructures are also projected to improve device performances. Cd-VI compounds, such as CdSe, CdS, and CdTe are others studied in hybrid systems. Vertically aligned NW structures based on these materials are fabricated by several groups using various methods, such as electrodeposition in AAO templates (Wang *et al.* 2007; Kong *et al.* 2006), electroless deposition based on solution–liquid–solid (SLS) mechanism (Kwak *et al.* 2009; Bozano *et al.* 1999), and gasphase VLS growth (Lee *et al.* 2009). Hybrid solar cell structure is then completed by spin-coating *photoactive polymers* onto the surface of NWs. Devices containing the vertically aligned II–VI Nanowires show a drastic improvement in energy-conversion efficiency over a polymer or NW only device (Kong *et al.* 2006; Lee *et al.* 2009).

Solution-phase-synthesized colloidal CdSe nanorods/poly(3-hexylthiophene) (P3HT) blend solar cell shows improved absorption in the spectral range, 300–720 nm, and with tunable absorption spectra by controlling the diameter (Bozano *et al.* 1999). The effect of aspect ratio on the performance is also characterized (Hyunh *et al.* 2002). Devices containing nanorods with higher aspect ratio demonstrated greater efficiencies, due to improved charge mobility in longer nanorods. Further, interface in hybrid system is critical for overall efficiency.

Wang *et al.* (2007) have demonstrated that the efficiency of a polymer/CdS solar cell can be increased through the use of pyridine as a solvent when spin coating the active layer of the device. This group has also shown that sample processed using pyridine demonstrates a more even dispersion of the nanotetrapods, which contributes to the increased efficiency. The studies of Wang *et al.* (2007) have revealed that samples processed with chlorobenzene achieved conversion efficiencies of 0.14% while those processed with pyridine showed efficiencies of 0.89%, which further increased to 1.17% through thermal annealing. The solvent effect and the thermal treatment are believed to create phase separation and conducting polymer networks, which lead to higher hole mobilities and thus lower series resistance.

For hybrid solar cells, vertical NW array offers: (a) improved light absorption, (b) direct conducting pathway to electrode contact, and (c) potential guidance for aligned polymer morphology, which leads to higher hole mobilities. As of now, efficiencies of the II-VI materials based devices are low compared to other classes of solar cell. However, there are benefits of using group II–VI materials that include low cost of production, general abundance of materials, as well as environment-friendly.

In general, the study of inorganic nanowire/polymer hybrid solar cells is not as much progressed as that of nanowire DSSCs, possibly due to low overall efficiencies. The best polymer BHJ cells could achieve 5% conversion efficiency (Reyes-Reyes *et al.* 2005; Ma *et al.* 2005) while DSSCs could record 7% even two decades ago (Oregan & Gratzel 1991). However, the polymer hybrid NW solar cells merit greater attention because, among all classes of NW solar cells, these devices can be easily manufactured at low cost.

A few prominent examples of Semiconductor NW/Polymer hybrid solar cells are explained.

6.3.1   InP nanowire – polymer (P3HT) hybrid solar cell

Novotny *et al.* (2008) have demonstrated an efficient carrier transport polymer hybrid system that does not require any intermediate step. This is achieved by growing NWs directly onto a metal oxide and enveloping the NWs with a conjugated polymer. In this device, like in bulk organic heterojunction solar cells, there are a very large number of junctions throughout the entire polymer matrix provided by the penetrating NWs, thus enhancing the probability of excitons dissociating at an interface. Furthermore, carriers created in the NWs have a direct pathway to the electrode and do not have to rely on carrier hopping. Thus, this hybrid device has a more efficient and defined pathway for carrier collection. Because the mobility in the semiconductor material is several orders of magnitude higher than that of the polymer, carrier transport will be much more efficient. Direct growth of NWs onto the electrode will thus improve device performance, reduce the problem of reproducibility of contacts to NWs, and provide a novel method of NW growth that does not use expensive substrates such as Si or InP. Indium phosphide (InP) nanowires can serve as electron fast-tracks and carry electrons released by the incident photons directly to the device's electrode boosting thin-film solar cell efficiency (Novotny *et al.* 2008).

n-InP nanowires were grown on 250–300 nm thick ITO-coated glass substrates, which enable to achieve an ohmic contact, thus providing an effective pathway for carriers (Figure 6.12(a) & (b)). The sputtered ITO layer covers the top, bottom, and sides of the glass substrate, providing a continuous layer around the glass. For growing n-InP NWs, a horizontal MOCVD reactor at a pressure of 100 torr was used. Trimethylindium (TMIn) and phosphine (PH3) were used as group III and V sources,

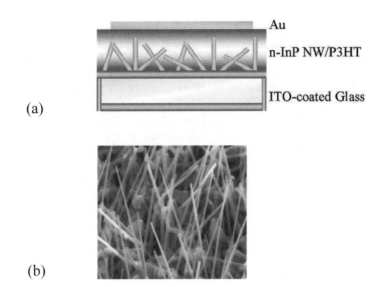

Au

n-InP NW/P3HT

ITO-coated Glass

(a)

(b)

Figure 6.12 (a) n-InP Nanowire/P3HT solar cell, (b) SEM image of n-type InP nanowire growth on ITO taken at a 45° tilt with scale bar of 500 nanometers
(*Source*: Novotny *et al.* 2008, Copyright © 2008 American Chemical Society, reproduced with the permission of Professor Paul Yu)

respectively, with hydrogen as the carrier gas. The growth temperature was 425°C with an input V/III ratio of 100. n-Nanowires were doped using a disilane source (25 ppm in hydrogen). The NWs nucleated and grew from indium droplets that formed on the surface. Details of this growth process are given in the earlier publication by Novotny and Yu (2005). Then, a layer of high hole mobility conjugate polymer, P3HT, also known as poly (3-hyxylthiophene) was deposited. The use of this high boiling point solvent has been shown to yield an increase in the mobility and overall device performance (Chang *et al.* 2004). The samples were then annealed at 120°C for 1.5 h under vacuum ($<10^{-6}$ Torr) followed by a sputter deposition of 200 nm of gold. After NW growth, it is dipped in ammonium sulfide to reduce surface recombination (Novotny *et al.* 2008).

This is the first demonstration of InP NW growth directly onto an electrode. In the conventional NW device fabrication, typical problems in the reproducibility of contacts to individual NWs have been experienced (Langford *et al.* 2006; Stern *et al.* 2006). This approach, however, allows for a large number of NWs to grow directly onto the electrode, thus improving the reproducibility of the device. Instead of relying on a single NW for the device, a very large number of NWs contribute to the carrier transport. Therefore, the overall performance of the device is based on the statistical average of the NWs rather than a single NW, leading to an enhanced reproducibility. Figure 6.12(b) is an SEM image, taken at a tilt of 45°, of as grown n-InP NWs on ITO. The diameters of these NWs range from 40 to 80 nm and the lengths range from 1 to 3 μm. There is no single growth direction on ITO due to the random orientation of the domains within the polycrystalline ITO layer. Studies have shown that similar InP NW growth can occur on various other poly substrates by relying on short-range ordering of the substrate (Kobayashi *et al.* 2006). The NWs grow away from the substrate, thus providing a structure that can penetrate several micrometers into the polymer matrix.

I-V measurements were performed to analyze the effects of the inclusion of NWs to the conjugated polymer, P3HT, which showed an increase of the forward bias current by 6–7 orders of magnitude. The origin of the observed p-n junction behaviour was the interface between the high hole mobility conjugated polymer and the n-doped NWs. The P3HT layer did not short the ITO and Au electrodes, as seen by the very low current of the P3HT only sample. However, care had to be taken to ensure that the polymer totally embedded the NWs such that the NWs could not short the two electrodes. The thickness of the P3HT was modified so that only P3HT was in contact to the Au electrode.

This research group has established that NW/polymer hybrid devices have very good diode characteristics. Growing InP NWs directly onto an ITO electrode has provided an ohmic contact for efficient carrier collection. The inclusion of NWs to P3HT has increased current through the device and provided efficient pathways for the carriers. High rectification ratios, low ideality factors, and improved reverse saturation currents are achieved as well as a photoresponse from this hybrid photodiode. It is shown that by optimizing the thickness of the polymer between the tips of the NWs to the Au electrode as well as the interface between the NW surface states and the polymer side chains, the overall efficiency of the device can be improved. A concept of great potential for the fabrication of low cost, high efficient PV cells is experimentally demonstrated which can emerge as a promising alternative to the current solar cell technology.

6.3.2 Microcrystalline silicon nanorods/P3HT hybrid solar cells

Ma *et al.* (2009) have studied another hybrid solar cell design, using silicon nanorods (SiNRs) array as an inorganic electron acceptor and 3-hexylthiophene (P3HT) as an organic electron donor.

The oriented microcrystalline silicon (μc-Si) nanorods grown on Si wafer, ITO glass and stainless steel substrates are successfully prepared by hotwire chemical vapor deposition with mixture of silane and hydrogen gases at low temperature. Reaction gases flux incidents with a glancing angle of 80° with respect to the substrate surface. The distance of 9.2 cm between the substrate and the filament provides a good directionality of incidence flux. The deposition pressure of 10^{-2} Pa prevents the possible gas phase reaction.

The morphologies and structures of the Si nanorods are characterized by field emission scanning electron microscopy and Raman scattering measurements. The optical properties of SiNRs/P3HT films on glass are analyzed by UV-V Absorption spectrometer and PL spectrometer.

The hybrid solar cells are prepared by spin coating P3HT into the Si nanorods array. Then ITO and metal (Ti and Ag) grids are evaporated onto the polymer to form the electrode. The performances of these SiNRs/P3HT devices are characterized by dark and illuminated (AM1.5, 100 mW/cm^2) I-V measurements.

It is found that the doping level has no remarkable influence on the morphology of SiNRs. The average diameters for intrinsic and n-doped SiNRs are 30~60 nm. The nanorods show inclined growth for the static substrate condition due to the shadowing effect. By rotating the substrate with a uniform rotation speed of 15 rpm, vertically oriented SiNRs are developed. Crystallized SiNRs are achieved under a low substrate temperature. This hybrid structure has achieved an open circuit voltage of 219 mV, current density of 3.12 mA/cm^2, FF of 0.31 and a power conversion efficiency of 0.2%.

6.3.3 TiO$_2$ nanotube arrays in DSCs

In this example, the use of nanotube arrays in Dye-sensitized solar cells to improve the performance is discussed. The disordered pore structure of TiO$_2$ anode presently used in DSC's fabrication, though offers a large surface area, has been a major factor limiting the progress of many approaches for further improvement of DSCs. Firstly, it induces a short electron diffusion length, order of 7–30 μm, which increase the recombination probability of the electrons with redox species limiting the effort of increasing the absorption of photons. Secondly, when solid-state electrolytes, ionic liquid, or other hole-transporting materials are used in DSCs, the TiO$_2$ porous film thickness has been limited to be less than 3 μm due to the ineffective infiltration of such materials into the disordered pores leading to insufficient dye adsorption and, hence, poor light absorption.

To overcome this problem, utilizing nanowire arrays that form direct conduction channels for electrons has been investigated (Hochbaum & Yang 2010). Primarily, unlike the mesoporous nanoparticle films, the average nanowire diameter is thick enough to support a depletion layer near the surface. This potential barrier can provide a strong driving force for exciton dissociation at the interface between the nanowire and the dye resulting in more efficient charge injection. Additionally, the rate of recombination is reduced because the electrons are swept away from surface by band bending. Further, the electron mobility in the nanowire is larger due to their directional and uninterrupted conduction channel compared to that in the nanoparticle films. The electron diffusion constant gets increased due to this directed transport, thus improving the charge collection and facilitating the production of optically thick cells that absorb more incident light. Figure 6.13 shows the schematic comparison of these two types of devices (Hochbaum & Yang 2010).

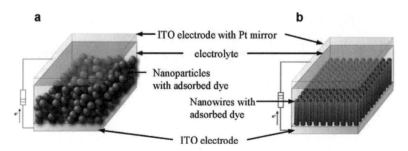

Figure 6.13 Schematic representation of a DSSC; (a) conventional cell with nanoparticle film electrode, and (b) nanowire DSS cell
(*Source:* Hochbaum & Yang 2010, reproduced with permission of Prof. Peidong Yang 2010, Copyright © 2010 American Chemical Society)

Several groups (for example, Law *et al.* 2006) have used radial (core-shell) nanowire heterostructures to utilize the charge transfer characteristics and surface stability of TiO_2 while retaining the fast electron transport of single-crystalline nanowires. ZnO nanowire arrays coated with TiO_2 films of varying thickness by 'atomic layer deposition' method were synthesized by Law and co-workers (2006). The radial NW structures have demonstrated a two-fold improvement in efficiencies which was attributed to improved charge injection and reduced recombination current (Palomares *et al.* 2003; Hochbaum and Yang 2010). Compared to ZnO, TiO_2 offers better material properties such as higher open circuit voltage, larger fill factor, and thus higher efficiency. But achieving model TiO_2 nanowire array has been a challenge. Recently, Feng and co-workers (2008) and Liu and co-workers (2008) have realized high-density, single-crystalline, vertically-aligned TiO_2 nanowire arrays. The performance of DSS cells fabricated with these arrays has been promising despite very low nanowire density offering reduced surface area for dye adsorption. Using ZnO 1-D nanostructures and TiO_2 nanotube arrays, rapid electron transport and, thus, large electron diffusion length (order of $100 \mu m$) and enhanced electron collecting efficiency have been observed in DSCs.

Recently, Xu *et al.* (2010) have developed a process for synthesizing vertically aligned top-end opened TiO_2 nanotube arrays directly on TCO. The process involves growing ZnO nanowires on TCO and then converting them into TiO_2 nanotubes.

Figure 6.14   ZnO nanowire and converted TiO_2 nanotube arrays. (a) ZnO nanowire arrays on ITO. Scale bar, 5 μm. (b) SEM of converted TiO_2 nanotube arrays with a closed top-end on ITO. Scale bar, 5 μm. (c) SEM top view of ZnO nanowires. Scale bar, 500 nm. (d) SEM top view of converted TiO_2 nanotubes with a closed top-end. Scale bar, 500 nm. (e) SEM top view of TiO_2 nanotube arrays with an open top-end. Scale bar, 5 μm. (f) TEM of TiO_2 nanotubes. Scale bar, 1 μm

(*Source*: Xu *et al.* 2010, reproduced with permission of Prof. Gao, Copyright © 2010 American Chemical Society)

All the synthesis steps utilize wet chemical processes, which facilitate low cost, low temperature, and easy scaling-up. The resulting TiO_2 nanotube arrays have a tube length of 10–11 μm, an interior diameter of 150–300 nm, a 50–100 nm thick wall, and a tube density of $1.6 \times 10^9/cm^2$.

The SEM images of ZnO nanowires and converted TiO_2 nanotubes are shown in Figure 6.14. These nanotube arrays are used in the fabrication of DSCs and have achieved an efficiency of 3.6% under 100 mW/cm² simulated sunlight. This result is nearly three times that of ZnO nanowire-based DSCs due to the significant increase in V_{oc}, FF, and internal surface area of the anode. It is also observed that the lifetime of photogenerated electrons in TiO_2 nanotubes is more than an order of magnitude larger than that in sintered TiO_2 nanoparticles (Xu *et al.* 2010).

6.4 THIRD-GENERATION CONCEPTS UNDER DEVELOPMENT

At Stanford University, under Global Climate & Energy Project (GCEP), research was undertaken with many research groups participating to develop very high efficiency PV solar devices using low cost processes based on three novel third-generation concepts: Hot carrier solar cells, Plasmonic photovoltaics, and Nanostructured material based solar cells. In addition to Stanford University, several major Institutions (CalTech; UNSW; University of Sydney; FOM-Institute, Amsterdam; IRDEP, Paris; IES-UPM, Madrid) are associated with this Project. These concepts, the approach to achieve targeted goals, and the progress realised are briefly explained. Several other researchers are also working on these concepts.

6.4.1 Hot Carrier solar cells

Hot Carrier solar cells represent one of the most promising third-generation PV concepts. They have a structure that is practically simple compared to other very high efficiency solar devices such as multijunction monolithic tandem cells. Further, the prospect to achieve high efficiencies is substantially more than the theoretical efficiency limit for traditional single-junction devices (31%) and also to significantly reduce the cost of solar electricity. These merits make their fabrication easy to find favour with 'thin film' deposition techniques, with their associated benefits such as low materials costs and energy usage and facility to use abundant, non-toxic elements. In this structure, the energy conversion is enhanced by reducing energy losses related to the absorption of solar photons with energy larger than the energy bandgap of the active PV material.

An ideal Hot Carrier cell would absorb a broad range of photon energies and derive a large fraction of the energy by extracting 'hot' carriers before they thermalise to the band edges. Therefore, the main feature of a hot carrier cell is to slow the rate of carrier cooling to allow hot carriers collected whilst they are still at elevated energies (hot), and thus enabling higher voltages to be achieved from the cell ensuing in higher efficiency. In addition, a Hot Carrier cell must only allow extraction of carriers from the device through contacts which accept only a very narrow range of energies (energy selective contacts or ESCs). This is necessary in order to prevent cold carriers in the contact from cooling the hot carriers, i.e., the increase in entropy on carrier extraction is minimized

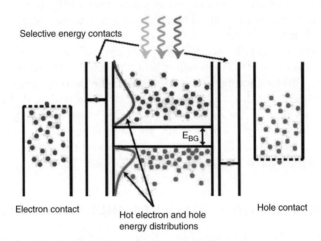

Selective energy contacts

E_{BG}

Electron contact

Hot electron and hole
energy distributions

Hole contact

Figure 6.15   Band diagram of the Hot Carrier cell. Energy distributions of hot electron and hole pop-
ulations created by photon absorption are shown together with SECs and cell electrodes
(*Source*: Conibeer *et al.* – Stanford University GCEP Report, September 2008 – August
2011, reproduced with the permission of Prof. Conibeer)

(Aliberti *et al.* 2010). The limiting efficiency for the hot carrier cell is 65% at 1 sun,
and 85% at maximum concentration – very close to the limits for an infinite number
of energy levels (Patterson *et al.* 2010; Clady *et al.* 2011; Takeda *et al.* 2010). Figure
6.15 is a schematic band diagram of a Hot Carrier cell illustrating the requirements.
To realize efficient hot carrier solar cell devices, the main requirements are the following: (a) Slowing of thermalization of photogenerated electrons and holes in the active
layer relative to radiative recombination rates. The major carrier cooling is due to the
interaction with phonons in the semiconductor lattice. On timescale of picoseconds,
hot carriers relax by the emission of optical phonons; and subsequent cooling to the
band edge occurs via acoustic phonon scattering (Klemens mechanism). It means the
approaches that interfere with optical to acoustic phonon scattering have the prospect
to significantly reduce carrier cooling rates. It has been shown that periodic networks
of quantum structures, such as semiconductor nanoparticles, incorporated into the
absorber material can considerably reduce the scattering between phonon modes. This
effect, known as 'phonon bottleneck' has been implemented by engineering a nanoparticle network embedded in an absorber matrix that is homogeneous on a scale greater
than a few tens of nanometers. Preparation of such a structure is compatible with
thin-film deposition techniques and has the prospects for high efficiencies.

(b) Developing energy selective contacts (ESCs) which have ability to extract
charges through a narrow allowed energy range in order to prevent cold carriers in
the contacts cooling the hot carriers to be extracted. Quantum mechanical resonant
tunneling structures are the most likely to satisfy the requirements of selective energy
transmission over a small energy range. These structures take the form of double barrier
resonant tunnel structures with the resonant energy level provided by a single quantum dot layer. The size, uniformity, and density of this QD layer have to be optimized
against overall cell performance.

(c) A working device would require integration of the structures used to tackle the above requirements without compromising their performance. Modelling and characterization studies and eventual integration into a combined device have been undertaken on the initial structures for the absorber layer with periodic quantum dot arrays and the selective energy contacts.

In the first device prototypes developed under the GCEP project, III-V compound semiconductors grown with metal-organic vapor phase epitaxy (MOVPE) and molecular beam epitaxy (MBE) techniques are used. After successful demonstration of these prototypes, studies are focused on strategies to transfer such devices to thin-film or self-assembly deposition technologies using abundantly available non-toxic materials such as silicon and tin (group IV materials). Time resolved spectroscopy is used throughout the project.

Reseach groups from UNSW (Conibeer and Green), Inst. for Solar Energy, Madrid (Marti and Luque), CNRS (Guillemoles) and U. Sydney (Timothy Schmidt) are involved in this major project. The recent GCEPAnnual report, '*Hot Carrier solar cell: Implementation of the Ultimate PV Converter – Annual Report April 2011*' (Conibeer *et al.* 2011) provides the progress achieved and the future plan of work. Several publications that appeared last year, and papers presented at International Meets are also listed in the report. For example, Green *et al.* 2010a & b, Konig *et al.* 2010, Conibeer 2010, Conibeer *et al.* 2010, so on present the summary of the present status and progress.

6.4.2 Plasmonic photovoltaics

Plasmonics are properly engineered metal structures which can confine incident light on a sub-micrometric scale. This feature of plasmonics was utilized to enhance solar light absorption in ultrathin semiconductor films (10's nm to 100's nm). By considerably reducing the thickness of the active layer, the application of plasmonics is expected to expand the range and quality of absorber materials that are suitable for inorganic thin-film PV devices. In particular, this would enable effective utilization of both low-dimensional semiconductor structures and thin films of earth-abundant, low-cost, and non-toxic absorbers with poor charge transport properties. Plasmonic structures which offer the possibility to construct *optically-thick but physically very thin* photovoltaic absorber layers could effect considerable changes in the design of high efficiency solar devices.

Recently, the use of metallic nanostructures for PV has received renewed attention with the availability of new nanofabrication tools and the growing understanding of their optical properties provided by the rapidly growing field of plasmonics (Brongersma & Kik 2007). Experimentally, efficiency enhancements have been achieved with respect to basic cells employing organics, a-Si, and GaAs respectively (Atwater & Polman 2010).

This research programme under GCEP has, therefore, three focal points: (1) design and realization of plasmonic structures to enhance solar light absorption in ultrathin film and low-dimensional inorganic semiconductor absorber layers; (2) synthesis of earth-abundant semiconductors in ultrathin films and lowdimensional (quantum wire and quantum dot) multijunction and multispectral absorber layers; and (3) investigation of carrier transport and collection in ultrathin plasmonic solar cells.

Figure 6.16 Plasmonic light trapping geometries for thin-film solar cells (a) Light trapping by scattering from metal nanoparticles at the surface of the solar cell, (b) Light trapping by the excitation of localized surface plasmons in metal nanoparticles embedded in the semiconductor, (c) Light trapping by the excitation of surface plasmon polaritons at the metal-semiconductor interface
(*Source:* Atwater 2010 & Polman 2010, reproduced with permission of Prof. Atwater)

Plasmonic structures can be obtained in atleast three different ways:

1   Metallic nanoparticles can be used as subwavelength scattering elements to couple and trap light from the sun into an absorbing semiconductor thin film by 'folding' the light into a thin absorber layer causing an enhancement of the effective optical path length (see Figure 6.16(a)),
2   Metallic nanoparticles can be used as subwavelength antennas in which the plasmonic near field is coupled to the semiconductor, enhancing its effective absorption cross section (i.e., causing the creation of electron-hole pairs in the semiconductor) (see Figure 6.16(b)), and
3   A corrugated metallic film on the back surface of a thin PV absorber layer couples sunlight into surface plasmon polariton modes supported at the metal-semiconductor interface as well as guided modes in the semiconductor layer, where the light is converted to photocarriers in the semiconductor (see Figure 6.16(c)).

Enhanced absorption offers several benefits to a solar cell. The reduced thickness implies that less absorber material is used in making a cell per generated Watt of electricity. This has very important technical and strategic consequences as photovoltaics scales up in manufacturing capacity from its current level of ~ 8 GW in 2008 to > 50 GW by 2020, and eventually to the TW scale. Extension of photovoltaics technology to the TW scale would be possible if the materials utilized in solar cells be abundant in the earth's crust and suitable to the fabrication of efficient photovoltaic devices.

Consider an extremely thin multi junction semiconductor structure shown in Figure 6.17. These types of structures use active layers prepared from semiconductor absorbers (e.g., Cu_2O, Zn_3P_2, Si, β-$FeSi_2$, $BaSi_2$, Fe_2O_3) and low-cost plasmonic metals (Al, Cu). For absorber layers with good surface passivation, the capacity to reduce the solar cell base thickness via plasmonic design yields an increased open circuit voltage, in addition to enhancing carrier collection. This development would open up new approaches to carrier collection from quantum wires and dots that do not rely on inter-dot or inter-wire carrier transfer prior to collection at the device contacts.

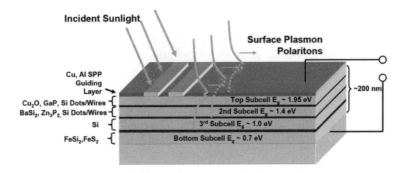

Figure 6.17   Schematic of a plasmonic multijunction solar cell with >45% efficiency potential (*Source:*Atwater *et al.* – Stanford University GCEP Report, September 2007 – August 2011, reproduced with permission of Prof. Atwater)

By appropriately designing the shape and dielectric environment, metal plasmonic layers can be used both for light concentration, as described, and for spectral filtering. The plasmon resonance can be broadly tuned across the visible and infrared frequency range. Thus, metallic contacts can be designed whose plasmon resonant spectral absorption features are well-matched to the PV absorber properties, e.g., making metallic contacts more transparent at photon frequencies at or above bandgap by designing the plasmon resonance frequency to lie below the bandgap frequency. This feature is utilized to use plasmonics as inter-cell ohmic contacts in the device.

The solar cell's performance under concentrated illumination is highly sensitive to the angle of incidence of incoming radiation. Hence, solar trackers are often used for tracking the sun across the sky. Such systems are, in general, bulky and energy-intensive, and require maintenance. Therefore, if the solar tracking functionality could be integrated into the solar cell design without the use of mechanical parts, it would be highly desirable. A planar metallic microcavity structure consisting of a metal–dielectric–metal (MDM) configuration can exhibit an omni-directional resonance, i.e., a resonance for which light can efficiently couple to cavity modes independent of the incidence angle for a carefully chosen metal-to-metal spacing (order of 100 nanometers). MDM structures are studied to develop built-in tracking systems that would enhance cell performance throughout the day and simultaneously decrease the cost of balance of system of the overall PV system.

This research undertaken by research groups under Atwater (Caltech), Brongersma (Stanford) and Polman (FOM Institute, Amsterdam) under GCEP included theory, numerical simulation, and experimental work to realize prototype tandem cells using plasmon enhanced light absorption and omnidirectional plasmonic absorbers. Full-field electromagnetic codes based on finite-difference methods are used to design plasmonic structures and define incidence angle and spectral dependence of the plasmonic light absorption efficiency.

Several prototype solar cells that feature surface plasmon polariton coupling to thin film absorbers as illustrated in Figure 6.18 are developed including Si vertical and lateral *p-n* junctions, Si nanocrystal and nanowire cells, and Si-based two junction tandem

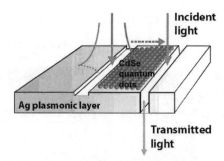

Figure 6.18   Schematic of a test structure consisting of a surface wave interferometer coated with CdSe quantum dots to test plasmonic light absorption. Absorption along the Ag surface is achieved through incident light conversion into surface plasmon polaritons that exhibit mode overlap with the absorber layer
(*Source:* Atwater *et al.* – GCEP Report, Sept. 2007 – August 2011, reproduced with permission)

cells. Test cells are used to investigate carrier transport and collection in ultrathin plasmonic solar cells and to assess the potential for plasmon enhanced light absorption and omnidirectional plasmonic absorbers in TF concentrators.

Very recently, enhanced solar cell absorption and photocurrent due to scattering by 100 nm Ag nanoparticles in a single 2.5 nm thick InGaN quantum well photovoltaic device based on a standard LED design was demonstrated. Nanoparticle arrays were fabricated on the surface of the device using an anodic alumina template masking process. The Ag nanoparticles increase light scattering, light trapping, and carrier collection in the III-N semiconductor layers leading to enhancement of the external quantum efficiency by up to 54%. Additionally, the short-circuit current in cells with 200 nm p-GaN emitter regions is increased by 6% under AM 1.5 illumination. AFORS-Het simulation software results were used to predict cell performance and optimize emitter layer thickness (Pryce *et al.* 2010; Atwater & Polman 2010).

Plasmonic backreflectors were studied on ultrathin a-Si solar cells that are integrated with plasmonic light trapping structures built into the metallic back contact. Substrate conformal imprint lithography (SCIL) was used to pattern a 150 nm thick sol-gel based resist that is applied by spin-coating over a 150 mm silicon wafer. To form the back contact, the patterned sol-gel layer is sputter-coated with 200 nm Ag (1%Pd), as illustrated in the SEM image in Figure 6.19(a). The metal holes are 225 nm in diameter, 240 nm in depth, and have a pitch of 513 nm after coating. The back contact for a flat reference cell was made by evaporating 200 nm of Ag on glass. Both cells were then processed side-by-side in the remaining steps to ensure identical deposition conditions. A 100 nm ZnO spacer layer was sputtered on top of the Ag contact, followed by standard n-i-p a-Si:H cell deposition using 13.56 MHz PECVD with an i-layer thickness of 500 nm.

An 80 nm ITO top contact was sputtered on top, which also serves as an antireflection coating. Finally a metal grid was evaporated over the ITO using a contact mask. The active area of the cell is 0.13 cm^2. Figure 6.19(b) shows a cross section of a cell after fabrication on top of the patterned cell, made using focused ion beam.

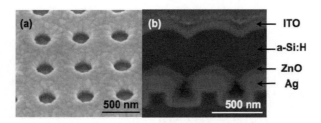

Figure 6.19 (a) SEM image of a nanoimprinted pattern of holes after overcoating with Ag, (b) SEM image showing a FIB cross section of a fully fabricated n-i-p a-Si:H solar cell grown on the patterned back contact
(*Source*: GCEP Annual Report, 2010, reproduced with permission)

The different layers can be clearly identified, and the holes are conformably coated with Ag and ZnO (Polman & Catchpole 2010).

The J-V measurements have shown short circuit current densities exceeding that of planar cells by 26% due to near-field coupling to waveguide modes of the a-Si:H layer, the fill factor increasing by 11%, from 0.55 to 0.61, and V_{oc} showing a slight decrease, by 2%. Combined, there is a significant increase in efficiency from 4.5% to 6.2% due to the patterned metal back contact. Though a-Si was used as a test bed for photonic nanopattern design, the approach is broadly applicable to other thin-film solar cell material systems. Spectral response measurements have shown that the enhance photocurrent is due to an increased absorption of the infrared portion of the solar spectrum, clearly demonstrating the light trapping effect of the nanostructured metal back contact (Polman & Catchpole 2010).

Brongersma and his group (2010) have studied the ability of aperiodic, nanometallic (i.e., plasmonic) nanostructures for light trapping because aperiodic structures enable more broadband coupling by providing more spatial frequencies (i.e., k-vectors) for coupling. To experimentally verify the anticipated performance improvement of their proposed light trapping layers, a rapid prototyping platform based on silicon-on-insulator (SOI) technology is developed and initial tests are performed on single structure and periodic metallic nanostructure arrays. Figure 6.20(a) shows a schematic cross section of the platform and its operation. The thin Si layer in SOI plays the role of the active absorbing layer of the solar cell. SOI wafers are readily available and allow a performance evaluation without the need to realize a complete PV cell. By lithographic means, hundreds to thousands of test devices can be generated on a single wafer. Schottky contacts or lateral pn-junctions have been utilized for efficient carrier extraction. Figure 6.20(b) shows an example of a photolithography mask developed for this purpose. Each mask consists of 100 dies per wafer and there will be 162 devices per die.

Different light trapping structures can be generated on each die by using focused ion beam milling, electron-beam lithography, or by chemical means. Photocurrent measurements are performed as a function of wavelength using a white light source coupled to a monochromator which have enabled to assess the spectral dependence of the photocurrent enhancement. The ability to effectively separate photo-generated carriers depends on the absorber microstructure, metallic nanostructure size, shape,

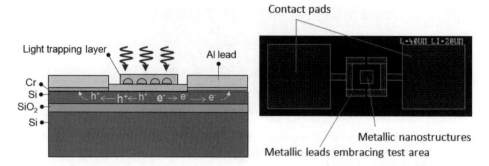

Figure 6.20 (a) Schematic of rapid prototyping. (b) Photolithography mask of a Platform based on SOI technology test structure
(*Source*: Brongersma *et al.*, GCEP Report, reproduced with permission of Prof. Brongersma)

spacing and interfaces. Scanning Electron Microscopy and Atomic Force Microscopy are used to characterize the size, distribution and surface morphology of nanostructured light trapping layers. These structural features are correlated to the spectral dependence of the absorption enhancements.

The results of these studies along with the Rapid prototyping Platform are recently published in Nature Nanotechnology (Barnard, Pala, & Brongersma, September 2011).

The detailed research results of these three research groups and their other collaborators are published recently as 'Plasmonic photovoltaics – GCEP Progress Report May 1, 2010 – April 28, 2011'. Several research publications that have appeared during 2007 – 2010 on the development of these solar cells of great potential are also listed.

6.4.3 Nanostructured materials for thin film solar cells

It is already seen that Multi junction solar cells achieve extremely high energy conversion efficiency compared to single-junction solar cells by reducing the thermodynamic losses associated with the absorption of photons with energy larger than the bandgap of the active layers thereby making a more efficient use of the whole solar spectrum. Triple-junction cells have theoretical efficiencies close to 50% and to date, III-V triple-junction laboratory cells have reached record efficiencies exceeding 40%. But, this technology as on today may have limited market demand because of high fabrication costs due to the low-rate and high-vacuum deposition processes used.

This limitation is addressed in this project undertaken by Salleo, Cui and Peumans (Stanford), by proposing a multijunction device concept where all layers can be sequentially deposited using *solution-based* processing techniques. The resulting structure would thus be free of complex lattice-matching requirements and compatible with low-cost and large-volume roll-to-roll fabrication technologies. Colloidal nanocrystals and nanowires are the preferred materials. Optical and electric properties of nanocrystalline semiconductors can be tuned by choosing the right composition and size to match the requirements of all subcells in the multijunction. Metal or

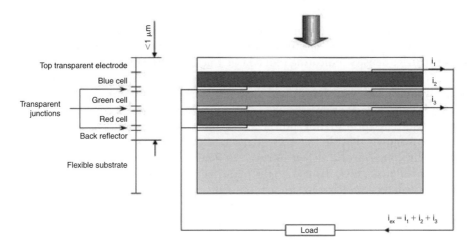

Figure 6.21 Schematic of the proposed multijunction thin-film (<1 μm thick) Photovoltaic. Individual cells are separated by thin transparent contact junctions for lateral photocurrent extraction
(*Source*: Stanford University GCEP Report, September 2007 – August 2011, reproduced with permission)

metal-oxide nanowires can be used to build highly conductive transparent electrodes to be inserted between each subcell. This design allows the lateral extraction of the photocurrent at each junction (Figure 6.21), thereby making each sub-cell independent and decoupled from the underlying structure. The current matching requirement of series-connected multijunctions is thus eliminated and the subcells can be deposited and optimized independently. A variety of colloidal semiconductor nanocrystals are investigated to fabricate the active layers of the multijunction device.

This research activity is based on recent advances that charge separation in nanocrystal solar cells is not driven by the presence of a built-in electric field as in the case of conventional thin film cells but rather requires a donor-acceptor heterojunction (similar to organic photovoltaics) where charge separation results from charge diffusion directed by different energy levels across the junction. Each cell, therefore, consists of a nanocrystal bilayer – forming the heterojunction – with bandgap and energy levels that must be carefully engineered for each separate cell by controlling quantum-confinement effects and material composition. Individual cells are separated by highly conductive transparent electrodes that serve as electric junctions and allow for lateral charge removal. This parallel connection design implies that the total photocurrent results from the addition of the current produced by the individual cells so that they can be optimized independently since no current matching is required.

Metal and highly doped metal-oxide nanowires are studied to develop transparent electrodes suitable for solution-processing and compatible to large-area fabrication. Early results have illustrated that solution-synthesized silver nanowire meshes can compete with traditional ITO electrodes. Optical and electrical network models are used to explore the limits of optical transmission and sheet resistance of random nanowire meshes and to optimize the wire coverage and aspect ratio. Highly-doped

ZnO films have the potential to better the state-of-the-art ITO. Hence, highly doped ZnO nanowires are synthesized and carrier density and mobility in the wires are measured. The carrier densities are of the order of 10^{20} cm^{-3} and mobilities of the order of 20–40 cm^2/V.s. Dopant incorporation, however, is up to one order of magnitude higher than the carrier density indicating that a large fraction of the dopants (Al or Ga) is not in substitution sites. In an effort to increase the dopant activation, the nanowires are thermally annealed, which has caused the free carriers to disappear. Anamolous X-ray diffraction experiments are performed and the effect of surfactant concentration (oleic acid) and solvent coordinating power are studied to understand this problem. The results have shown that decrease in carrier concentration upon annealing at 400°C is caused by creation of a defect intrinsic to ZnO with acceptor-like character. The formation of defects is confirmed by the photothermal deflection spectroscopy, which reveals the emergence of broad sub-gap absorption upon annealing.

Non-aqueous synthesis and doping processes are investigated and optimized to control the morphology and aspect ratio of the nanowires. Various techniques are being explored to further enhance the conductivity of both the metal and metal oxide-based films by aligning the nanowires during processing. A process to combine ZnO nanostructures and Ag nanowires to produce a transparent electrode with low sheet resistance and high haze are successfully developed. In order to deposit this new electrode, a spray-coating set-up capable of depositing materials over relatively large areas (10×10 cm^2) at high speed is built and optimal deposition conditions are determined.

In addition to nanowire synthesis, a new air-stable ink rolling (AIR) process has been developed to form CuInS$_2$ (CIS) absorber thin films. The process consist of three steps: (a) Ink deposition process using roller-bar; (b) Oxide bilayer formation after heating at 370°C in air; and (c) Film after sulfurization in sulfur vapor at 525°C. The initial cells show the short-circuit current, J$_{sc}$ (18.49 mA/cm^2) comparable to high efficiency CuInS$_2$ solar cells, but power conversion efficiency (2.15%), FF, and V$_{oc}$ are low. This is most likely due to the low shunt resistance which might be due to film cracking induced by mechanical stresses created during sulfurization, a chemical transformation involving structural and volumetric changes. Further development of AIR process and integration with other materials are planned on CIS solar cells.

Simulation tools to optimize the design of the multijunction and all processing parameters (layer number, thickness, and morphology, band-edges and work-function position) need to be developed to study a variety of fabrication and operation conditions. For more details, the report on 'High efficiency, low cost TF solar cells' and other publications may be consulted.

The fabrication of MJ solar cells using solution processes, integrating ZnO/Ag electrodes in solar cells, optimization of device design are the challenges before the research groups.

Lateral nanoconcentrator nanowire multijunction PV cells

The objective of this research project undertaken by Profs. Wong, Peumans, Brongersma, and Nishi at Stanford University is to develop a novel type of multijunction solar cell that uses lateral arrays of semiconductor nanowires of various bandgaps as the elements that convert optical energy into electrical energy. In contrast to conventional stacking multijunction cells, the NWs of varying bandgap will not be connected in series. Instead, a specially-designed nanostructured metal film is used to split the incident broadband solar spectrum and localize spectral energy in different lateral spatial locations (spectral splitting and concentration) coinciding with the location of the NWs of the optimized bandgap. The same nanostructured metal film also allows for current extraction from each nanowire separately such that photocurrent matching is not required. This approach allows to use a wide range of bandgaps (depending on the performance of the lateral metal spectral splitter and concentrator) without requiring current matching. It also allows wide choice of materials which can approach the ideal performance limits due to the less spectral mismatch losses. This removes the most important efficiency bottleneck of multijunction cells such that efficiencies >45% may be achieved over a wide range of spectral conditions. The NWs can be grown by the vapor-liquid-solid (VLS) method and the sol-gel approach, both of which have the potential for low cost manufacturing since epitaxial growth conditions are easily met at moderate temperatures over the short length scale of a NW. To demonstrate the lateral multijunction principle, NWs of materials that span the solar spectrum, including Si, Ge, III-V materials (e.g., GaP) and other abundant, non-toxic, low-cost elements such as TiO_2 are used.

The details of this research can be found from GCEP Progress Reports 2009 & 2011.

6.5 CRYSTALLINE SILICON ON GLASS (CSG) SOLAR CELLS

The silicon wafer based technology has limitations in reaching very low costs required for large-scale photovoltaic applications; similarly, a few fundamental difficulties exist with thin-film technologies for large-scale production and utilization as will be seen in later pages.

Crystalline silicon on glass (CSG) solar cell technology involves deposition of a polycrystalline silicon layer on glass which requires small quantity of material. CSG technology was developed combining the advantages of standard silicon wafer-based technology such as durability, good electronic properties and environmental reliability with the advantages of thin-films, specifically low material use, large monolithic structure and a desirable glass superstrate configuration. The challenge in the development process, however, is to match the different preferred processing temperatures of silicon and glass, and to obtain strong solar absorption in low-absorbing silicon of around 1.4 μm thickness, the thinnest active layer of the prominant thin-film technologies. A durable silicon thin-film technology with a possible lowest likely manufacturing cost of these competing technologies is now available. The confirmed efficiency of 8–9% for small pilot line modules is achieved, which is moving on to 12–13% range.

Figure 6.22(a) shows a schematic of the crystalline silicon on glass structure developed at UNSW with its features (Green et al 2004). The CSG structure is a thin layer of silicon deposited onto borosilicate glass coated with an antireflection coating of silicon nitride and a layer of half-micron glass beads (not shown) that give the glass the textured surface necessary for light trapping. The silicon is deposited in amorphous form using PECVD, followed by solid-phase crystallisation, rapid thermal annealing, and hydrogen passivation.

The thin-film research group at UNSW has focussed on three different solar cell types: cells deposited by PECVD (PLASMA cells); by e-beam evaporation (EVA cells); and on the AIC (aluminium induced crystallization) seed layers (ALICE cells).

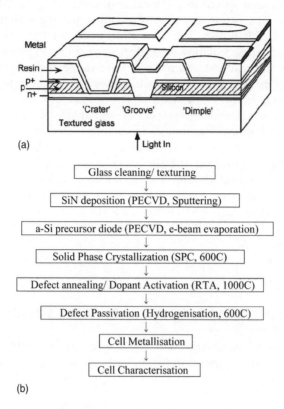

(a)

(b)

Figure 6.22 (a) Schematic representation of CSG solar cell (*Source*: Green *et al.* 2004, reproduced with permission of Prof. Martin Green, Copyright © 2004 Elsevier). (b) Process sequence of the four types of poly-Si thin-film on glass solar cells under development in UNSW (ARC-UNSW Annual report 2008)

Figure 6.22(b) shows the fabrication sequence followed at UNSW. The first working cells were produced in collaboration with CSG Solar AG from Si films deposited in a newly commissioned e-beam evaporation system (MANTIS cells).

The substrates used are 3.3 mm thick planar borosilicate glass (Borofloat33). While EVA and ALICE cells are fabricated on planar glass, PLASMA cells are fabricated on textured glass. The texture on the glass aims to reduce reflection losses and enhance light trapping in the silicon film. The texture consisting of an irregular array of sub-micron sized dimples is prepared by AIT method (Chuangsuwanich *et al.* 2004; Widenborg *et al.* 2004). In the AIT method, a thin (~100 nm) Al film is deposited onto the glass, followed by thermal annealing at about 600°C for 30 min to allow Al to react with the glass surface, and subsequent wet-chemical etching of the reaction products. Then, SiN film (refractive index ~2.1 at 633 nm) of about 75nm thick is deposited at about 400°C glass temperature by PECVD (direct or remote plasma) or at 200°C by reactive sputtering. This SiN film acts both as an AR coating and a barrier layer for contaminants from the glass.

For EVA, MANTIS and PLASMA cells, an a-Si n^+pp^+ or p^+nn^+ precursor diode is then typically deposited by either EVA or PECVD on planar or AIT glass respectively, followed by SPC at 600°C for 25 hrs in a nitrogen atmosphere.

The process for ALICE cells involves a preparation of a thin (~100 nm) poly-Si seed layer on the SiN. The seed layer is formed by AIC of an intrinsic a-Si layer at about 500°C (AIC seed) deposited by dc magnetron sputtering or PECVD (Widenborg & Aberle 2002; Goldschmidt et al. 2003). Then, deposition of a pp^+ or nn^+ a-Si structure onto the heavily doped seed layer is performed, followed by crystallisation of the a-Si structure by SPE, and post-deposition treatments (rapid thermal annealing, hydrogenation, metallisation) similar to those for EVA cells. The RTA process uses a lamp-based system that rapidly heats the samples to the desired temperature (900–1000°C), maintains at that temperature for a certain time, and then lowers the temperature in a controlled way to values below 200°C. Hydrogenation is done at a maximum sample temperature in the range of 500–620°C, using a remote plasma tool.

The MANTIS cells were made using proprietary CSG technology described elsewhere (Basore 2006; Keevers et al. 2007) from the n^+pp^+ a-Si precursor diodes deposited by UNSW in the Mantis evaporator onto SiN coated planar glass.

The main advantage of EVA, MANTIS, and PLASMA cells is their simplicity, since no seed layer is used in the fabrication process. However, the grain size of the resulting poly-Si films is small (typically 1–2 μm) and hence their electronic properties (effective carrier lifetime, lateral conductivity, etc) are significantly affected by grain boundary effects. The seed layer ALICE solar cells offer an advantage that the properties of the seed layer and the remainder of the solar cell can be independently optimised. This allows the realisation of larger, up to 20 μm poly-Si grains and/or more heavily doped materials.

Metal contacting of the entire module is achieved in one step using a sputtered layer of aluminium that is insulated from the silicon by a layer of white resin. Contacts occur through vias that are defined in the resin layer using inkjet printing. The metal layer is scribed using a laser to create an interdigitated contacting pattern that monolithically interconnects the adjacent cells in series. This contacting scheme leaves 99% of the silicon area active for photogeneration (Basore 2003, 2006a).

The salient feature of the CSG fabrication approach is the small number of layers involved, and the simplicity arising from the separation of the active layer deposition and interconnection steps.

The high quality of light-trapping within the device provided by the combination of the textured glass surface and rear reflector enables the use of 1.4 μm very thin silicon layer.

Another unique feature made possible by the use of the silicon's conductance is the 'strip interconnects' scheme shown in more detail in Figure 6.23.

Each metal strip extends across only two adjacent cells and is isolated from all other strips by intervening silicon. Unlike in other thin-film technologies where a shunt would short-circuit metal to the conducting TCO seriously affecting the module performance, the effect of such a shunt in a CSG module will be isolated automatically to a small surrounding region by the resistance of the intervening silicon and have no detectable impact on the module performance. The several merits of CSG photovoltaic

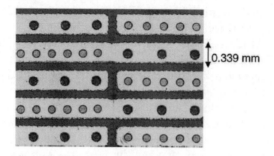

Figure 6.23 Fault-tolerant metallisation scheme showing inkjet defined 'dimple' and (darker) 'crater' contact holes. The 'groove' separating adjacent cells can also be discerned in the centre of the image perpendicular to the rows of dimples and craters
(*Source*: Green *et al.* 2004; Reprinted with permission of Prof. Martin Green, Copyright © 2004 Elsevier)

technology make it possibly the most promising thin-film solar cell option so far developed.

(i)   It requires minimal material because of very effective light-trapping. As the technology does not require a thick TCO layer to provide lateral conductance, it involves the least active material of all the traditional thin-films. Further, the key materials are silicon and its oxide and nitride rather than the complex and often metastable phases required by other thin-film technologies (Green 2003c).

(ii)   The device processing is simple and elegant. The entire active device structure is formed during a single deposition cycle, and a proprietary inkjet patterning approach is used for defining contact areas. This separation of the device deposition and interconnection stages is unique amongst thin-film approaches.

(iii) The ability to strike into the much larger active matrix display field, where deposition equipment costs reportedly are dropping at a rate of 21% per year (Buechel, 2002).

(iv)   There is rapid improvement in the performance by about 1% per year from the current efficiency of 8–9% towards a value that appears to lie in the 12–13% range. The commercial module on the market has the efficiency range near the top of present thin-films, whereas 'champion' research modules (Green 2003c) and the likely performance of delivered product (Green 2003a) is higher than most wafer-based modules.

The greatest advantage, however, lies in its exceptional durability. It appears that the technology not only can undercut the present wafer based standard in costs, but offers new levels of stability by avoiding common failure modes in conventional modules such as wafer breakage, fatigue of the metal interconnect strips joining the cells, discolouring or delaminating of encapsulants, or degradation of the solder joints between interconnects and cell metallization (Green *et al.* 2004).

On alumina substrates, IMEC, Belgium, using the AIC approach, has the world's best p-type cell efficiency of 8% (Gordon *et al.* 2008). To further improve poly-Si cell efficiency, the material quality of the seed layer (van Gestel 2007) needs to be improved

Figure 6.24 CSG module produced by CSG Solar, Germany
(*Source*: USNW Annual Report 2009, reproduced with permission of Prof. Martin Green)

which requires a deeper understanding of the AIC mechanism. On the other hand, the p-type seed layer can be converted to an n-type layer by an overdoping process, making it possible to fabricate efficient (p) a-/(n) poly-Si heterojunction structures, which can offer higher open-circuit voltages compared with the standard (n) a-/(p) poly-Si structure (Jensen *et al.* 2002). Furthermore, a higher minority carrier lifetime and a higher tolerance towards metal impurities (Macdonald and Geerligs 2004) are also expected in n-type poly-Si. This research group has recently studied n-type poly-Si solar cells made by overdoping of AIC seed layers with solid phosphorus spin-on dopant followed by epitaxial thickening with low-pressure chemical vapor deposition (Qiu *et al.* 2009). These cells recorded an efficiency of 5.0% which is worldwide the best efficiency so far for n-type poly-Si solar cells based on the AIC approach. The cell has the following 1-Sun parameters: $J_{sc} = 16.8 \text{ mA/cm}^2$, $V_{oc} = 462 \text{ mV}$, $FF = 0.645$, Efficiency $= 5.0\%$. These cell parameters are reached without doping profile control, optimized emitter, optimized hydrogen passivation and light trapping, all of which are expected to be addressed in their future research.

6.5.1 Production of CSG modules

The highest solar conversion efficiency for CSG that has been independently validated is 9.8% for a mini module masked to an aperture area of 96 cm^2 (Basore 2006a).

In the late 1990s, Pacific Solar Pvt Ltd, a spin-off company of UNSW, successfully transferred the PECVD-based SPC approach to borosilicate glass sheets (Borofloat33 from Schott AG, Germany). Major breakthroughs have been achieved at Pacific Solar in the areas of light trapping (novel glass texture, Ji & Shi 2002) and cell metallisation and interconnection (point contacts) (Basore 2006a, 2002).

The best efficiency obtained so far with the CSG technology is 10.4%, with a 94-cm^2, 20-cell mini-module with a fill factor of 72.1%, J_{sc} of 29.5 mA/cm^2, and an average cell V_{oc} of 492 mV (Keevers 2007). The Jsc is remarkably high for a silicon film thickness of merely 2.2 μm, confirming that CSG devices feature excellent light trapping properties (Aberle 2009).

Another major manufacturer is CSG Solar AG, started in 2004 in Germany (Basore 2006b), with a rated capacity of 10 MW$_p$/year. Silicon deposition is conducted in a KAI-1200 PECVD unit from Oerlikon Solar, using 1.4 m^2 glass sheets. The large-area CSG modules were available in late 2006, and by mid-2007, the module efficiencies were in the 6–7% range and improving steadily (Keevers et al. 2007). Typical CSG module is shown in Figure 6.24. A second KAI-1200 machine has been installed in 2007, doubling the rated factory capacity to 20 MWp/year. For a factory producing 20 MWp/year of 8% efficient CSG modules, the expected module fabrication costs are about 120 €/m2 or, equivalently, about 1.50 €/Wp (Basore 2004).

REFERENCES

Aberle, A.G. (2009): Thin film Solar Cells, *Thin Solid Films*, **517**, 4706–4710.

Aliberti, P., Feng, Y., Takeda, Y., Shrestha, S.K., Green, M.A., & Conibeer, G. (2010a): Investigation of theoretical efficiency limit of hot carriers solar cells with a bulk indium nitride absorber, *Journal of Applied Physics*, **108**, 094507.

Aliberti, P., Shrestha, S.K., Teuscher, R., Zhang, B., Green, M.A., & Conibeer, G. (2010b): Study of Silicon Quantum Dots in a SiO2 Matrix for Energy Selective Contacts Applications, *Sol. Energy Mater. and Sol. Cells*, **94**, 1936–1941.

Andrä, G., Pietsch, M., Stelzner, T., Falk, F., Ose, E., Christiansen, S., Scheffel, A., & Grimet, S. (2007): Silicon Nanowires for Thin Film Solar Cells, *Proc. 22nd EU PV Solar energy Conf.* Milan, Italy, pp. 481–483.

Arango, A.C., Johnson, L.R., Bliznyuk, V.N., Schlesinger, Z., Carter, S.A., & Hörhold, H.H. (2000): Efficient titanium oxide/conjugated polymer photovoltaics for solar energy conversion, *Adv. Mater.* **12**, 1689–1692.

ARC, Photovoltaic Center of Excellence, UNSW, Annual Report 2008, available at UNSW_Photovoltaics_An_Report_2008.pdf

ARC, Photovoltaic Center of Excellence, UNSW, Annual report 2009, available at ARC_PV_Anrep_UNSW_2009.pdf

Arici, E., Hoppe, H., Schäffler, F., Meissner, D., Malik, M.A., & Sariciftci, N.S. (2004): Hybrid solar cells based on inorganic nanoclusters and semiconductive polymers, *Thin Solid Films*, **451–452**, 612.

Atwater, H.A. & Polman, A. (2010): Plasmonics for Improved Photovoltaic Devices, *Nature Materials*, **9**, 205–213.

Atwater, H.A. (2010): Plasmonic photovoltaics, Kavli Award Lecture, *MRS Spring Meeting*, San Fransisco, CA, March 2010.

Atwater, H.A. (2010): Plasmonic Photovoltaics (Invited lecture), *American Physical Society March Meeting*, Portland, OR, March 19, 2010.

Atwater, H.A., Brongersma, M., & Polman, A.: Stanford University GCEP report, September 2007 – August 2011.

Bailey, C.G., Hubbard, S.M., Forbes, D.V., & Raffaelle, R.P (2009): Evaluation of strain balancing layer thickness for InAs/GaAs quantum dot arrays using high resolution x ray diffraction and photoluminescence, *Appl. Phys. Lett.* **95**, 203110—203113.

Barnard, E.S., Pala, R.A., & Brongersma, M. (2011): Photocurrent mapping of near-field optical antenna resonances, *Nature Nanotechnology*, **6** (9), Published online, 21st August 2011, DOI:10.1038/NNA No.2011.131.

Barnham, K.W.J., & Duggan, G. (1990): A new approach to high-efficiency multi-bandgap solar cells, *J. Appl. Phys*, **67**, 3490–3493.

Basore, P.A. (2002): Pilot production of thin-film crystalline silicon on glass modules, *Proc. 29th IEEE PV Specialists Conf.*, May 2002, New Orleans, LA. pp. 49–52.

Basore, P.A. (2003): Large-area deposition for crystalline silicon on glass modules *Proc. 3rd WC on PV Energy Conversion*, Osaka, Japan, May 2003, pp. 935–938.

Basore, P.A. (2004): Simplified processing and improved efficiency of crystalline silicon on glass modules, *Proc. 19th EU PV Solar Energy Conf.*, Paris, France, June 2004, p. 45.

Basore, P.A. (2006a): CSG-2: Expanding the production of a new polycrystalline silicon PV technology, *Proc. 21st EU PV Solar Energy Conf.*, September 2006, Dresden, Germany, p. 544.

Basore, P.A. (2006b): CSG-1: Manufacturing a New Polycrystalline Silicon PV Technology, *Proc. 4th WC on PV Energy Conversion*, Waikoloa, HI, May 2006, pp. 2089–2093.

Baxter, J.B., & Aydil, E.S. (2005): Nano-wire based dye-sensitized solar cells, *Appl. Phys. Lett.* **86**, 053114.

Baxter, J.B., Walker, A.M., van Ommering, K., & Aydil, E.S. (2006): Synthesis and characterization of ZnO nanowires and their integration into dye-sensitized solar cells, *Nanotechnology*, **17**, S304.

Beek, W.J.E., Wienk, M., & Janssen, R.A.J. (2004): Efficient hybrid solar cells from zinc oxide nanoparticles and a conjugated polymer, *Adv. Mater.* **16**, 1009–1013.

Beek, W.J.E., Wienk, M., & Janssen, R.A.J. (2006): Hybrid solar cells from regioregular polythiophene and ZnO nanoparticles, *Adv. Funct. Mater.* **16**, 1112–1116.

Boettcher, S.W., Spurgeon, J.M., Putnam, M.C., Warren, E.L., Turner-Evans, D.B., Kelzenberg, M.D., Maiolo, J.R., Atwater, H.A., & Lewis, N.S. (2010): Energy-conversion properties of vapor–liquid–solid-grown silicon wire-array photocathodes, *Science*, **327**, 185–187.

Bozano, L., Carter, S.A., Scott, J.C., Malliaras, J.J., & Brock, P.J. (1999): Temperature- and field-dependent electron and hole mobilities in polymer light-emitting diodes, *Appl. Phys. Lett.* **74**, 1132–1134.

Brongersma, M.L. & Kik, P.G. (Eds), Surface Plasmon Nanophotonics, Series: *Springer Series in Optical Sciences*, 2007, **131**, VII, 271.

Brongersma, M.L., Atwater, H.A., & Polman, A. (2011): Plasmonic Photovoltaics, *GCEP Progress Report*, Available at 2.2.8_Brongersma_Web_Public_2010PlasmonicPV.pdf

Buechel, A.R. (2002): KAI PECVD systems in AM-LCD: Cross fertilization from displays to solar, *PV NET Workshop Proceedings*, JRC, ISPRA, pp. 120–134.

Cao, H., Wang, G., Zhang, S., Zhang, X., & Rabinovich, D. (2006): Growth and optical properties of wurtzite-type CdS nanocrystals, *Inorg. Chem*, **45**, 5103–5108.

Catchpole, K. (2008): Plasmonic Photovoltaics (Invited lecture), *MRS Spring meeting*, March 2008, San Fransisco, CA.

Catchpole, K. & Polman, A. (2008): Design Principles for particle Plasmon enhanced Solar cells, *Appl. Phys. Lett.* **93**, 191113.

Chadi, D.J. (1994): The problem of doping in II–VI semiconductors, *Annu. Rev. Mater. Sci*, **24**, 45–62.

Chang, J.F., Sun, B., Breiby, D.W., Nielsen, M.M., Sölling, T.I., Giles, M., McCulloch, I., & Sirringhaus, H. (2004): Enhanced mobility of poly(3-hexylthiophene) transistors by spin-coating from high-boiling-point solvents, *Chem. Mater.* **16**, 4772–4776.

Cho, Y.H., Green, M.A., Cho, E.-C., Huang, Y., Trupke, T., & Conibeer, G. (2005): Silicon quantum dots in SiNx matrix for third-generation photovoltaics, *Proc. 20th EU PV Solar Energy Conf*, June 2005, Barcelona, Spain, pp. 47.

Cho, E.C., Park, S., Hao, X., Song, D., Conibeer, G., Park, S.C., & Green, M.A. (2008): Silicon Quantum Dots/Crystalline Silicon Solar Cells, *Nanotechnology* **19**, 245201.

Chuangsuwanich, N., Widenborg, P.I., Campbell, P., & Aberle, A.G. (2004): Light Trapping Properties of Thin Silicon Films on AIT-Textured Glass, *Tech. Digest 14th International Photo voltaic Science and Engineering Conference*, Bangkok, p. 325.

Clady, R., Tayebjee, M.J.Y., Aliberti, P., Konig, D., Ekins-Daukes, N.J., Conibeer, G., Schmidt, T.W., & Green, M.A. (2011): Interplay between Hot phonon effect and Interval-ley scattering on the cooling rate of hot carriers in GaAs and InP, *Progress in Photovoltaics*, March 2011.

Colombo, C., Heibeta, M., Gratzel, M., & Fontcuberta i Morral, A. (2009): Gallium arsenide p-i-n radial structures for photovoltaic applications, *Appl. Phys. Lett.* **94**, 173108.

Conibeer, G., Green, M., Cho, E.-C., Konig, D., Huang, S., Song, D., Hao, X., Perez-Wurfl, I., Gentle, A., Gao, F., Park, S., Flynn, C., So, Y., Zhang, B., Di, D., Campbell, P., Huang, Y., & Puzzer, T. (2009): *Proc. EU Mater. Res. Society Spring Meet*, BP2-10.

Conibeer, G., Patterson, R., Huang, L., Guillemoles, J.-F., Konig, D., Shrestha, S., & Green, M.A. (2010): Modelling of Hot carrier Solar cell absorbers, *Sol. Energy Mater. Sol. Cells*, **94**, 1816–1821.

Conibeer, G., Green, M.A., Guillemoles, J.-F., Schmidt, T., Marti, A., & Luque, A. (2011): Project: Hot Carrier solar cell: Implementation of the Ultimate PV Converter, Annual Report April 2011, Available at 2.2.5_Conibeer_Public_2011hotcarriercells.pdf

Conibeer, G., Green, M.A., Marti, A., Luque, A., Guillemoles, J.-F., & Schmidt, T.: Stanford University GCEP Report, September 2008 – August 2011.

Cooke, Mike (2011): Built-in charge boosts QD Solar cell efficiency, News item in *Semiconduc-torToday*, May 2011.

Cress, C.D., Hubbard, S.M., Landi, B.J., Raffaelle, R.P., & Wilt, D.M. (2007): Quan-tum dot solar cell tolerance to alpha-particle irradiation, *Appl. Phys. Lett.* **91**, 183108, doi:10.1063/1.2803854.

Cui, Y., Lauhon, L.J., Gudiksen, M.S., Wang, J.F., & Lieber, C.M. (2001): Diameter-controlled synthesis of single-crystal silicon nanowires, *Appl. Phys. Lett.* **78**, 2214–2216.

Cui, Y., Wei, Q., Park, H., & Lieber, C.M. (2001): Nanowire nano-sensors for highly sensitive and selective detection of biological and chemical species, *Science*, **293**, 1289–1292.

Cuevas, A., Samundsett, C., Kerr, M.J., Macdonald, D.H., Mäckel, H., & Altermatt, P.P. (2003): Back junction solar cells on n-type multicrystalline and Cz-Si wafers, *Proc. 3rd WC on PV Energy Conversion*, May 2003, Okasa, Japan, pp. 913–918.

Czaban, J.A., Thompson, D.A., & LaPierre, R.R. (2008): GaAs core – shell nanowires for photovoltaic applications, *Nano Lett.* **9**, 148–154.

Dong, A., Wang, F., Daulton, T.L., & Buhro, W.E. (2007): Solution – liquid – solid (SLS) growth of ZnSe – ZnTe quantum wires having axial heterojunctions, *Nano Lett.* **7**, 1308–1313.

Dong, Y., Tian, T.J. Kempa, & C.M. Lieber (2009): Coaxial group III nitride nanowire photovoltaics, *Nano Lett.* **9**, 2183–2189.

Desnica, U.V. (1998): Doping limits in II–VI compounds—Challenges, problems and solutions, *Prog. Cryst. Growth Charact. Mater*, **36**, 291–357.

Dubrovski, V.G., Sibirev, N.V., Cirlin, G.E., Soshnikov, I.P., Chen, W.H., Larde, R., Cadel, E., Pareige, P., Xu, T., Grandidier, B., Nys, J.P., Stievenard, D., Moewe, M., Chuang, L.C. & Chang-Hasnain, C. (2009): Gibbs-Thomson and diffusion-induced contributions to the growth rate of Si, InP, and GaAs nanowires, *Phy. Rev, B: Condensed Matter*, **79**, 205316–7.

Egan, R.J., Young, T.L., Evans, R., Schubert, U., Keevers, M., Basore, P.A., Wenham, S.R., & Green, M.A (2006): Silicon deposition optimization for peak efficiency of CSG modules, *Proc. 21st EU PV Solar Energy Conf*, September 2006, Dresden, Germany, 2CV.3.33, pp. 874–876.

Fanfair, D.D. & Korgel, B. (2007): Twin-related branching of solution grown ZnSe Nanowires, *Chem. Mater.*, **17**, 4416–4425.

Fang, H., Li, X., Song, S., Xu, Y., & Zhu, J. (2008): Fabrication of slantingly aligned silicon nanowire arrays for solar cell applications, *Nanotechnology*, **19**, 255703–6.

Feng, X.J., Shankar, K., Varghese, O.K., Poulose, M., Latempa, T.J., & Grimes, C.A. (2008): Vertically aligned single crystal TiO$_2$ nanowire arrays grown directly on transparent conducting oxide coated glass: synthesis details and applications, *Nano. Lett.*, 8, 3781.

Garnett, E., & Yang, P. (2008): Silicon nanowire radial p-n junction solar cells, *J. Amer. Chem. Soc.* 130, 9224–9225.

Garnett, E. & Yang, P. (2010): Light trapping in Silicon nanowire solar cells, *Nano Lett.*, 10, pp. 1082–1087.

Goldschmidt, J.C., Roth, K., Chuangsuwanich, N., Sproul, A., Vogl, A., & Aberle, A.G. (2003): Electrical and optical properties of polycrystalline silicon seed layer made on glass by solid phase crystallisation, *Proc. 3rd WC on Photo voltaic Energy Conversion*, May 2003, Osaka, Japan, p. 1206.

Goto, H., Nosaki, K., Tomioka, K., Hara, S., Hiruma, K., Motohisa, J., & Fukui, T. (2009): Growth of core–shell InP nanowires for photovoltaic application by selective-area metal organic vapor phase epitaxy, *Appl. Phys. Expr.* 2, 035004.

Green, M.A. (2003b): *Third Generation PV: Advanced Solar Energy Conversion*, Berlin, Springer.

Green, M.A. (2003a): Crystalline and thin-film silicon solar cells: state of the art and future potential, *Solar Energy*, 74, 181–192.

Green, M.A. (2003c): Thin-film Photovoltaics, In: Goswamy, D.Y. (Ed.) '*Advances in Solar energy – Annual review of R&D*', American Solar Energy Society, New York, pp. 187–214.

Green, M.A., Basore, P.A., Chang, N., Clugston, D., Egan, R., Evans, R., Hogg, D., Jarnason, S., Keevers, M., Lasswell, P., Sullivan, J.O., Schubert, U., Turner, A., Wenham, S.R. & Young, T. (2004): Crystalline Si on glass (CSG) thin film solar cell modules, *Solar energy*, 77, 857–863.

Green, M.A., Cho, E.C., Huang, Y., Pink, E., Trupke, T., & Lin, A., Fangsuwannarak, T., Puzzer, T., Conibeer, G., & Corkish, R. (2005): All-Si tandem cells based on 'artificial' semiconductor synthesized using Si quantum dots in a dielectric matrix, *Proc. 20th EU PV Solar Energy Conf*, June 2005, Barcelona, Spain, pp. 3–7.

Green, M.A., Conibeer, G., König, D., Shrestha, S., Huang, S., Aliberti, P., Treiber, L., Patterson, R., Veettil, B., Hsieh, A., Luque, A., Marti, A., Linares, P., Cánovas, E., Antolín, E., Fuertes Marrón, D., Tablero, C., Hernández, E., Guillemoles, J.-F., Huang, L., Schmidt, T., Clady, R., & Tayebjee, M. (2010a): Recent Progress with Hot Carrier Solar Cells, *25th EU PV Solar EnergyConf*, Valencia, Spain, Sept. 2010.

Green, M.A., Conibeer, G., König, D., Shrestha, S., Huang, S., Aliberti, P., Treiber, L., Patterson, R., Puthen-Veettil, B., Hsieh, A., Feng, Y., Luque, A., Marti, A., Linares, P., Cánovas, E., Antolin, E., Fuertes Marrón, D., Tablero, C., Hernandez, E., Guillemoles, J.-F., Huang, L., Le Bris, A., Schmidt, T., Clady, R., & Tayebjee, M. (2010b): Hot Carrier Solar Cells: Challenges and Recent Progress, *35th IEEE PV Specialists Conf*, Honolulu, HI, June 2010.

Greene, L.E., Law, M., Yuhas, B.D., & Yang, P. (2007): ZnO-TiO$_2$ core–shell nanorod/P3HT solar cells, *J. Phys. Chem. C*, 111, 18451–18456.

Greenham, N.C., Peng, X., & Alivisatos, A.P. (1996): Charge separation and transport in conjugated polymer/semiconductor-nanocrystal composites, *Phys. Rev. B*, 54, 17628–17637.

Gunawan, O. & Guha, S. (2009): Characteristics of vapor–liquid–solid grown silicon nanowire solar cells, *Sol. Energy Mater. Sol. Cells*, 93, 1388.

Guo, Y., Zhang, Y., Liu, H., Lai, S.-W., Li, Y., Li., Y., Hu, W., Wang, S., Che, C.-M., & Zhu, D. (2009): Assembled organic/inorganic p-n junction interface and photovoltaic cell on a single nanowire, *J. Phys. Chem. Lett.* 1, 327–330.

Gur, I., Fromer, N.A., Geier, M.L., & Alivisatos, A.P. (2005): Air-stable All-inorganic nao-crystal solar cells processed from Solution, *Science*, 310, 462–465.

Hao, Y., Pei, J., Wei, Y., Cao, Y., Jiao, S., Zhu, F., Li, J., & Xu, D. (2010): Efficient semiconductor-sensitized solar cells based on poly(3-hexylthio-phene) @CdSe @ZnO core – shell nanorod arrays, *J. Phys. Chem. C*, **114**, 8622–8625.

Hochbaum, A.I. & Yang, P. (2010): Semiconductor nanowires for Energy conversion, *Chem. Rev.*, **110**, 527–546.

Hsueh, T.J., Chang, S.J., Hsu, C.L., Lin, Y.R., & Chen, I.C. (2007): Highly sensitive ZnO nanowire ethanol sensor with Pd adsorption, *Appl. Phys. Lett.* **91**, p. 053111.

Hu, M.S., Chen, H.-L., Shen, C.-H., Hong, L.-S., Huang, B.-R., Chen, K.H., & Chen, L.-C. (2006): Photosensitive gold-nanoparticle-embedded dielectric nanowires, *Nature Materials*, **5**, 102–106.

Hu, L., & Chen, G. (2007): Analysis of Optical Absorption in Silicon Nanorwire Arrays for Photovoltaic Applications, *Nano Lett.*, **7**, 3249.

Huang, C., Yang, Y.J., Chen, J.-Y., Wang, C.-H., Chen, Y.-F., Hong, L.S., Liu, C.-S., & Wu, C.-Y. (2010): p-Si nanowires/SiO2/n-ZnO heterojunction photodiodes, *Appl. Phys. Lett.* **97**, 013503-3.

Hubbard, S.M., Cress, C.D., Bailey, C.G., & Raffaelle, R.P. (2008): Effect of strain compensation on quantum dot enhanced GaAs solar cells, *Appl. Phys. Lett.* **92**, 123512, doi:10.1063/1.2903699.

Hubbard, S.M., Bailey, C., Polly, S., Cress, C., Andersen, J., Forbes, D., & Raffaelle, R. (2009a): Nanostructured photovoltaics for space power, *J. Nanophoton*, **3**, 031880–031816.

Hubbard, S.M., Bailey, C., Polly, S., Aguinaldo, R., Forbes, D., & Raffaelle, R. (2009b): Characterization of quantum dot enhanced solar cells for concentrator PV, *Proc. 34th IEEE PV Specialists Conf.* Philadelphia, PA, June 2009, 1, pp. 1–6.

Hubbard, S.M. & Raffaelle, R. (2010): Boosting solar cell efficiency with QD-based nanotechnology, SPIE Digital library, Spie News, Feb. 2010, at http??:spie.org/x39022.xml?ArticleID= x39022.

Hull, K.L., Grebinski, J.W., Kosel, T.H., & Kuno, M. (2005): Induced branching in confined PbSe nanowires, *Chem. Mater.*, **17**, 4416–4425.

Huynh, W.U., Dittmer, J.J., & Alivisatos, A.P. (2002): Hybrid nanorod polymer solar cells, *Science*, **295**, 2425.

Jensen, N., Hausner, R.M., Bergmann, R.B., Werner, J.H., & Rau, U. (2002): Optimization and Characterization of Amorphous/Crystalline Silicon Heterojunction Solar Cells, *Prog. Photovolt: Res. Appl.* **10**, 1–13.

Kanama, D. & Kawamoto, H. (2008): R & D trends of solar Cell for high Efficiency, *Quaterly Review*, No. 28, July 2008, available at STTQr2804R&Dforhighefficiencysolarcells 2008.pdf.

Kang, Y., Park, N.G., & Kim, D. (2005): Hybrid solar cells with vertically aligned CdTe nanorods and a conjugated polymer, *Appl. Phys. Lett.* **86**, 113101.

Kang, Y., & Kim, D. (2006): Well-aligned CdS nanorod/conjugated polymer solar cells, *Sol. Energy Mater. and Sol. Cells*, **90**, 166–174.

Kayes, B.M., Atwater, H.A. & Lewis, N.S. (2005): Comparison of the device physics principles of planar and radial p-n junction nanorod solar cells, *J. Appl. Phys*, **97**, 114302.

Keevers, M.J., Turner, A., Schubert, U., Basore, P.A., & Green, M.A. (2005): Remarkably effective hydrogenation of crystalline silicon on glass modules, *Proc. 20th EU PV Solar energy Conf*, June 2005, Barcelona, Spain, pp. 1305–1308.

Keevers, M.J., Young, T.L., Schubert, U. & Green, M.A. (2007): 10% Efficient CSG Minimodules, *Proc. 22nd EU PV Solar Energy Conf*, Sept. 2007, Milan, Italy, pp. 1783–1790.

Kelzenberg, M.D., Turner-Evans, D.B., Kayes, B.M., Filler, M.A., Putnam, M.C., Lewis, N.S., & Atwater, H.A. (2008): Photovoltaic measurements in single-nanowire silicon solar cells, *Nano Lett.* **8**, 710–714.

Kelzenberg, M.D., Putnam, M.C., Turner-Evans, D.B., Lewis, N.S., & Atwater, H.A. (2009): Predicted Efficiency of Si Wire Array Solar Cells, *Proc. 34th IEEE PV Specialists Conf*, Philadelphia, PA, June 7–12, 2009, pp. 391–396.

Kelzenberg, M.D., Boettcher, S.W., Petykiewicz, J.A., Turner-Evans, D.B., Putnam, M.C., Warren, E.L., Spurgeon, J.M., Briggs, R.M., Lewis, N.S., & Atwater, H.A. (2010): Enhanced absorption and carrier collection in Si wire arrays for photovoltaic applications, *Nature Materials*, 9, 239–244.

Kempa, T., Tian, J.B., Kim, D.R., Hu, J., Zheng, X., & Lieber, C.M. (2008): Single and tandem axial p-i-n nanowire photovoltaic devices, *Nano Lett.* 8, 3456–3460.

Kim, J.Y., Kim, S.H., Lee, H.H., Lee, K., Ma, W., Gong, H., & Heeger, A.J. (2006): New architecture for high-efficiency polymer photovoltaic cells using solution-based titanium oxide as an optical spacer, *Adv. Mater.* 18, 572–576.

Kobayashi, N.P., Wang, S.Y., Santori, C., & Williams, R.S. (2006): *Appl. Phys. A: Mater. Sci. Process*, 85, 1.

Konagai, M. (2011): Present status and Future prospects of Silicon thin-film solar cells, *Japanese J. Appl. Phys.*, 50, 030001, online March 20, 2011, doi:10.1143/JJAP.50.030001.

Konig, D., Casalenuovo, K., Takeda, Y., Conibeer, G., Guillemoles, J., Patterson, R., Huang, L.M., & Green, M.A. (2010): Hot Carrier Solar Cells: Principles, Design and Materials, *Physica E: Low Dimens. Syst. & Nanostructures*, 42, 2862.

Kurokawa, Y., Miyajima, S., Yamada, A., & Konagai, M. (2006): Preparation of Nanocrystalline Silicon in Amorphous Silicon Carbide Matrix, *Jpn. J. Appl. Phys.* 45, L1064–1066.

Kurokawa, Y., Tomita, S., Miyajima, S., Yamada, Y., & Konagai, M. (2008): Observation of the photovoltaics effect from the solar cells using silicon quantum dots superlattice as a light absorption layer, *Proc. 33rd IEEE PV Specialists Conf*, San Diego, Calif. May 2008, pp. 1–6.

Kwak, W.C., Kim, T.G., Lee, W., Han, S.-H., & Sung, Y.-M. (2009): Template free liquid-phase synthesis of high-density CdS nanowire arrays on conductive glass, *J. Phys. Chem. C*, 113, 1615–1619.

Kymakis, E., & Amaratunga, G.A.J. (2002): Single-wall carbon nanotube/conjugated polymer photovoltaic devices, *Appl. Phys. Lett.* 80, 112–114.

Laghumavarapu, R., El-Emawy, M., Nuntawong, N., Moscho, A., Lester, L.F., & Huffaker, D.L. (2007): Improved device performance of InAs/GaAs quantum dot solar cells with GaP strain compensation layers, *Appl. Phys. Lett.* 91, 243115.

Langford, R.F., Wang, T.X., Thornton, M., Heidelberg, A., Sheridan, J.G., Blau, W., & Leahy, R.J. (2006): Comparison of different methods to contact to nanowires, *J. Vac. Sci. Technol.*, 24, 2306–2311.

Law, M., Greene, L.E., Johnson, J.C., Saykally, R., & Yang, P. (2005): Nanowire dye-sensitized solar cells, *Nature Materials*, 4, 455.

Law, M., Greene, L.E., Radenovic, A., Kuykendall, T., Liphardt, J., & Yang, P. (2006): ZnO-Al_2O_3 and $ZnO-TiO_2$ Core – Shell Nanowire Dye-Sensitized Solar Cells, *J. Phys. Chem, B.* 110, 22652.

Lee, J.C., Lee, W., Han, S.-H., Kim, T.G., & Sung, Y.-M. (2009): Synthesis of hybrid solar cells using CdS nanowire array grown on conductive glass substrates, *Electrochem. Commun.*, 11, 231–234.

Lew, K.-K., Pan, L., Bogart, T.E., Dilts, S.M., Dickey, E.C., Redwing, J.M.Y., Wang, Y., Cabassi, M., Mayer, T.S., & Novak, S.W. (2004): Structural and electrical properties of trimethylboron-doped silicon nanowires, *Appl. Phys. Lett.* 85, 3101–3103.

Li, T.T., & Cuevas, A. (2009): Effective surface passivation of crystalline silicon by rf sputtered aluminum oxide, *Physica Status Solidi – Rapid Res. Letts*, 3, 160–162, Published online June 2009, DOI: 10.1002/pssr.200903140.

Li, Y., Xiang, J., Qian, F., Gradečak, S., Wu, Y., Yan, H., Blom, D.A., & Lieber, C.M. (2006): Dopant-free GaN/AlN/AlGaN radial nanowire heterostructures as high electron mobility transistors, *Nano Lett*, **6**, 1468–1473.

Liu, B., Boercker, J.E., & Aydil, E.S. (2008): Oriented single crystalline titanium dioxide nanowires, *Nanotechnology*, **19**, 505604.

Luque, A. & Marti, A. (1997): Increasing the Efficiency of Ideal Solar Cells by Photon Induced Transitions at Intermediate Levels, *Phys. Rev. Lett.* **78**, 5014–5017.

Ma, D.D.D., Lee, C.S., Au, F.C.K., Tong, S.Y., & Lee, S.T. (2003): *Science*, Small-diameter Silicon nanowire Surfaces, **299**, 1874–1877.

Ma, Y., Liu, F., Zhu, M., Liu, J., Yang, Y., & Li, Y. (2009): Microcrystalline silicon nanorods/ P3HT hybrid solar cells, *34th IEEE PV Specialists Conf*, Philadelphia, PA, June 7–12, 2009.

Macdonald, D., & Geerligs, L.J. (2004): Recombination Activity of Interstitial Iron and Other Transition Metal Point Defects in p- and n-Type Crystalline Silicon, *Appl. Phys. Lett.* **85**, 4061–4063.

Marti, A., Antolin, E., Stanley, C.R., Farmer, C.D., Lopez, N., Diaz, P., Canovas, E., Linares, P.G., & Luque, A. (2006): Production of photocurrent due to intermediate-to-conduction-band transitions: a demonstration of a key operating principle of the intermediate-band solar cell, *Phys. Rev. Lett.* **97**, 247701.

Miller, A.J., Hatton, R.A., & Ravi Silva, P.S. (2006): Interpenetrating multiwall carbon nanotube electrodes for organic solar cells, *Appl. Phys. Lett.* **89**, 133117–133119.

Novotny, C. & Yu, P.K.L. (2005): Vertically aligned, catalyst-free InP nanowires grown by metalorganic chemical vapor deposition, *Appl. Phys. Lett.* **87**, 203111–203113.

Novotny, C.J., Yu, E.T., & Yu, P.K.L. (2008): InP nanowire/polymer hybrid photodiode, *Nano Lett.* **8**, 775–779.

Okada, Y. (2006): Application to self assembler quantum dot super-grid and highly effective solar battery, *JACG Magazine*, **33**, No. 2 (Japanese).

Okada, Y., Oshima, R. & Ryoshi, J. (2007): The next generation solar battery that introduces quantum nano structure, *Appl. Phys*, **76**, No. 1 (2007) (Japanese).

Oshima, R., Nakamura, Y., Takata, A. & Okada, Y. (2007): Multi-stacked InAs/GaNAs Strain-compensated Quantum Dot Solar Cells, *Proc. 22nd EU PV Solar Energy Conversion*, September 2007, Milan, Italy.

Ouyang, L., Maher, K.N., Yu, C.L., McCarty, J., & Park, H. (2006): Catalyst-Assisted Solution-Liquid-Solid Synthesis of CdS/CdSe Nanorod Heterostructures, *J. Amer. Chem. Soc.*, **129**, 133–138.

Palomares, E., Clifford, J.N., Haque, S.A., Lutz, T., & Durrant, J.R. (2003): Control of charge recombination dynamics in dye sensitized solar cells by the use of conformally deposited metal oxide blocking layers, *J. Am. Chem. Soc.* **125**, 475.

Patterson, R., Kirkengen, M., Puthen Veettil, B., Konig, D., Green, M.A., & Conibeer, G. (2010): Phonon lifetimes in model quantum dot superlattice systems with applications to the hot carrier solar cell, *Sol. Energy Mater. and Sol. Cells*, **94**, 1931–1935.

Peiro, A.M., Ravirajan, P., Govender, K., Boyle, D.S., O'Brien, P., Bradley, D.D.C., Nelson, J., & Durrant, J.R. (2006): Hybrid polymer/metal oxide solar cells based on ZnO columnar structures, *J. Mater. Chem.* **16**, 2088–2096.

Peng, K.-Q., Yan, Y.J., Gao, S.P., & Zhu, J. (2002): Vapor-Liquid-solid growth of silicon nanowires and its application in Nanosensor, hybrid organic and nanostructure solar cell, *Adv. Mater.* **14**, 1164.

Peng, K.-Q., Xu, X., Wu, Y., Yan, Y., Lee, S.T., & Zhu, J. (2005): Aligned single crystalline Si nanowire arrays for photovoltaic applications, *Small*, **1**, 1062–1067.

Peng, K.-Q., Wang, X., Wu, X.-L., & Lee, S.-T. (2009): Platinum nanoparticle decorated silicon nanowires for efficient solar energy conversion, *Nano Lett.* **9**, 3704–3709.

Peng, K.-Q., Wang, X., Li, L., Wu, X.L., & Lee, S.T. (2010): High performance silicon nanohole solar cells, *J. Amer. Chem. Soc*, **132**, 6872–6873.

Plass, R., Pelet, S., Krueger, J., Gratzel, M., & Bach, U. (2002): Quantum Dot Sensitization of Organic-Inorganic Hybrid Solar Cells, *J.Phys. Chem. B*, **106**, 7578–7580.

Polman, A. (2010): Plasmonic Solar Cells (Invited lecture), *MRS Fall meeting*, Nov–Dec, 2010, Boston, MA.

Ponce, F.A. & Bour, D.P. (1997): Nitride-based semiconductors for blue and green light-emitting devices, *Nature*, **386**, 351–359.

Poortmans, J. (2008): Processing and Characterization of Efficient Thin-Film Polycrystalline-Silicon Solar Cells, *Proc. 33rd IEEE PV Specialists Conf*, May 2008, San Diego, CA. 4922763.

Pryce, I., Fischer, A.J., Koleske, D.D., & Atwater, H.A. (2010): Plasmonic nanoparticle enhanced photocurrent in GaN/ InGaN/GaN Quantum well solar cells, *Appl. Phys. Lett.* **96**, 153501.

Ravirajan, P., Peiro, A.M., Nazeeruddin, M.K., Graetzel, M., Bradley, D.D.C., Durrant, J.R., & Nelson, J. (2006): Hybrid polymer/zinc oxide photovoltaic devices with vertically oriented ZnO nanorods and an amphiphilic molecular interface layer, *Phys. Chem. B*, **110**, 7635.

Qiu, Y., Tüzün, O., Gordon, I., Venkatachalam, S., Slaoui, A., Beaucarne, G. & Poortmans, J. (2009): n-type thin film polycrystalline silicon solar cells using a seed layer approach, *34th IEEE PV Specialists Conf*, Philadelphia, PA, June 7–12, 2009, available at Mpublic88_0703001626.pdf.

Sablon, K.A., Little, T., Mitin, V., Sergeer, A., Vagidov, N., & Reinhardt, K. (2011): Strong enhancement of solar cell efficiency due to QD with Built-in-Charge, *Nano Lett.* **11**(6), 2311–2317.

Salleo, A., Cu, Y., & Peumans, P. (2011): 'High efficiency, low cost TF solar cells' *GCEP Annual report*, available at '2.2.6_salle_public_2011.pdf'

Schneider, J. & Evans, R. (2006): Industrial Solid phase crystallization of Silicon, *Proc. 21st EU PV Solar Energy Conf*, September 2006, Dresden, Germany, 2CV.4.32, pp. 1032–1034.

Sharps, P., Cornfeld, A., Stan, M., Korostyshevsky, A., Ley, V., Cho, B., Varghese, Diaz, T.J., & Aiken, D. (2008): The future of high efficiency, multi-junction space solar cells, *Proc. 33rd IEEE PV Specialists Conf.* May 2008, San Diego, CA, 1, pp. 1–6.

Shu, Q., Wei, J., Wang, K., Zhu, H., Li, Z., Jia, Y., Gui, X., Guo, N., Li, X., Ma, M., & Wu, D. (2009): Hybrid heterojunction and photoelectrochemistry solar cell based on silicon nanowires and double-walled carbon nanotubes, *Nano Lett.* **9**, 4338–4342.

Soci, C., Zhang, A., Xiang, B., Dayeh, S.A., Aplin, D.P.R., Park, J., Bao, X.Y., Lo, Y.H., & Wang, D. (2007): ZnO nanowire UV photodetectors with high internal gain, *Nano Lett.* **7**, 1003–1009.

Stelzner, T., Pietsch, M., Andrä, G., Falk, F., Ose, E., & Christiansen, S. (2008): Silicon nanowire-based solar cells, *Nanotechnology*, **19**, 295203-4.

Stern, E., Cheng, G., Klemic, J.F., Broomfield, E., Turner-Evans, D., Li, Zhou, C., Reed, M.A. (2006): *J. Vac. Sci. Technol.* B **24**, 231.

Sun, Ke, Kargar, A., Park, N., Madsen, K.N., Naughton, P.W., Bright, T., Jing, Y., & Wang, D. (2011): Compound Semiconductor Nanowire solar cells, *IEEE Journal of Selected topics in Quantum Electronics*, **17**(4), 1033–1049.

Takeda, Y., Motohiro, T., König, D., Aliberti, P., Feng, Y., Shrestha, S., & Conibeer, G. (2010): Practical factors lowering conversion efficiency of hot carrier solar cells, *Appl. Phys. Express*, **3**, 104301.

Tian, B., Zheng, X., Kempa, T.J., Fang, Y., Yu, N., Yu, G., Huang, J., & Lieber, C.M. (2007): Coaxial silicon nanowires as solar cells and nanoelectronic power sources, *Nature*, **449**, 885–889.

Torchynska, T.V. & Polupan, G.P. (2002): "III–V material solar cells for space application, *Semiconductor Phys. Quantum electronics, Opto-electronics*, **5**, 63–75.

Tsakalakos, L. (2007): Strong broadband optical absorption in silicon nanowire films, *J. Nanophotonics*, **1**, 013552.

Tsakalakos, L., Balch, J., Fronheiser, J., Korevaar, B.A., Sulima, O., & Rand, J. (2007): Silicon nanowire solar cells, *Appl. Phys. Lett.* **91**, 233117-3.

Wagner, R.S., & W.C. Ellis (1964): Vapor–liquid–solid mechanism of single crystal growth, *Appl. Phys. Lett.* **4**, 89.

Wagner, D.K. & Shealey, J.R. (1984): Graded band-gap AlGaAs solar cells grown by MOVPE, *J. Cryst. Growth*, **68**, 474–476.

Wang, L., Liu, Y., Jiang, X., Qin, D., & Cao, Y. (2007): Enhancement of photovoltaic characteristics using a suitable solvent in hybrid polymer/multiarmed CdS nanorods solar cells, *J. Phys. Chem. C*, **111**, 9538–9542.

Widenborg, P.I., Chuangsuwanich, N. & Aberle, A.G.: Glass texturing, AU2004228064A1.

Widenborg, P.I., & Aberle, A.G. (2002): Surface morphology of poly-Si films made by aluminium-induced crystallisation on glass substrate, *J. Crystal Growth*, **242**, 270.

Xu, C., Shin, P.H., Cao, L., Wu, J., & Gao, D. (2010): Ordered TiO2 nanotube arrays on TCO for DSS Cells, *Chem. Mater.*, **22**, 143–148.

Yamaguchi, M., Yamamoto, A., Uchida, N., & Uemura, C. (1986): A new approach for thin film InP solar cells, *Solar cells*, **19**, 85–86.

Yang, Q., Tang, K., Wang, C., Qian, Y., & Zhang, S. (2002): PVA-assisted synthesis and characterization of CdSe and CdTe nanowires, *J. Phys. Chem. B*, **106**, 9227–9230.

Yoon, J., Baca, A.J., Park, S.I., Elvikis, P., Geddes, J.B., Li, L., Kim, R.H., Xiao, J., Wang, S., Kim, T.H., Motala, M.J., Ahn, B.Y., Duoss, E.B., Lewis, J.A., Nuzzo, R.G., Ferreira, P.M., Huang, Y., Rockett, A., & Rogers, J.A. (2008): Ultrathin silicon solar microcells for semitransparent, mechanically flexible and microconcentrator module designs, *Nature Materials*, **7**, 907–918.

Zacharias, M., Heitmann, J., Scholz, R., Kahler, U., Schmidt, M., & Blasing, J. (2002): Size-controlled highly luminescent silicon nanocrystals: A SiO/SiO$_2$ superlattice approach, *Appl. Phys. Lett.* **80**, 661–663.

Zhang, D., Liu, Z., Li, C., Tang, T., Liu, X., Han, S., Lei, B., & Zhou, C. (2004): detection of NO2 down to ppb levels using individual and multiple In2O3 nanowire devices, *Nano Lett.*, **4**, 1919–1924.

Zheng, G.F., Patolsky, F., Cui, Y., Wang, W.U., & Lieber, C.M. (2005): Multiplexed electrical detection of cancer markers with nanowire sensor arrays, *Nat. Biotechnol.*, **23**, 1294–1301.

Zhou, C. & Reed, M.A. (2006): *J. Vac. Sci. Technol.*, B, **24**, 231.

Zhu, J. & Cui, Y. (2010): Photovoltaics: More solar cells for less, *Nature Materials*, **9**, 183–184.

Chapter 7

Policies and incentives

7.1 INTRODUCTION

The photovoltaics are utilized in three broad segments which are generally referred to as (i) additive systems, (ii) ground-mounted systems, and (iii) integrated systems. The additive systems account for major share in Germany with more than 8 giga-watts of PV installed by early 2010. Spain dominates currently in ground-mounted installations as a result of grid-input tariff paid since 2008. France is considered as flourishing market for integrated systems (BIPV) because nearly 95% of the installed PV systems are integrated into building roofs and facades. The on-grid PV capacity in France amounts to 91.2 MW at the end of 2008. This is due to a high grid input remuneration rate given for integrated systems as well as attractive sunny climate (Sun & Wind Energy 2010).

These examples clearly establish that in any country, proper environment in terms of legislation/policy is an essential prerequisite for the utilization of appreciable volumes of PV systems. Furthermore, attractive rates of remuneration push the market most effectively towards building integrated PV. Appropriate and encouraging legislation and proper feed-in tariff (FIT) in countries such as Japan and European countries – Germany, France, Spain, Italy and Austria – have encouraged more widespread application of photovoltaics for power generation. In this chapter, policies and incentives that are currently in vogue in a few selected countries are discussed.

The financial incentives are offered to electricity consumers to install and operate solar-electric generating systems. The incentives are also designed for energy producers to move away from conventional fossil fuels to Solar or other renewable energy sources. They are also intended to encourage the PV industry to achieve the economies of scale needed to compete where the cost of PV-generated electricity is above the cost from the existing grid. These policies result in the promotion of energy independence, high tech job creation and more importantly, reduction of carbon dioxide emissions which cause global warming.

The government legislation must also guarantee a fixed, premium rate for solar electricity fed into the national grid. The power companies are obliged by the government legislation to buy the solar electricity and the additional costs are passed onto the customers.

7.2   INCENTIVE MECHANISMS

There are basically four types of incentive mechanisms, in general, offered to help install and generate electric power through renewable sources including solar. These incentives are sometimes operated in combination.

1   *Investment subsidies*: In this programme, the authorities refund part of the cost of installation of the system. With investment subsidies, the financial burden falls upon the taxpayer. The investment subsidy is easy to administer. Investment subsidies are paid out based on the declared capacity of the installed system and are independent of its actual power yield over time, and even its poor durability and maintenance.

2   *Feed-in Tariffs (FITs)*: This is 'most commonly' implemented programme. FITs typically include three key provisions: guaranteed grid access, long-term contracts for the electricity produced, and purchase prices based on the cost of generation. The electricity utility buys PV electricity from the producer under a multiyear contract at a guaranteed rate. The need for a feed-in tariff comes from the fact that it is far more expensive to produce energy from green sources than it is from fossil fuels. With feed-in tariffs, the initial financial burden falls upon the consumer. Feed-in tariffs reward the number of kilowatt-hours produced over a long period of time. But, since the rate is fixed by the authorities, it may result in an apparent over payment of the owner of the PV installation. The main argument in favour of feed-in tariffs is the encouragement of quality. The price paid per kWh under a feed-in tariff exceeds the price of grid electricity.

   The Feed-in tariff system has already been in place in many countries; for example, Germany, Israel, USA, Spain and Australia have the system for some time now and have been instrumental in the success and growth of photovoltaics.

3   *Net metering*: 'Net metering' refers where the price paid by the utility (for the PV power fed back into the grid) is the same as the price charged. It is often achieved by having the electricity meter spin backwards as electricity produced by the PV installation exceeds the amount being used by the owner of the installation, which is fed back into the grid.

4   *Solar Renewable Energy Certificates* (SRECs): SRECs allow for a market mechanism to set the price of the solar generated electricity subsidy. In this mechanism, a renewable energy production or consumption target is set, and the utility (more technically the Load Serving Entity) is obliged to purchase renewable energy or face a penalty (Alternative Compliance Payment or ACP). The producer is credited for an SREC for every 1,000 kWh of electricity produced. If the utility buys this SREC and retires it, they avoid paying the ACP. In principle, this system delivers the cheapest renewable energy, since all the solar facilities are eligible and can be installed in the most economic locations.

7.3   POLICIES IN SELECTED COUNTRIES

Japan

Japan has become a leading PV nation world-wide despite wide open fields for the installations of large-scale PV are rare. When Japan signed Kyoto Protocol, it has formulated a goal to install 4.8 GW by 2010. The principles underlying Japan's Energy Policy are: (i) security of Japanese energy supply, (ii) economic efficiency, and (iii) harmony with environment. This Energy policy is supported by clear objects such as promoting energy conservation measures, developing and introducing diverse sources of energy and so on.

In Nov. 2008, a Plan was announced to support the Government's 'Action Plan for Achieving a Low-carbon Society' whose targets are to increase the amount of installations of solar power generation systems 10-fold by 2020 and 40-fold by 2030, and to bring down the current price of the solar power generation system to 50% within three to five years.

The Comprehensive Immediate Policy Package (2008) also cites the promotion of the installation of solar power generation systems in homes, businesses and public facilities.

Implementation of Photovoltaics: The Japanese residential PV implementation programme has been the longest scheme implemented from 1994 to October 2005. During this period, the average price for 1 kWp in the residential sector fell from 2 million yen/kWp in 1994 to 670,000 yen/kWp in 2004. This scheme has been responsible for the expansion of Japan's PV market for 12 years. New installations have decreased since 2005, mainly due to closure of the scheme. At the end of 2008, total cumulative installed capacity was 2.15 GW, less then half of the original 4.8 GW goal for 2010.

In order to stimulate the home market, METI announced the introduction of investment subsidy for residential PV systems in 2009. This new measure along with the New National Energy Strategy (2006) and the Action Plan for Promoting the Introduction of Solar Power Generation (2008) to revitalise the Japanese market confirm the political will of the government for promoting renewable energies.

To develop an independent and sustainable new energy business/industry that includes the whole value chain from raw material production, cell, module and BOS component manufacturing to the establishment of business opportunities in overseas markets, various support measures for PV are declared:

1   Strategic promotion of technological developments as a driving force for competitiveness such as promotion of technological development to overcome high costs, and development of PV systems to facilitate grid-connection and creation of the environment for its implementation;
2   Accelerated demand creation by developing several support measures besides subsidies, and help to create new business models;
3   Developing competitiveness to establish a sustainable PV industry such as establishment of standards, codes and an accreditation system to contribute to the availability of human resources, as well as securing performance, quality and safety, and improving the awareness for Photovoltaic systems; and promotion of international co-operation.

In addition, over 300 local authorities have introduced measures to promote the installation of PV systems. For example, Tokyo Metropolitan Government plans to support the installation of 1 GW of PV systems in 40,000 households in 2009 and 2010; the Federation of Electric Power Companies of Japan (FEPC) announced that they intend to install PV plants with a cumulative installed capacity of 10 GW by 2020.

PV Roadmap 2030⁺:

In 2004, the New Energy and Industrial Technology Development Organization (NEDO) presented the PV Roadmap 2030 (PV2030) for the technological development of PV power generation toward the target set for 2030, namely, PV systems to supply 10% of the annual total electricity consumed in Japan, which is currently about 1×10^{12} kWh. The target was therefore to establish PV systems with a cumulative installed capacity of 100 GW by 2030 (this target was reduced to 54 GW in 2008). To accomplish this objective, all aspects related to PV power generation technologies were reviewed and numerical targets for PV power generation were set for each element of development. At that time, these numerical targets appeared to be very difficult to achieve. However, the rapid expansion of PV production and markets globally, the accelerated growth of energy demand in Asia, changed outlook towards climate change and the greenhouse gas reductions in Japan have led to a revision in 2009 and the target year was changed from 2030 to 2025, resulting in the revised roadmap, PV2030⁺.

Figure 7.1 shows PV2030⁺ presented by NEDO (NEDO website). The milestones in this roadmap are the reduction of electricity cost to that feasible for business use [14 yen/(kW h)] by 2017 and to that feasible for industrial use [7 yen/(kW h)] by 2025. To realize these electricity costs, numerical targets were set for the manufacturing cost, lifetime, and conversion efficiency of PV modules; for example, a conversion efficiency of 25% is targeted for practical PV modules by 2025. Several breakthroughs must be achieved to attain the targets in PV2030⁺. Today, the R&D of a wide range of both organic and inorganic materials is ongoing because current material technology is expected to make significant breakthroughs.

The review aims at further expanding PV utilisation and maintaining the international competitiveness of Japan's PV industry. 2030⁺ also emphasizes 'making PV power generation one of the key technologies that would play a significant role in reducing CO_2 emissions by 2050, so that it can contribute not only to Japan, but also to the global society'.

In PV2030⁺, the target year has been extended from 2030 to 2050 and a goal is set to cover 5 to 10% of domestic primary energy demand with PV power generation by 2050. The document assumes that Japan can supply roughly one-third of the required overseas market volumes. To improve economic efficiency, the concept of 'realiszing Grid Parity' was unchanged, and so also, the generation cost targets. In addition, PV2030⁺ aims to achieve generation cost of below 7 yen/kWh in 2050. Regarding the technological development, an acceleration to realise these goals is aimed to achieve the 2030 target already in 2025, five years ahead of the schedule set in PV2030. For 2050, ultra-high efficiency solar cells with 40% efficiency and even higher conversion efficiency will be developed.

NEDO PV Programme

New Energy Development Organisation (NEDO) is responsible for the Research Programme for renewable energies. The current PV programme in the frame of Energy and Environment Technologies Development Projects has three main areas: (1) New Energy Technology Development, (2) Introduction and Dissemination of New Energy and Energy Conservation, and (3) International Projects. One of the top priorities is obviously the cost reduction of solar cells and PV systems. NEDO supports efforts

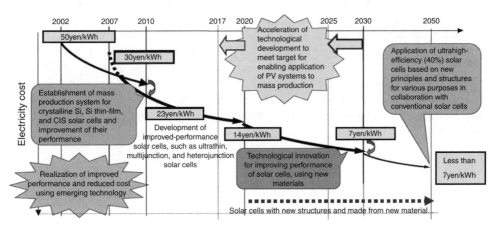

Cost reduction scenario and PV system development

Table for "Roadmap for development of PV systems, PV2030+"

Target year (completion of development)	2010 or after	2020(2017)	2030(2025)	2050
Generation cost	Equivalent to residential use (23yen/kWh)	Equivalent to business use (14yen/kWh)	Equivalent to industrial use (7yen/kWh)	Equivalent to general-purpose power source use (Less than 7yen/kWh)
Module conversion efficiency (R&D level)	16% for practical modules (20% for R&D-level cells)	20% for practical modules (25% for R&D-level cells)	25% for practical modules (30% for R&D-level cells)	40% for ultrahigh-efficiency modules
Annual production for Japanese market (GW/year)	0.5 ~ 1	2 ~ 3	6 ~ 12	25 ~ 35
Annual production for overseas market (GW/year)	~ 1	~ 3	30 ~ 35	~ 300
Major purposes	Individual houses, public facilities	Individual houses, condominiums, public facilities, offices, etc.	Individual houses, condominiums, public facilities, consumer business use, charging of electric cars	Consumer use, general purpose, industrial use, transportation use, off-grid power source

Figure 7.1 Japan's Roadmap, PV2030+
(*Source*: Konagai 2011, reproduced with permission, Copyright © 2011, The Japan Society of Applied Physics)

in certain technology fields that have the potential for an early practical application, including fullscale production, commercialisation and market competitiveness by 2015.

To maintain the competitiveness of Japan's technology development, NEDO provides subsidies (a subsidy ratio of 50%) for projects that address the following challenges: (i) Enhanced production technologies for thin-film silicon solar cells (including super large area cell production and high-speed film production) and light-weighting technology, (ii) Slicing techniques for ultra thin polycrystalline silicon solar cells, and (iii) Selenisation process optimisation techniques for CIS thin-film solar cells. Japan's resolve to cover nearly 50% of its residential power load from BIPV systems in 2030, has spurred large-scale commercialization resulting in lowering of the system prices. Industrially prefabricated houses with an integrated PV system are readily available in the market, and are offered at prices around 8% higher than those of a traditionally built home. Major companies like Sanyo and Kaneka are developing innovative roof-top BIPV systems.

China

China has emerged as a major producer of solar PV modules as well as a market leader in recent years. As the Renewable Energy Law endorsed in February 2005, the Chinese Government set a target for renewable energy to contribute 10% of the country's gross energy consumption by 2020, a huge increase from the current 1%. The main features of the Law which came into effect in 2006 are:

(a) Responsibility for implementing and managing renewable energy development, including resource surveys lays with Energy Authorities of the State Council;
(b) Availability of renewable energy development fund to support R&D and resource assessment;
(c) Various types of grid-connected renewable energy power generation would be encouraged;
(d) Grid enterprises shall purchase the power produced with renewable energy sources within the coverage of their power grid, and provide grid-connection service;
(e) The grid-connection price of renewable energy power generation shall be determined by the price authorities, and the excess shall be shared in the power selling price within the coverage of the grid. However, no specific rate was set for electricity from PV installations.

The 2006 Report on the Development of the PV Industry in China, prepared by the National Development and Reform Commission (NDRC), the Global Environment Facility (GEF) and World Bank (WB), estimates a market of 130 MW in 2010. The report states that the imbalance between solar cell production and domestic market development impedes not only the sustainable development of energy sources in China, but also the healthy development of the PV industry.

The document released by National Outlines for Medium and Long-term Planning for Scientific and Technological Development (2006-2020) emphasised the development of (i) PV cell technology with high cost-effect ratio and its utilization, and (ii) solar power generation technology and the study of solar powered buildings.

China's laboratory Photovoltaic cell has currently achieved a top efficiency of 21%, the commercialised PV components and normal commercialized cells respectively showed an efficiency of 14–15% and 10–13%. The NBR Programme of China (the '973 Programme) stresses on 'Basic research of mass hydrogen production using solar energy'.

PV Production costs: China has reduced the production cost of solar cells, and the price has gradually declined since the year 2000 from 40 RMB/Wp (4.40 €/Wp); and in July 2009, the National Energy Administration has set a subsidized price for solar power at 1.09 RMB/kWh (0.112 €/kWh). According to the Energy Research Institute, this cost doesn't allow Chinese companies to work profitably; the companies need between 1.3 and 1.5 RMB/kWh (0.134 and 0.155 €/kWh) to become profitable and to accelerate the domestic market growth.

When the PV electricity generation cost declines to around1 RMB/kWh (0.103 €/kWh) in 2010/11, this would be within the cost price of conventional power generation.

In July 2009, Chinese government has announced subsidies for PV demonstration projects through a programme called 'Golden Sun'. The subsidies are 50% of total investment in PV power generation systems and power transmission facilities in on-grid projects, and 70% for independent projects. The Government earmarked a budget that should allow about 500 MW of PV installations.

Under a new plan called, 'Energy Revitalization Plan', the investment in new energies would reach more than RMB 3 trillion (€309 billion), and investments in smart-grids exceed RMB 4 trillion (€436 billion) by the next decade.

PV Resources and Utilisation: China's continental solar power potential is estimated at 1,680 billion toe (equivalent to 19,536,000 TWh) per year. One percent of China's continental area, with 15% conversion efficiency, could supply 29, 304 TWh of solar energy which is 189% of the world-wide electricity consumption in 2001.

Despite the Renewable Energy Law coming into effect in January 2006, the impact on PV installations is limited because no tariff has been set for PV.

'Strategic Status of Photovoltaics in China' says that the national target for the cumulative capacity of PV systems set was 500 MW in 2010. The predictions of the PV Market in China for 2020 were rather optimistic; and the cumulative installed capacity was given as 30 GW and included 12 GW in the frame of the Chinese Large-Scale PV Development Plan, a project which was scheduled to start in 2010. However, the actual growth of PV installations was far below the set targets. Therefore, the 2007 China Solar PV Report authored by the China Renewable Energy Industry Association, Greenpeace China, European PV Industry Association, and WWF, reduced the market predictions to 300 MW as cumulative installed capacity in 2010. For 2020, two scenarios are given. The low target scenario predicts 1.8 GW, in line with the old Government policy, whereas a high target of 10 GW would be possible if strong support mechanisms were to be introduced.

According to 'China's Solar Future' report of May 2009, China faces a rapidly increasing demand for energy, and the country is building a massive PV industry, representing all facets of the supply chain, from polysilicon feedstock, ingots and wafers to cells and modules. The report recommends an accelerated adoption of PV generated electric power in China to reach global average level of PV power generation by 2014. The main policy recommendations in the report include: Establishing clear targets for PV installation by adjusting current national targets and achieve global average level by 2014; Enacting clear and easy-to-administer PV incentive policies that are suitable for China's unique situations, using both market and legal mechanisms to encourage private investment in PV; Immediately implementing a State financed direct investment subsidy model at central and local levels, and effectively implementing feed-in tariff programmes stipulated in the Renewable Energy Law.

PV subsidy programme
In March 2009 the Chinese government has announced a solar subsidy programme: For 2009 the subsidy is 20 RMB/Wp (2.06 €/Wp) for BIPV and 15 RMB/Wp (1.46 €/Wp) for roof top applications. The document neither mentions a cap on individual installations nor a cap for the total market. The subsidy is paid as a 70% down payment and 30% after the final acceptance of the project.

All systems >50 kW which use modules with efficiencies of >14% (polycrystalline), >16% (monocrystalline), or >6% (thin-film) are eligible.

In addition to the solar subsidy programme, the Golden Sun Programme – for pilot cities to support the use of renewable energies in buildings was announced.

In July 2009 the new Chinese energy stimulus plan revised the 2020 targets for installed solar capacity to 20 GW.

European Union

The political structure of the European Union, with 27 member States is quite diverse and there is no unified approach towards the development and utilization of renewable energies. However, at a EC meeting in Brussels in March 2007, the Council endorsed a binding target of a 20% share of renewable energies in the overall EU energy consumption by 2020 and a 10% binding minimum target to be achieved by all Member States for the share of Biofuels in overall EU transport petrol and diesel consumption.

In order to meet the new targets, the EC formulated an overall coherent framework for renewable energies resulting in the Directive on the 'Promotion of the Use of Energy from Renewable Sources'. This Directive exceeds the targets set within the White Paper 'Energy for the Future: Renewable Sources of Energy' and the Green Paper 'Towards a European Strategy for the Security of Energy Supply'. The goals were that renewable energies should provide 12% of the total and 21% of electric energy in the European Union by 2010, in order to meet the obligations of CO_2-reductions pledged in the Kyoto Protocol and to lower the dependence on energy imports systems. The electricity generation from the PV systems would then be in the order of 2.4 to 3.5 TWh, depending under which climatic conditions these systems are installed.

The Directive admits that '. . . Due to widely varying potentials and developments in different Member States regarding renewable energies, a harmonization seems to be very difficult to achieve in the short term. In addition, short term changes to the system might potentially disrupt certain markets and make it more difficult for Member States to meet their targets. Nevertheless, the merits and demerits of harmonisation towards the different current systems have to be analysed and monitored, also notably for the medium to longer term development'.The Commission considers co-operation between countries as wll as optimisation of the impact of national schemes for renewable energy sources to be appropriate.

While the new Directive indicated the overall percentage of renewable energies for the different Member States as well as the indicative trajectory how to reach it, the decision to utilize the kind of technologies to reach the national targets is left to the States.

Market and implementation in the European Union
The market conditions for Photovoltaics differ substantially from country to country. This is due to different energy policies and public support programmes for renewable

energies, especially PV, as well as the varying grades of liberalisation of domestic electricity markets. Between 2001 and 2008, installations of PV systems in the EU increased more than ten times and reached 9.5 GW cumulative installed capacity at the end of 2008. A total of about 24 GW of new power capacity was constructed in the EU last year of which 4,590 MW (19%) is PV.

The scenario in countries where PV related activity is fairly significant is outlined:

Spain: In 2008, Spain was the biggest market due to the almost five-fold increase from 560 MW in 2007 to about 2.7 GW in 2008. This was more than twice the expected capacity and was due to an exceptional race to install systems before the Spanish Government introduced a cap of 500 MW on the yearly installations in the autumn of 2008. This tremondous expansion during 2006–2008 was the result of Government's approval of the Plan de Energías Renovables en España (PER) for 2005–2010, which stipulated that 12.1% of Spain's overall energy requirements and 30.3% of total electricity consumption should be covered with renewable energies by 2010. The generous feed-in tariffs set by the Royal Decree 436/2004, helped the development of the Spanish PV market. In 2007 the Royal Decree 661/2007 with an increased cap of 1.200 MW for PV installations was passed which triggered the plan to install multi-megawatt free-field solar PV power generating systems. This progress had led to the revision of the solar PV legislation in 2008, and the new Decree set considerably lower FITs for new systems and restricts the market to 500 MW per year with the provision that two thirds are rooftop mounted. As of 2009, FIT for Building mounted PV are: for $\leq$20 kWp, 0.34 €/kWh, and for >20 kWp, 0.32 €/kWh, and for Ground mounted, 0.32 €/kWh. These feed in tariffs are capped at approximately 500 MWp/year, of which 241 MW ground-mounted, 233 MWp building-mounted.

Germany: Germany was the second largest single market, with around 1.5 GW. Germany now produces over 14 per cent of its energy from renewable sources, which has been attributed to the generous and comprehensive feed-in tariff system implemented by the German government.

In 2000, the German government introduced the first large-scale feed-in tariff system, under 'The German renewable Energy Sources Act (EEG)' which is reviewed on a regular basis and the 2010 version is currently in force. This Act has resulted in explosive growth of PV installations in Germany. The FIT was over 3 times the retail price or 8 times the industrial price. The principle behind the German system is a 20 year flat rate contract. The value of new contracts is programmed to decrease each year, in order to encourage the industry to pass on lower costs to the end users. The model of EEG has been adapted by many countries around the world. EEG has been amended several times and has triggered an unprecedented boom in solar electricity production. This success is largely due to the creation of favourable political framework conditions. Grid operators are legally obliged to pay producers of solar electricity a fixed remuneration for solar generated electricity fed into the grid, depending on the size and type of the system, as well as the year of installation. The tariffs differ to account for the different costs of rooftop or ground-mounted systems in accordance with the size of the system and system cost reductions over time. Since the EEG guarantees the FIT-payments for 20 years, it provides sustained planning security for investors in PV systems. In view of the unexpectedly high growth rates, the depreciation was accelerated and a new category (>1000 kWp) was created with a lower tariff. The facade premium was abolished.

Peak power dependent FiT in ct/kWh

Type		2004	2005	2006	2007	2008	2009	2010	Jul 2010	Oct 2010	2011
Rooftop mounted	up to 30 kW	57.4	54.53	51.80	49.21	46.75	43.01	39.14	34.05	33.03	28.74
	between 30 kW and 100 kW	54.6	51.87	49.28	46.82	44.48	40.91	37.23	32.39	31.42	27.34
	above 100 kW	54.0	51.30	48.74	46.30	43.99	39.58	35.23	30.65	29.73	25.87
	above 1000 kW	54.0	51.30	48.74	46.30	43.99	33.00	29.37	25.55	24.79	21.57
Ground mounted	contaminated grounds	45.7	43.4	40.6	37.96	35.49	31.94	28.43	26.16	25.37	22.07
	agricultural fields	45.7	43.4	40.6	37.96	35.49	31.94	28.43	–	–	–
	Other	45.7	43.4	40.6	37.96	35.49	31.94	28.43	25.02	24.26	21.11

Since 2009, there are additional tariffs for electricity immediately consumed rather than supplied to the grid with increasing returns, if more than 30% of production, in general, is consumed on site. This is to incentivise a demand side management and help develop solutions to the intermittency of solar power. In July 2010, the EEG was amended to reduce the tariffs by a further 16% in addition to the normal annual depreciation, due to the sharp drop in the prices of PV-panels in 2009. Most recently in 2011, the changes occurred, when part of the degression foreseen for 2012 was brought forward to mid 2011 as a response to surprisingly high installations during 2010. In 2011 the additional costs may rise to €0.035/kWh. The contract duration is fixed for 20 years with constant remuneration. Feed-in tariffs will be lower in value in future years (decreasing by 9% default and a maximum of 24% per year). Degression will be accelerated or slowed down by three percentage points for every 1,000 MWp/annum divergence from the target of 3,500 MWp/a.

Italy: In Italy, the Ministry for Industry issued a decree on 5 August 2005 that provides the legal framework for the system known as 'Conto Energia'. This has led to a steep rise in applications in the second half of 2005 and early 2006; but, the increase in the new systems capacity has been moderate in 2006. At the end of the first quarter of 2006, applications with more than 1.3 GW were submitted to the 'implementing body' Gestore del Sistema Elettrico (GRTN SpA.), nearly 2.6 times more than the 500 MW cap up to 2012. The actual installations in 2006 were only 12.5 MW, far less than the predicted 50 to 80 MW. In February 2007, a Decreto Interministeriale was issued, which changed the national projected target for cumulative installed PV systems from 2,000 MW in 2015 to 3,000 MW in 2016. This led to a steep growth in PV installations and 50.2 MW were installed in 2007 and 127 MW in 2008. In June 2009, more than 500 MW of PV systems were connected to the grid and the total was expected to reach ~900 MW in 2009. The following incentive tariffs are from the decree of 19 Feb 2007.

System size (kWp)	Free-standing (€-ct/kWh)	Semi-integrated (€-ct/kWh)	Integrated (€ct/kWh)
1 to 3	0.40	0.44	0.49
3 to 20	0.38	0.42	0.46
Greater than 20	0.36	0.40	0.44

The contract duration is 20 years with constant remuneration. A new government decree issued in March 2011, allows lower tariffs (exact amounts not announced) for new installations from June 2011.

France: In France, revised feed-in tariffs came into force in July 2006 and resulted in a moderate growth of the French PV market. In 2006 and 2007, just 7.6 MW and 12.8 MW respectively were installed, despite attractive and cost competitive feed-in tariff for building integrated PV. In 2008, the PV installations rose to 44.3 MW. In November 2009, the Government announced a new programme to substantially increase the solar generated power 400 times by 2020 to a total installed capacity of 5.4 GW.

The French feed-in tariff offers a uniquely high premium for building integrated systems. As of 2010, the FIT are: (a) BIPV on houses, hospitals, and schools: €0.58/kWh; (b) BIPV on other buildings: €0.50/kWh; (c) Semi-integrated (PV panels located on buildings do not have any architectural function): €0.42/kWh; (d) Ground-mounted PV in DOM-TOM and Corsica: €0.40/kWh; (e) Ground-mounted PV < 250 kWp in mainland: €0.314/kWh; (f) Ground-mounted PV > 250 kWp in mainland: €0.314–0.3768/kWh according to the administrative region. The contract duration of 20 years is linked to inflation. Additional investment subsidies are available as tax credits. Accelerated depreciation of PV systems is possible for enterprises. These tariffs remain valid until 2012 when they will be reviewed.

Each year, starting from 2012, the new contracts will be 10% lower than the previous year. Only 1500 kWh/kWp per year are brought from any fixed installation in mainland (2200 for tracking). In DOM-TOM and Corsica, the caps for fixed and tracking installation are respectively 1800 and 2600. The National Target is 160 MW by 2010 and 450 MW by 2015. The tax credit allowed for income tax payer is 50% reimbursement on equipment cost.

Austria: Austria passed the second amendment of Eco electricity law in August 2008 which provides for an investment subsidy limited to €2.1 million for all PV electricity systems larger than 5 kW capacity. The provision in the first amendment (2006) remained changed, and €18 million were available to support new PV systems (capacity >5 kW) in 2009 through the Austrian Climate & Energy Fund. The subsidy for free field and roof-mounted systems upto 5 kWp was 2,500 €/kWp and €3,200/kWp for building integrated systems. Since, there has been a great rush on this incentive, while increasing the budget, reduction of incentive rates are also planned. Still the payments for BIPV would be higher than for 'additive' installations. A second incentive programme 'BIPV for pre-fabricated Homes' was launched towards the end of 2009, in which a private home buyer satisfying 'Passive House' criteria, receives a single subsidy of 2,600 €/kW for installations not exceeding 5 kW. The programme is seen as a development plan for the domestic BIPV market, which in turn strengthens the Austrian industry and create new jobs. 33% of all homes built in 2008 in Austria are prefabricated homes. According to a study, a cost saving upto 16% compared to retrofit installation was estimated (Sun & Wind Energy 2010).

Greece: Greece has introduced a new feed-in-tariff scheme in January 2009, given in the Table that remain unchanged until August 2010 and are guaranteed for 20 years. However, if a grid connection agreement is signed before that date and if the system is finalised within the next 18 months, the unchanged FIT is applied. The degression of the tariffs for new systems will be 5% each half year. A 40% grant is still available on top of the new FITs for most of the systems where minimum investment eligible for grants is €100,000.

System size (kWp)	Mainland (€/kWh)	Island (€/kWh)
Less than 100 kWp	0.45	0.50
>100 kWp	0.40	0.45

The new contract prices are expected to reduce 1% per month starting 2010. Special programme with higher FIT but no tax rebates is planned to drive 750 MWp installations of BIPV. Investment subsidies such as Tax rebates and grants (40%) are available.

A new incentives programme is also introduced in June 2009 which covers rooftop PV systems up to 10 kWp (both for residential users and small companies). The new FIT is set at 0.55 €/kWh and is guaranteed for 25 years, as well as being adjusted annually for inflation (25% of last year's Consumer Price Index). An annual degression of 5% is foreseen for newcomers as of 2012. In addition to the FIT, small residences are eligible for a 20% tax deduction capped at €700 per system. Residential users do not have to be registered as 'business' with the tax authorities and are exempted from any tax (with the exception of the 19% VAT paid for the initial investment). Small companies are also exempted from any tax as long as they keep the income from PV as untaxed reserves.

To become eligible for this FIT, a residence has to meet part of its hot water needs by solar thermal. The programme, though approved for the mainland grid only, the islands with autonomous grids would be covered in a second phase. PV facades are not eligible for the new support scheme; but, a PV facade on a commercial building could benefit from the old FIT regime (i.e., 0.45 €/kWh for 20 years). These measures are expected to spur the Greek PV market which has been sluggish over the last years with just 18.5 MW installed capacity at the end of 2008.

Czech Republic: In Czech Republic, as of 2010 feed-in tariffs are 12.25 CZK/kWh for ≤30 kWp and 12.15 for >30 kWp. The contract duration is 20 years with yearly increase (within range 2–4%) linked to inflation. New contract prices are changed by 5% yearly. However, due to unexpected rise of number of installations in 2009, new bill is proposed allowing 25% change.

Slovakia: Slovakia, in July 2009, published new legislation (Slovakia's Law Code) supporting renewable energy sources and efficient co-production of heat and power. Under this law energy companies that generate electrical energy from renewable sources enjoy a price guarantee for fifteen years. The guarantee involves purchase prices set by the Regulatory Office for Network Industries (URSO) and obligatory purchase of this energy for the electrical energy transmission grid. URSO determines the price for electricity from renewable energy sources by taking into consideration the type of the renewable energy source, the technology used, the installed capacity and the date of launching of the facility. As of December 2010, the FIT is for PV systems <100 KWp, 0.43 €/kWh, and for Systems >100 KWp, 0.425 €/kWh. The contract duration is 15 years with a decrease of 10% in 2011.

Bulgaria: In Bulgaria as of April 1, 2010, the incentives offered are: (a) for ≤5 kWp, 792.89 Leva/MWh (about 0.405 €/kWh), and (b) for >5 kWp, 728.29 Leva/MWh (about 0.372 €/kWh). Also, remuneration for 25 year contract, with possible next year changes set related to 2 components (electricity sales price in the previous year and RES component) are available.

United Kingdom: In the United Kingdom, the Energy Act of November 2008 includes specific provisions for the implementation of feed-in tariffs in the UK by 2010. The FIT legislation fixes an above market rate for utility companies to buy electricity from renewable energy producers. It means that if the retail price of fossil fuel electricity were 15 p/kWh, then the rate for renewable electricity could be up to 60 p/kWh. In this case, the 45p difference per kWh would be spread across every customer of the relevant utility company.

As of November 2010, the UK Government has introduced a feed-in tariff for small scale (up to 5 MW) renewables from 1 April 2010, with a review in 2012 for changes from 1 April 2013.

From April 2010, the FIT offers a fixed payment per kWh generated and a guaranteed minimum payment of 3p per kWh exported to the market (assumed 50% only) or entitled to opt out with own Power Purchase Agreement (PPA). Tariffs will not be index-linked to the RPI.

1   Projects up to 5 MW will be eligible, including off-grid installations,
2   All renewable technologies are eligible for the FIT from April 2010,
3   FIT is offered for a 20 year period, with the exception of solar PV projects which get upto 25 years,
4   FIT designed with the aim of delivering 2% of the UK's energy from small scale projects by 2020,
5   Support will degress in line with expected technology cost reductions, where appropriate,
6   Support levels will be reviewed periodically and in response to sudden changes in technology costs. However, projects continue to receive the levels of support offered at their registration,
7   Projects below 50 kWp must be installed by MCS accredited installers. 50 kWp to 5 MWp projects will be subject to accreditation similar to the current RO process,
8   Projects installed in the interim period between the announcement of the FIT (15 July 2009) and the start of the scheme (April 2010) are eligible to receive the tariff with some conditions on the support period. However, any non-domestic projects that receive grant funding from central government have to return the grant before they can receive FIT payments,
9   Regardless of technology, projects installed prior to 15 July 2009 are eligible to receive generation payments currently being auctioned of 9 p/kWh and export payments of 5 p/kWh, provided they were previously receiving support under the RO scheme,
10 Projects up to 50 kWp in size can no longer claim the RO; existing installations are automatically transferred to the FIT. New and interim period projects between 50 kW and 5 MW are given a one-off choice between claiming support under the FIT or the RO. Existing projects between 50 kW and 5 MW in size will remain under the RO, with no opportunity to transfer to the FIT,
11 No further capital/financial support for the up-front capital costs of projects,

12   Feed-in-tariff rates for PV grid connected are the following: 43.3 p/kwh for
      projects <4 kWp; 37.9 p/kWh for projects, >4 kWp to 10 kWp; 32.8 p/kWh for
      projects, >10 KWp to 100 kWp; 30.7 p/kWh for projects, >100 kWp; and Stand
      alone, 30.7 p/kWh.

From 1 August 2011, tariff rate for >50 KWp is 19.0 p/kwh.

Despite the fact that the European PV production grew by almost 60% and reached 1.9 GW, the excellent Spanish market growth and the stable large German market demand did not change the role of Europe as an importer of solar cells and/or modules. Further capacity expansions and technology progress are necessary to secure a leading role of the European PV industry in the future. The support measures for Photovoltaics in the European Union Member States and Switzerland are listed in the PV Research Report 2009, prepared by European Commission, Joint Research Center, and Institute for Energy (August 2009).

PV Research in Europe

In addition to the 27 national programmes for market implementation, research and development, the EU has been funding research (DG RTD) and demonstration projects (DG TREN) with the Research Framework Programmes (FP) since 1980. This is of particular importance, as research for PV in a number of Member States is closely linked to EU funds. A large number of research institutions, covering from basic material research to industry process optimisation, are contributing to the progress of PV. The EU's R&D activities are organized under FPs.

In addition, there were Marie Curie Fellow-ships and the 'Intelligent Energy – Europe' (EIE) Programme. The CONCERTO Initiative was a Europe-wide initiative proactively addressing the challenges of creating a more sustainable future for Europe's energy needs.

The PV Technology Platform established with an aim to mobilise all the concerned sharing a long-term European vision for Photovoltaics has developed the European Strategic Research Agenda for PV for the next decade(s) and recommends for its implementation to ensure that Europe maintains industrial leadership.

The following impacts are expected from the research activities: Through technological improvements and economies of scale, the cost of grid-connected PV electricity in Europe is expected to reach 0.10–0.25 €/kWh by 2020. Research and development should lead to reduced material consumption, higher efficiencies and improved manufacturing processes, based on environmental-friendly processes and cycles.

Some of the Projects selected were: (i) Multi-approach for high efficiency integrated and intelligent concentrating PV modules/systems, (ii) Heterojunction solar cells based on a-Si/c-Si (iii) Large grained, low stress multi-crystalline silicon thin-film solar cells on glass by a novel combined diode laser and solid phase crystallisation process; (iv) Intermediate Band Materials and Solar Cells for Photovoltaics with High Efficiency and Reduced Cost. This project pursues the manufacturing of intermediate band materials and solar cells based on the strategies: Insertion of transition elements into III-V semiconductor matrices; Use of quantum dot systems to artificially engineer intermediate band solar cells; Development of intermediate band materials and solar cells based on InGaN; and Insertion of transition elements into thin-film polycristalline hosts; (v) The project on 'Dye Sensitised Solar Cells' aims to develop materials and manufacturing procedures for DSCs with long life-time and increased module efficiencies (7% target); (vi) Next generation Solar Cells and Module Laser Processing Systems; (vii) Ultra Thin Solar Cells for Module Assembly; (viii) New Applications for CPV's: A fast Way to Improve Reliability and Technology Progress; (ix) Metamorphosis of Power Distribution: System Services from Photovoltaics; (x) Efficiency and material issues for thin-film Photovoltaics; (xi) Manufacturing and product issues for thin-film Photovoltaics; (xii) Further development of very thin wafer based c-Si photovoltaics; (xiii) Development of novel materials, device structures and fabrication methods suitable for thin-film solar cells and TCOs, including Organic photovoltaics; (xiv) Development of new concentrator modules and field performance evaluation of CPV systems.

On 22 November 2007, the EC announced the European Strategic Energy Technology Plan (SET-PLAN) [EC 2007a] in which PV was identified as one of the key technologies. With the help of the SET-Plan, the European PV Industry Association (EPIA) plans to develop the solar PV sector so that up to 12% of European electricity is generated with PV systems. This corresponds to around 420 TWh of electricity or 350 GWp of Photovoltaic system installations. To realise this grand object, new PV installations would have to increase from around 1.6 GW per annum in 2007 to 4 GW per annum in 2010 and 80 GW per annum in 2020. By then, electricity generation costs with PV systems will have reached grid parity in most of Europe. The EPIA suggests the following measures to realize this vision: (i) the responsibilities to be divided between industry and academic research; and (ii) research issues to be categorized as short term and medium/long term.

In addition, other issues such as necessary policy framework, securing human resources and a general awareness campaign are listed. The prerequisites needed are (i) Co-operation with other renewable technologies, (ii) Interaction with utilities and grid operators, (iii) Internalisation of external costs, (iv) Liberalised utility market, and (v) Fair and transparent electricity rate structure. One of the important conditions to reach the 12% target mentioned above is a favourable political framework (EU and national) both in the precompetitive phase, as well as in the phase when Grid parity is reached and beyond. The necessary supportive national policies needed for the precompetitive phase are reasonable-feed in tariffs (7–8% ROI) with no caps, investment security, waiving of administrative barriers, priority access to the grid, and support of building codes; and for the grid parity phase and beyond are investment security, waiving of administrative barriers, priority access to the grid and grid regulation, and support of building codes.

United States

In February 2009, the American Recovery and Reinvestment Act (ARRA) became effective. The main solar provisions in the Law are: (i) creation of a Department of Treasury (DOT) Grant Programme, (ii) improvement to the investment tax credit by eliminating ITC penalties for subsidised energy financing, (iii) a new DOE Loan Guarantee Programme, and (iv) creation of tax incentives for manufacturing by offering accelerated depreciation and a 30% refundable tax credit for the purchase of equipment used to produce solar material and components for all solar technologies.

Earlier, Clean Renewable Energy Bonds (CREBs) were created under the Energy Tax Incentives Act of 2005, for funding State, local, tribal, public utility and electric cooperative projects.

The Energy Improvement and Extension Act of 2008 extended the CREBs and changed some programme rules. The ARRA of 2009 expanded funding to $2.4 billion (€1.7 billion) of new allocations. Low interest financing, some as low as 0.75%, is provided to approved projects.

On 27 May 2009, the US President announced to spend over $467 million (€333.6 million) from the ARRA to expand and accelerate the development, deployment, and use of geothermal and solar energy throughout the country. The DOE will provide $117.6 million (€84 million) in Recovery Act funding to accelerate the widespread commercialization of solar energy technologies across America. $51.5 million (€36.8 million) will go directly for PV Technology development and $ 40.5 million (€28.9 million) will be spent on Solar Energy deployment, where projects will focus on non-technical barriers to solar energy deployment.

In September 2004, the US Photovoltaic Industry has published their PV Roadmap through to 2030 and beyond, 'Our Solar Power Future' (Sei 2004). The main goal of this Roadmap is that 'solar provides half of all new US electricity generation by 2025'. The Industry Association has advocated effective policies sustained over time to increase solar power production and implementation and recommended the following provisions for market expansion, research and development, and incentives:

(i)   *Market Expansion:*

    (a)   Enact a residential and commercial tax credit that augments current State and Federal support. The first 10 kW installed would receive a 50% tax credit capped at $3 per watt. Any system above 10 kW would be eligible for a 30% tax credit capped at $2 per watt. Decreasing the caps by 5% per year will encourage a steady decline in prices and ease the transition to a market without tax credits;

    (b)   Modify the wind tax credit for solar so that it can be used together with the existing 10% investment tax credit;

    (c)   Establish uniform net metering and interconnection standards to give solar power owners simple, equitable access to the grid and fair compensation;

    (d)   Boost Federal Government procurement of solar power to $100 million per year to build public-sector markets for solar power; and

    (e)   Support State public benefit charge programmes and other State initiatives to advance solar power and build strategic alliances with public and private organizations to expand solar markets

(ii)   *Research and Development:*

    (a)   Increase R&D investment to $250 million/annum by 2010;

    (b)   Strengthen investments in crystalline silicon, thin-film, and balance-of-systems components, as well as new system concepts that are critical to the industry now – reducing the gap between their current cost and performance and their technical potential;

    (c)   Support higher-risk, longer-term R&D for all system components that can leap-frog beyond today's technology to new levels of performance and reduce installed system costs;

(d)   Enhance funding for facilities and equipment at Centres of excellence, Universities, National labs (Sandia National Laboratories and the National Renewable Energy Laboratory) – as well as the Science and Technology Facility at NREL – to shorten by 50% the time between lab discoveries and industry use in manufacturing and products; and

(e)   Develop partnerships among industry, universities, and national laboratories to advance PV manufacturing and product technologies.

(iii)   *Incentives:*

Many State and Federal policies and programmes have been adopted to encourage the development of markets for PV. These consist of direct legislative mandates (such as renewable content requirements) and financial incentives (such as tax credits). Financial incentives typically involve appropriations or other public funding, whereas direct mandates typically do not. In both cases, these programmes provide important market development support for PV. Amongst the types of incentives (given below), investment rebates, loans and grants are the most commonly used – at least 39 States in all regions of the country, have such programmes in place.

Most common mechanisms are:

(a)   personal tax exemptions (Federal Government, 21 States + Puerto Rico)
(b)   corporate tax exemptions (Federal Government, 24 States + Puerto Rico)
(c)   sales tax exemptions for renewable investments (27 States + Puerto Rico)
(d)   property tax exemptions (35 States + Puerto Rico)
(e)   buy-down programmes (19 States + District of Columbia, Virgin Islands, 234 utilities, 8 local)
(f)   loan programmes and grants (Federal Gov., 40 States + Virgin Islands; 69 utilities, 17 local, 7 private)
(g)   industry support and production incentives (Federal Government, 24 States + Puerto Rico, 33 Utilities, 9 private).

All different support schemes and information on State, local, utility, and selected federal incentives that promote renewable energy are given in the Database of State Incentives for Renewable Energy, DSIRE (Dsi 2009). For more details, it is recommended to visit the DSIRE web-site http://www.dsireusa.org/ and the corresponding interactive tables and maps.

Solar Energy Technologies Programme (SETP)

The aim of this SETP or Solar Programme is to develop cost-competitive solar energy systems for America. The current Multiannual work-programme runs from 2008 to 2012 (DOE 2008). More than $170 million (€121.4 million) are spent each year for research and development on the two solar electric technologies, Photovoltaics and Concentrated solar power, which are considered to have the greatest potential to reach cost competitiveness by 2015. The greatest R&D challenges that this programme lists are the reduction of costs, improvement of system performance, and the search for new ways to generate and store energy captured from the sun.

The SETP also aims to ensure that the new technologies are accepted in the market place. Hence, necessary work is done to remove many non-technical market barriers, such as updating codes and standards that aren't applicable to new technologies, improving interconnection agreements among utilities and consumers, and analyzing utility value capacity credits for utilities. Such activities should help consumers, businesses, and utilities to make more informed decisions when considering renewable energy, and they also facilitate the purchase of solar energy. The SETP conducts its key activities through four sub-programmes: (i) photovoltaics; (ii) concentrated Solar Power; (iii) systems Integration; and (iv) market transformation.

The SETP supports the DOE 2006 Strategic Plan (DOE 2006), which has identified five strategic themes including energy security. In addition, it supports the research and development provisions and broad energy goals outlined in the National Energy Policy Act 2005, and the Energy Independence and Security Act (EISA). In both acts, the Congress has expressed strong support for decreasing dependence on foreign energy sources and decreasing the cost of renewable energy generation and delivery. This support along with the availability of financial incentives is important for achieving the SETP goals.

The Solar Programme lists economic targets for PV, which was determined by an analysis of key markets. They are set based on assessments of the Levelised Costs of Energy (LCOE) for solar technologies to be competitive in these markets.

According to SETP, the residential and commercial price targets are based on current retail electricity prices and take into consideration the rather optimistic projection of the Energy Information Administration (EIA) that electricity prices will remain fairly constant (in real terms) through 2025. With these assumptions, the Programme predicts that meeting the solar market cost goals will result in PV installations of 5–10 GW by 2015 and 70–100 GW by 2030 in USA.

The PV technology roadmaps were developed in 2007 by NREL, Sandia National Laboratories, DOE, and experts from universities and industry (DOE 2008a). This work was done, in part, to support activities within the Solar America Initiative. These technology roadmaps summarise the current status and future goals for the specific technologies. The Roadmaps for Intermediate-Band PV, Multiple-Exciton-Generation PV and Nano-Architecture PV are yet to be finalised.

Solar Technology Research Plan

The U.S. strategy for overcoming the challenges and barriers to massive manufacturing, sales, and installation of PV technology is to achieve challenging targets throughout the development pipeline. Specific broad R&D efforts toward achieving these goals include: PV Systems & Module Development, PV Materials & Cell Technologies, Testing & Evaluation, and Grid/Building Integration.

The PV sub-programme's R&D activities include the following:

I.   Research on 'New Devices and Processes' focuses on two areas:
  1   Next Generation PV: to develop innovative photovoltaic cells and/or processes by 2015; potential areas of interest included, but were not limited to:
    (a)   Photovoltaic devices – Organic, crystalline, non-single crystal devices, photoelectrochemical, advanced multijunction, low-dimensional structures, optimised interfaces, transport properties, and cross-cutting issues;

    (b)   Hybrid PV concepts – Hydrogen generation, powered electrochromics, and storage; and

    (c)   Manufacturing – Low-cost techniques, environmental/recycling issues and novel manufacturing processes.

    The PV device and manufacturing process activities in this area are expected to produce prototype PV cells and/or processes by 2015, with full commercialization by 2020-2030.

2   Photovoltaic Technology Pre-Incubator: The new project is aimed to help small solar businesses transition from concept verification of a solar PV technology to the development of a commercially viable PV prototype by 2012. The goals of the project include promoting grid parity for PV technologies, transitioning innovative PV technologies into the prototype stage, and developing prototype PV concepts with manufacturing costs of less than $1/watt.

II.   Prototype Components and Systems: The Solar America Initiative's research in component and system prototypes emphasizes development of prototype components and systems produced at pilot-scale. The display of cost, reliability, or performance benefits is required.

III.   Systems Development and Manufacturing: These R&D activities are intended for collaboration and partnership among industry and university researchers on components and systems that are ready for mass production and capable of delivering electricity at Solar America Initiative target costs.

Very High Efficiency Solar Cell Programme

In 2005 the US Defence Advanced Research Projects Agency initiated the Very High Efficiency Solar Cell (VHESC) Programme to develop 50% efficient solar cells over the next years. The aim of the Programme is to reduce the average load of 20 pounds (ca 9 kg) that an average soldier has to carry to power the portable technology gadgets used.

In the first phase, many Universities and Industrial units have participated. In July 2007 DARPA announced the start of the second phase of the programme by funding DuPont-University of Delaware VHESC Consortium to transition the lab-scale work to an engineering and manufacturing prototype model. For this purpose, DARPA awarded the consortium $12.2 million as part of a three-year, multi-phase programme that could total up to $100 million. DuPont is managing the consortium of proposed companies and scientific institutions dedicated to the optimisation of the VHESC solar cells for efficiency and cost.

The recent report from Greentech Solar (GreentechMedia.com, news item dated 21st Sept., 2011), the PV market consumption in 2010 is 887 MW, and is projected to go to 1.8 GW in 2011. The report sees USA positioning to nearly double its global market share in 2011 supporting a greater diversity of installation types than has been seen previously in any demanding situation.

India

India has one of the highest potential yields of solar power globally which makes it attractive to consider solar energy generation. Government of India released country's National Action Plan on Climate change in June 2008. The Plan identifies eight missions and one of them is National Solar mission. The Jawaharlal Nehru National Solar

Mission (JNNSM) was officially announced by Prime Minister of India on 12 January 2010. This programme aims to install 20,000 MW of solar power by 2020. The first phase of this programme aims to install 1000 MW by paying a tariff fixed by the Central Electricity Regulatory Commission (CERC) of India. While in spirit this is a feed in tariff, there are several conditions on project size and commissioning date. Tariff for solar PV projects is fixed at Rs. 17.90 (US$ 0.397/kWh), and will be reviewed periodically by the CERC. The mission further aims to ensure large-scale deployment of solar generated power for both grid connected as well as distributed commercial energy services, which is estimated to cost US$ 9 billion. Money will be spent on incentives for production and installation as well as for research and development. Further, the Plan offers financial incentives and tax holidays for utility companies.

The Indian Renewable Energy Development Agency (IREDA) under the Ministry of New and Renewable Energy (MNRE) provides revolving fund to financing and leasing companies offering affordable credit for the purchase of PV systems in India. State Utilities are mandated to buy green energy via a Power Purchase Agreement from Solar Farms.

The MNRE has launched a new scheme (Jan 2008) for installation of Solar Power Plants. For the projects, a Generation-based subsidy is available up to Indian Rs. 12/kWh (€0.21/kWh) from the Ministry, in addition to the price paid by the State Utility for 10 years with a cap of 50 MW. Several State governments announced generation-based subsidies for upto 12 years with caps ranging from 50 MW to 500 MW, the prominent states being Haryana, Punjab, Gujarat, Rajasthan, Karnataka, Tamilnadu, West Bengal and Orissa.

The State Electricity Regulatory Commissions are setting up preferential tariffs for Solar Power: For example, Rajasthan, Rs. 15.6 (€0.27) per kWh (proposed); West Bengal, Rs. 12.5 (€0.22) per kWh (proposed); Punjab, Rs. 8.93 (€0.15) per kWh. 80% accelerated depreciation, concessional duties on import of raw materials, and excise duty exemption on certain devices are allowed.

Taiwan

In 2002 the Renewable Energy Development Plan was approved which aimed for the generation of 10% or more of Taiwan's total electricity generation through renewable energy sources by 2010. This plan led to concerted efforts by all the concerned to develop renewable energy and to aggressively adopt its use. In 2004, Taiwan enacted 'Measures for Subsidising Photovoltaic Demonstration Systems', as part of its National Development Plan 2008. This programme provides subsidies that cover up to 50% of the installation costs for PV systems.

The adopted support scheme foresees a maximum investment subsidy of NT$ 150,000/kWp (3,225 €/kWp) but only up to 50% of installation costs. Administration Agencies, public schools and hospitals, suitable for demonstration projects, are eligible for 100% investment subsidies for systems under 10 kWp. In addition, for all renewable energies, 2 NT$/kWh (0.043 €/kWh) are paid to approved applicants for 10 years, and this can be extended up to 20 years.

Other support measures for renewable energies are a 13% tax credit for investment in energy conservation, as well as renewable energy utilisation equipment, a 2-year accelerated deprecation and low interest loans.

The Solar Energy Development Project has several long-term goals. It is planned that a total of 7.5 million residents should utilise solar energy by 2030. Industrial and commercial use should be about half that of residential use. Public utilities are expected to have the same solar power generating capacity as the industrial and commercial sectors, and independent solar power generating systems will be set up in mountains and on off-shore islands. By 2020, the goal is to take the island's solar power generating capacity to 4.5 GW (1.2 GW PV).

In July 2008, the government has earmarked solar energy and light emitting diodes (LED) for active development in the coming years.

The Government provides subsidies to manufacturers involed in R&D and offers incentives to solar energy consumers. Material suppliers are expanding operations and increasing their investments in the field. About a dozen manufacturers have planned to invest in fabricating thin-films for solar cells and eight of them will set up their own plants to process the products.

The Government – backed Industrial Technology Research Institute (ITRI) has drawn up an R&D strategy to take up research in ranging from efficiency increase in the various wafer based and thin-film solar cells to concentrator concepts and novel devices in order to lower module costs to around 1 $/Wp between 2015 and 2020. However, the main focus is on the industry support to increase production capacities and improved manufacturing technologies.

For 2008 and 2009, the Government has allotted NT$ 1 billion (€21.5 million) for subsidies to consumers who buy solarpower systems. The Government plans to subsidise half of the installation cost for solar devices; and households which install solar PV systems would be offered an electricity rate of 2.1 NT$/kWh (0.045 €/kWh). A national target to double the cumulative capacity installations to 31 MW was set for 2010.

The new law (Renewable Energy development Act of 2009) provides enhancement of incentives for the development via a variety of methods, including the acquisition mechanism, incentives for demonstration projects and the loosening of regulatory restrictions. The goal was to increase Taiwan's renewable energy generation capacity by 6.5 GW to a total of 10 GW within 20 years.

It is expected that the law would attract investment of at least NT$30 billion (€645 million) per year, create at least 10,000 jobs and generate output value of NT$100 billion within two years. The industry had recommended to set a price floor of 8 NT$/kWh (0.172 €/kWh) for green energy which would give firms a reasonable profit margin.

Canada

Only Ontario offers significant incentive. In 2006 the Ontario Power Authority introduced the Renewable Energy Standard Offer Program which was replaced with the 2009 Feed-In Tariff program for renewable energy (FIT). The FIT program was further divided into the MicroFIT program for projects less than 10 kW, designed to encourage individuals and households to generate renewable energy. The program was launched in September 2009 and the tariffs were fixed then. The solar projects ≤10 kW received $0.802; however, as of 13 August 2010, ground mounted systems would receive a lower tariff than rooftop mounted systems. Feed-In tariff rates for the

Ontario Power Authority's FIT and MicroFIT Programmes, for renewable generation capacity of 10 MW or less, connected at 50 kV: Solar Photovoltaic:

(a)   Rooftop ≤10 kW, $0.802/kWh CDN;
(b)   Ground Mounted ≤10 kW, $0.642/kWh CDN;
(c)   Rooftop >10 ≤ 250 kW, $0.713/kWh CDN;
(d)   Rooftop >250 ≤ 500 kW, $0.635/kWh CDN
(e)   Rooftop >500 kW, $0.539/kWh CDN;
(f)   Ground Mounted, 2 > 10 kW, $0.443/kWh CDN

REFERENCES

PV Research Report – Research, Solar cell Production, and Market Implementation of PV (2009) prepared by European Commission, Joint Research Center & Institute of Energy in August 2009 is the prime source of information.

Feed-in Tariffs: Wikipedia, free encyclopedia, at http://en.wikipedia.org/wiki/feed-in_tariff. http://solarfeedintariff.co.uk/the-feed-in-tariff/.

China unveils subsidies of 50 per cent on large solar power projects". BusinessGreen.com. 22 July 2009. http://www.businessgreen.com/business-green/news/2246509/china-unveils-subsidies-per%20.

http://www.pv-tech.org/news/_a/anwell_produces_its_first_a-si_thin-film_module_using_in-house_technology/.

PV Tech.org. 21 Sept 2009, at http://www.pvtech.org/news/_a/tianwei_solarfilms_ramps_production_as 70mw_supply_deal_signed/.

CDF (2003): China Development Forum 2003, 15–17 November 2003, Background Reports for China's National Energy Strategy 2000 to 2020, at http://www.efchina.org/documents/Draft_Natl_E_Plan0311.pdf.

CHI (2007): China Solar PV report, China Renewable energy Industry Assoc., Greenpeace China, EPIA, and WWF, China Environmental Sciences Press, Sept, 2007.

China Industry Brief, 10 Sept 2009 at http://www.chinabriefing.com/news/2009/09/10/china-industry-sept-10.html.

SEM (2009a): SEMI China White paper, 'China's solar future' 2009 at http://www.pvgroup.org.

Policy framework: The Renewable Energy Sources Act (EEG) Bundesverband Solarwirtschaft. 2 Feb 2011. http://www.neocoop.eu/Pictures/conto_energia_2007_en.pdf.

National Action Plan of India (2009).

SEM (2009): SEMI India White paper, 'The Solar PV landscape in India – An Industry Perspective' at http://www.pvgroup.org.

Mendonça, M. (2007): Feed-in Tariffs: Accelerating the Deployment of Renewable Energy, London: EarthScan.

NREL 2010, at www.nrel.gov/docs/fy10osti/44849.pdf.

Germany, Renewable Energy Sources Act (2000): 'Act on Granting Priority to Renewable Energy Sources,' Federal Ministry for the Environment, Nature Conservation and Nuclear Safety (BMU), at http://www.wind-works.org/ FeedLaws/Germany/GermanEEG2000.pdf.

Lipp, J. (2007): 'Lessons for effective renewable electricity policy from Denmark, Germany and the United Kingdom', Energy Policy, 35, Issue 11, pp. 5481–5495.

Klein, A. et al. (Fraunhofer ISI) (October 2008): Evaluation of Different Feed-in Tariff Design Options: Best Practice Paper for the International Feed-in Cooperation, 2nd Edition. Berlin, Germany: BMU; at: http://www.feed-in-cooperation.org/wDefault_7/wDefault_7/download-files/research/best_practice_ paper_2nd_edition_final.pdf.

REN21 Global Status Report (2010) at http://www.ren21.net/REN21Activities/Publications/GlobalStatusReport/tabid/5434/Default.aspx.

Ikk (2008): Osamu Ikki, PV Activities in Japan, **12**, July 2008.

NED (2007): NEDO Brochure, Energy and Environ. Technologies, Dec, 2007.

NDR (2006): Report on the development of the PV industry in China, China RED Project office, August 2008.

Pho (2007): PV Technology Platform: Stratagec research Agenda for PV Solar Energy Conversion technology, June 2007; Luxembourg: Office for the official publications of European communities.

CEU (2007): Council of the European Union, March 2007, presidency conclusions 7224/07.

EC (2007): Communication from the Commission to the Council and the European parliament; Renewable Energy roadmap – renewable energy in the 21st century; building a more sustainable future.

European Commission (COM) 2008: Commission Staff Working Document, Brussels, 57, 23 January 2008; at: http://ec.europa.eu/energy/climate_actions/doc/2008_res_working_document_en.pdf.

Epi (2009): European Union PV Industry association, Global market Outlook for PV until 2013.

Ewe (2009): EU Wind energy association annual report 2008.

EC (2009): *Directive 2009/287/EC (April 2009)*, Official Journal of the European Communities, L140/16, June 2009.

International Energy Agency (2008): *Deploying Renewables: Principles for Effective Policies*, ISBN 978-92-64-04220-9.

Sys (2009): Photovoltaic Energy Barometer, Systemes Solaires, le journal du photovoltaique no 1–2009, March 2009, ISSN 0295-5873.

Ges (2009): Gestore Servizi Elettrici, Press Release, 22 June 2009. http://www.gse.it/media/Comunicati Stampa/Comunicati%20Stampa/Comstampa_PV500MW. pdf.

MEE (2008): Ministre de l'Ecologie, de l'Energie, du Developpement durable et de la Mer, Press Release, 17 November 2008 http://www.developpement-durable.gouv.fr/article.php3?id_article=3903.

Zweibel, K. (2007): A Solar Grand Plan, *Scientific American*, 16 Dec 2007.

Konagai, M. (2011): Present Status and future prospects of Silicon solar cells, *Japan J. Appl. Phys.*, **50**, p 03001.

UNEP (2009): UNEP and New Energy Finance, 2009, Global Trends in Sustainable Energy Investment 2009, ISBN 978-92-807-3038.

Union of Concerned Scientists (2009): *Successful Strategies: Renewable Electricity Standards, Fact Sheet*, February 2009, http://www.ucsusa.org/.

DOE (2008): *U.S. Department of Energy, Solar Energy Technologies Programme* (Solar Programme): 2008–2012 Multi-Year Programme Plan.

DOE (2008): *Solar America Initiative*, at http://www1.eere.energy.gov/solar/solar_america/publications.html#technology_ roadmaps.

Annexure

The best measurements for cells and submodules are summarised in Table A.1.

Table A.1 Confirmed terrestrial cell and submodule efficiencies measured under the global AM1.5 spectrum (1000 W/m²) at 25°C (IEC 60904-3: 2008, ASTM G-173-03 global)

Classification[a]	Effic.[b] (%)	Area[c] (cm²)	V_{oc} (V)	J_{sc} (mA/ cm²)	FF[d] (%)	Test centre[e] (and date)	Description
Silicon							
Si (crystalline)	25.0 ± 0.5	4.00 (da)	0.706	42.7	82.8	Sandia (3/99)[f]	UNSW PERL[11]
Si (multicrystalline)	20.4 ± 0.5	1.002 (ap)	0.664	38.0	80.9	NREL (5/04)[f]	FhG-ISE[12]
Si (thin film transfer)	16.7 ± 0.4	4.017 (ap)	0.645	33.0	78.2	FhG-ISE (7/01)[f]	U. Stuttgart (45 μm thick)[13]
Si (thin film submodule)	10.5 ± 0.3	94.0 (ap)	0.492[g]	29.7[g]	72.1	FhG-ISE (8/07)[f]	CSG Solar (1–2 μm on glass; 20 cells)[14]
III–V Cells							
GaAs (thin film)	26.1 ± 0.8	1.001 (ap)	1.045	29.6	84.6	FhG-ISE (7/08)[f]	Radboud U. Nijmegen[15]
GaAs (multicrystalline)	18.4 ± 0.5	4.011 (t)	0.994	23.2	79.7	NREL (11/95)[f]	RTI, Ge substrate[16]
InP (crystalline)	22.1 ± 0.7	4.02 (t)	0.878	29.5	85.4	NREL (4/90)[f]	Spire, epitaxial[17]
Thin Film Chalcogenide							
CIGS (cell)	19.4 ± 0.6[h]	0.994 (ap)	0.716	33.7	80.3	NREL (1/08)[f]	NREL, CIGS on glass[18]
CIGS (submodule)	16.7 ± 0.4	16.0 (ap)	0.661[g]	33.6[g]	75.1	FhG-ISE (3/00)[f]	U. Uppsaia, 4 serial cells[19]
CdTe (cell)	16.7 ± 0.5[h]	1.032 (ap)	0.845	26.1	75.5	NREL (9/01)[f]	NREL, mesa on glass[20]
Amorphous/Nanocrystalline Si							
Si (amorphous)	9.5 ± 0.3[i]	1.070 (ap)	0.859	17.5	63.0	NREL (4/03)[f]	U. Neuchatel[21]
Si (nanocrystalline)	10.1 ± 0.2[j]	1.199 (ap)	0.539	24.4	76.6	JQA (12/97)	Kaneka (2 μm on glass)[22]
Photochemical							
Dye-sensitized	10.4 ± 0.3[k]	1.004 (ap)	0.729	22.0	65.2	AIST (8/05)[f]	Sharp[23]
Dye-sensitized (submodule)	**8.4 ± 0.3[k]**	**17.11 (ap)**	**0.693[g]**	**18.3[g]**	**65.7**	**AIST (4/09)**	**Sony, 8 serial cells[3]**
Organic							
Organic polymer	5.15 ± 0.3[k]	1.021 (ap)	0.876	9.39	62.5	NREL (12/06)[f]	Konarka[24]
Organic (submodule)	**2.05 ± 0.3[k]**	**223.5 (ap)**	**6.903**	**0.502**	**59.1**	**NREL (1/09)**	**Plextronics[4,25]**
Multijunction Devices							
GaInP/GaAs/Ge	32.0 ± 1.5[j]	3.989 (t)	2.622	14.37	85.0	NREL (1/03)	Spectrolab (monolithic)
GaInP/GaAs	30.3[j]	4.0 (t)	2.488	14.22	85.6	JQA (4/96)	Japan Energy (monolithic)[26]
GaAs/CIS (thin film)	25.8 ± 1.3[j]	4.00 (t)	–	–	–	NREL (11/89)	Kopin/Boeing (4 terminal)[27]
a-Si/μc-Si (thin submodule)[i,j]	11.7 ± 0.4[j,l]	14.23 (ap)	5.462	2.99	71.3	AIST (9/04)	Kaneka (thin film)[28]

[a]CIGS = CuInGaSe₂; a-Si = amorphous silicon/hydrogen alloy; [b]Effic. = efliciency; [c](ap) = aperture area; (t) = total area; (da) = designated illumination area; [d]FF = fill factor; [e]FhG-ISE = Fraunhofer Institut für Solare Energiesysteme; JQA = Japan Quality Assurance; AIST = Japanese National Institute of Advanced Industrial Science and Technology; [f]Recalibrated from original measurement; [g]Reported on a 'per cell' basis; [h]Not measured at an external laboratory; [i]Stabilised by 800 h, 1 sun AM1.5 illumination at a cell temperature of 50°C; [j]Measured under IEC 60904-3 Ed. 1: 1989 reference spectrum; [k]Stability not investigated; [l]Stabilised by 174 h. 1 sun illumination after 20 h, 5 sun illumination at a sample temperature of 50°C.

Below are the references (for the above data):

3. Morooka, M., Noda, K. (2008): 88th Spring Meeting of The Chemical Society of Japan, Tokyo, 26 March 2008.
4. Tipnis, R. (2009): "Printed Solar Power: From Lab to Market", Nortech Advanced Energy Speaker Series, February 3, 2009 (available at http://www.nortech.org/Docs/Ritesh%20Tipnis%20Plextronics.pdf).
11. Zhao, J. *et al.* (1998): Appl. Phys. Lett, 73: pp. 1991–1993.
12. Schultz, O. *et al.* (2004): Prog. in Photovolt: Res. and Appl; 12, pp. 553–558.
13. Bergmann, R.B. *et al.* (2001): Tech. Digest, PVSEC-12, Chefju Island, Korea, 11–15.
14. Keevers, M.J. *et al.* (2007): 22nd European Photovoltaic Solar Energy Conference, Milan.
15. Bauhuis, G.J. *et al.* (2005): 20th EU PV Solar Energy Conference, Barcelona, pp. 468–471.
16. Venkatasubramanian, R. *et al.* (1997): Conf. Record, 25th IEEE PV Specialists Conference, pp. 31–36.
17. Keavney, C.J. *et al.* (1990): Conf. Record, 21st IEEE PV Specialists Conference, pp. 141–144.
18. Repins, I. *et al.* (2008): IEEE Photovoltaics Specialists Conference Record, p. 33.
19. Kessler, J. *et al.* (2000): Proc. 16th EU PV Solar Energy Conference, Glasgow, pp. 2057–2060.
20. Wu, X. *et al.* (2001): Conf. Proc. 17th EU PV Solar Energy Conference, Munich, pp. 995–1000.
21. Meier, J. *et al.* (2004): Thin Solid Films, 451–452: pp. 518–524.
22. Yamamoto, K. *et al.* (1998): MRS Spring Meeting April, 1998, San Francisco.
23. Chiba, Y. *et al.* (2005): Tech. Digest, 15th Intl. PV Science and Engineering Conf., pp. 665–666.
24. See http://www.konarka.com.
25. Laird, D. *et al.* (2007): SPIE Proc., Vol. 6656, No. 12.
26. Ohmori, M. *et al.* (1996): Tech. Digest, Intl. PVSEC-9 Miyasaki, Japan, pp. 525–528.
27. Mitchell, K. *et al.* (1998): Conf. Record, 20th IEEE PV Specialists Conf., pp. 1384–1389.
28. Yoshimi, M. *et al.* (2003): Conf. Record, 3rd WC on PV Energy Conversion, pp. 1566–1569.

(*Source*: Green, M.A., Emery, K., Hishikawa, Y., & Warta, W. (2009): Solar cell efficiency tables (Version 34) Prog. Photovolt: Res. Appl. **17**, 320–326).

Subject index